ULTRASOUND
Its Chemical, Physical, and Biological Effects

ULTRASOUND

Its Chemical, Physical, and Biological Effects

Edited by

Kenneth S. Suslick

VCH

Kenneth S. Suslick
University of Illinois at
Urbana-Champaign
Noyes Laboratory
505 S. Mathews Avenue
Urbana, Illinois 61801

Library of Congress Cataloging-in-Publication Data

Ultrasound: its chemical, physical, and biological effects / edited by Kenneth S. Suslick.
p. cm.
Includes bibliographies and index.
ISBN 0-89573-328-5
1. Ultrasonic waves. 2. Biological chemistry. I. Suslick.
QC244.U454 1988
534.5'5--dc 19

87-28050
CIP

Printed in the United States of America.

ISBN 0-89573-328-5 VCH Publishers
ISBN 3-527-26645-3 VCH Verlagsgesellschaft

Distributed in North America by:

VCH Publishers, Inc.
220 East 23rd Street, Suite 909
New York, New York 10010

Distributed Worldwide by:

VCH Verlagsgesellschaft mbH
P.O. Box 1260/1280
D-6940 Weinheim
Federal Republic of Germany

Contributors

Anthony A. Atchley Department of Physics, Naval Postgraduate School, Monterey, California 93943

Philip Boudjouk Department of Chemistry, North Dakota State University, Fargo, North Dakota 58105

Lawrence A. Crum Physical Acoustic Research Group, Department of Physics and Astronomy, University of Mississippi, University, Mississippi 38677

Leon A. Frizzell Bioacoustics Research Laboratory, Electrical and Computer Engineering Department, University of Illinois, Urbana, Illinois 61801

James A. Rooney Jet Propulsion Laboratory, California Institute of Technology, Pasadena, California 91109

C. M. Sehgal Biodynamics Research Unit, Department of Physiology and Biophysics, Mayo Clinic/Foundation, Rochester, Minnesota 55905

Andrew Shoh Ultrasonic Research, Ridgefield, Connecticut 06877

Kenneth S. Suslick School of Chemical Sciences, University of Illinois at Urbana-Champaign, Urbana, Illinois 61801

G. R. ter Haar Physics Department, Institute of Cancer Research, Royal Cancer Hospital, Sutton, Surrey, SM 25PX United Kingdom

R. E. Verrall Department of Chemistry, University of Saskatchewan, Saskatoon, Saskatchewan, S7N 0W0 Canada

Preface

Heard melodies are sweet, but those unheard
  Are even sweeter

Ode on a Grecian Urn
John Keats, 1819

In 1894, the destroyer *H. M. S. Daring* failed to meet specifications: its speed and efficiency were inexplicably low. Sir John Thornycroft and Sidney Barnaby observed severe vibration and excessive slippage of the ship's screw propeller. After four sets of blades had been replaced, a solution to the problem was found by simply increasing the surface area of the propeller and decreasing its angular velocity. Their description of the observation of associated bubble formation on the moving propeller was the first report of the phenomenon known as cavitation,[1] which occurs both during turbulent flow and during ultrasonic irradiation of liquids. In attempting to explain such observations, Lord Rayleigh[2] described in 1917 the first mathematical model for the collapse of cavities in incompressible liquids and predicted enormous local temperatures (10,000K) and pressures (10,000 atm) during such collapse. The chemical[3] and biological[4] effects of ultrasound were first reported by Loomis 10 years later.

With this historical backdrop, the origins of the impact that ultrasound has had on modern science and technology are clear. The industrial uses of ultrasound in chemical and materials processing is substantial. Its use in clinic medicine is commonplace. Its application to cleaning of virtually anything has become quite conventional. In spite of this, there has been almost a complete lack of review materials on the underlying principles from which such effects originate. El' piner's monograph[5] was one of the first to attempt such a complete coverage. It is our hope that this volume serves the wide breadth of fields that have a substantial interest in ultrasonic phenomena.

This book is intended for chemists of all kinds, for materials scientists and engineers, for acoustical physicists and engineers, and for biologists and medical researchers who make use of ultrasound in their own work. The range of backgrounds of the ten authors who wrote these chapters is as diverse as the intended readership. Even so, there is a continuity in the underlying physical phenomena that ultrasound can create. The cross-fertilization which this volume hopes to provide should be fruitful to each of the areas of specialization represented herein.

Early in the study of acoustics, Minnaert observed that scientists

> have hardly ever investigated the sounds of running water. As a matter of fact we know very little about the murmur of the brook, the roar of the cataract, or the humming of the sea.[6]

Nonetheless, the authors of this volume have all contributed to a much fuller understanding of the effects of ultrasound on physical, chemical, and biological systems. Their efforts, cooperation, and tolerance have been greatly appreciated. The National Institute of Health and the Sloan Research Foundation are gratefully acknowledge for fellowship sponsorship, and the National Science Foundation for its generous research support.

Finally, no work of this size occurs without consequences in one's other endeavors. For her patience and kindness, I thank Adele most of all.

Kenneth S. Suslick
Urbana, Illinois

REFERENCES

1. Thornycroft, J.; Barnaby, S. W. *Inst C. Eng.* **1895**, *122*, 51.
2. Lord Raleigh, *Phil. Mag., Ser. 6* **1917**, *34*, 94.
3. Richards, W. T.; Loomis, A. L. *J. Am. Chem. Soc.* **1927**, *49*, 3086.
4. Wood, R. W.; Loomis, A. L. *Phil. Mag., Ser. 7* **1927**, *4*, 417.
5. El'piner, I. E. "Ultrasound: Physical, Chemical, and Biological Effects" Sinclair, F. L., trans.; Consultants Bureau: New York, 1964.
6. Minnaert, M. *Phil. Mag., Ser. 7* **1933**, *16*, 235.

Contents

1

Acoustic Cavitation and Bubble Dynamics

ANTHONY A. ATCHLEY AND LAWRENCE A. CRUM

Physical Acoustics Research Group,
Department of Physics and Astronomy,
University of Mississippi

INTRODUCTION

Statement of Purpose and Definition of Terms

"Cavitation" refers to the formation and the subsequent dynamic life of bubbles in liquids. These bubbles can be either gas or vapor filled and form in a wide variety of liquids under a wide range of conditions. Cavitation occurs in water, organic solvents, biological fluids, liquid helium, and molten metals, as well as many other fluids. It may be hydrodynamic, thermal, or acoustic in origin. The emphasis of this chapter is on acoustic cavitation. The section dealing with nucleation, however, may be applied to cavitation of different origins. It is traditional when describing cavitation to classify it as one of two types—transient or stable. Transient cavitation involves large-scale variations in the bubble's size (relative to its equilibrium size) over a time scale of a few acoustic cycles, and this rapid growth usually terminates in a collapse of varying degrees of violence. Stable cavitation, on the other hand, usually involves small-amplitude (compared to the bubble radius) oscillations about an equilibrium radius. Stable cavitation in most instances results in little appreciable bubble growth over a time scale of thousands of acoustic cycles. This classification of cavitation is not strict, however. Stable cavitation can lead to transient cavitation, and the collapse of a transient cavity can produce smaller bubbles that undergo stable cavitation.

The topic of this chapter, acoustic cavitation, is but one of several possible mechanisms through which ultrasound can interact with a liquid medium. Some of the other mechanisms are the topics of other chapters of this book. Acoustic cavitation can affect a liquid through two possible avenues. The first is the bubble itself. The liquid is disrupted by the inhomogeneous presence of the bubble. The second avenue through which cavitation affects a fluid is bubble dynamics. The bubble's interior and the liquid immediately surrounding the bubble are regions that undergo continual change. The bubble–liquid interface continually changes shape and size; liquid molecules diffuse into and out of the bubble; the concentration of gas in the surrounding liquid varies; acoustic streaming occurs in the liquid in the vicinity of the bubble, often resulting in severe shear stresses; the interior pressure and temperature fluctuate rapidly; the bubble radiates acoustic energy as it oscillates; thermal and viscous damping hinder the bubble oscillations. Each of these processes manifests itself differently, yet they all lead to changes in the properties of the liquid, although some of the changes are relatively small. Therefore, it is wholly appropriate to begin any study of the chemical, physical, and biological effects of ultrasound with a general analysis of acoustic cavitation.

The goal of this chapter is to introduce the broad field of acoustic cavitation. Although considerable information is contained in this chapter, it is by no means comprehensive in either breadth or depth. We refer the reader to numerous books, reviews, and articles for more in-depth discussions.

We begin the chapter with a brief historical review of the field of cavitation. The field has a long and colorful history and it has figured prominently in such diverse areas as ship design and elementary particle physics.

The consideration of the physical aspects of cavitation begins with a discussion of the origin of all cavitation processes: the nucleation phase. In terms of liquid properties, cavitation may result from either homogeneous or heterogeneous nucleation. "Homogeneous nucleation" refers to the formation of a bubble in a liquid free of any inhomogeneities such as solid particulates or organic micelles. "Heterogeneous nucleation" refers to the growth of a preexisting gas pocket or microbubble into a macroscopically observable bubble. Models for both kinds of nucleation are discussed and the predictions of these models are compared to experimental results. We also introduce the concept of a cavitation threshold or the minimum pressure amplitude necessary for the onset of either transient or stable cavitation.

Following this discussion, we digress somewhat to consider experimental techniques commonly used in transient cavitation research. This section begins with a brief overview of the basic components of any adequate experimental apparatus. The discussion is intended to inform the researcher of the various aspects of experimental work to which consideration should be given rather than to serve as a "complete guide" to experimental techniques (which it certainly is not). The section ends with a description of a versatile experimental apparatus that we have developed in our laboratory.

In the next section, the concept of rectified diffusion is discussed. If a bubble is caused to pulsate by an acoustic field, gas will diffuse in and out of the bubble during each cycle. There exists a threshold for which more gas diffuses in than out over an acoustic cycle and the bubble is caused to grow as a result of the sound field. This concept is called rectified diffusion and the words are an adequate description of the phenomenon. It is an important concept because it is the commonly observed phenomenon associated with intense sound fields in liquids. Rectified diffusion is a slow process, however, requiring thousands of cycles; if the sound field is sufficiently intense it plays a minor, secondary role to the rapid growth and collapse of a cavity during a single cycle, which is termed transient cavitation. Because of rectified diffusion, however, bubbles are dynamic systems, changing in size and so showing a variety of nonlinear responses to the action of the sound field.

Concluding the chapter, we discuss bubble dynamics. We cover the equations governing stable bubble oscillations and show how they lead to nonlinear effects, noise generation, and chaos. We also discuss the dynamics of a transient cavity, placing emphasis on the collapse phase. Of particular importance to the discussion of transient cavitation are the temperatures and pressures developed within the cavity during collapse. The final part of this chapter is devoted to posing questions about the important topics in cavitation research at the present. The questions not only summarize the material presented in the chapter but also serve to bring the reader up to data with current research activities.

Historical Perspective

Perhaps the first type of cavitation observed was the formation of bubbles in liquids supersaturated with gas. This form of cavitation is familiar to all of those who have ever opened or poured a carbonated beverage. Tomlinson[1] discussed a series of 12 experiments performed with sodawater and various solids. He contended that if the solids were "chemically clean," no bubbles would form on them. He went on to point out that if a solid, initially chemically clean, came into contact with a dirty cloth, unclean air, or dust particles, copious bubbles formed on it when it was immersed in the sodawater. Tomlinson concluded that chemically clean solids are perfectly wetted, whereas dirty solids are not. From this observation he argued that if a solid were dirty it would have a stronger adhesion force for the gas than for the liquid and that it was this asymmetry that caused bubble formation. Although this line of reasoning is incorrect, the results of his extensive experimentation with effervescence support the explanation offered by his contemporary, M. Gernez.[2] Gernez proposed that the outgassing of supersaturated gaseous solutions was caused by gas pockets embedded in the fissures found on the surface of every solid, regardless of its degree of finish. (It is therefore Gernez who is the initial proponent of the crevice model of the cavitation nucleus, a model to be discussed extensively here.) Effervescence is a form of cavitation that requires gaseous diffusion to "pump up" gas-containing nuclei until they grow large enough to rise to the surface of the liquid.

At about the same time that Tomlinson and Gernez were doing their work, another genre of cavitation was being investigated. This other kind of cavitation has its origin with the conception and development of hydrodynamics. Not until Daniel Bernoulli[3] set down the guidelines for this new branch of science in 1738 was it understood that a negative pressure could be produced in a liquid. In the following years Euler and d'Alembert debated over the consequences of negative pressures.[4] Euler believed that the application of a negative pressure could result in a rupture of the liquid, while d'Alembert refused to accept this view.

The rise of cavitation as a topic for scientific research began with the development of high-powered, high-rpm steam turbines in the mid 1800s. With this advance came the means of moving an object (such as a propeller) through a fluid rapidly enough so that the object lost contact with the fluid. In fact, the term "cavitation" was first used by Sir John Thornycroft in his observations of propeller inefficiency in torpedo boat destroyers.[5] The most obvious concern to which this new phenomenon drew attention was that the state of the art propeller design was poor. A second and, as was soon evident, much greater concern was propeller erosion. In fact, this problem was so critical that in 1915 the British Admiralty appointed a special subcommittee to investigate this phenomenon. In 1917, Lord Rayleigh[6] became involved. His typically straightforward and ingenious solution of the equations governing vapor cavity dynamics is still quite useful today.

Concurrent to this work the first attempts to measure the tensile strengths of liquids were being undertaken. In about 1850, Berthelot began to measure the tensile strength of water and found it to be on the order of 50 bar (maximum).[7] (Note that in this chapter we shall use the traditional unit of pressure, the bar, which is approximately 1 atm. In SI units, 1.0 bar = 0.1 MPa. Further, since it has been traditional to use cgs units for the other quantities as well, we shall do so in this chapter.) His method consisted of heating a liquid, which was sealed in an evacuated glass capillary tube, until the expanding liquid completely filled the tube. The temperature at which the tube was filled was recorded and the liquid was then allowed to cool. Because the liquid adhered to the sides of the tube, this subsequent contraction resulted in the liquid undergoing a tensile stress, eventually ending in the rupture of the liquid. The temperature at which rupture occurred was also recorded. From a knowledge of the filling and rupturing temperatures, along with the thermal expansion coefficient for the liquid being tested, Berthelot was able to calculate a tensile strength. Many other measurements of the tensile strength of water through these static means have been made over the intervening 130-odd years, and their results (ranging from as low as about 5 to as high as about 200 bar) are comparable to Berthelot's. A detailed discussion of these works is not given here. The point is that for the first time investigators were intentionally putting liquids under stress and considering the consequences. The application of a tensile stress to a liquid is the basis for present-day acoustic cavitation studies, which are actually studies of the "dynamic" tensile strength, as opposed to these earlier "static" measurements.

Acoustic cavitation was first observed during the period from 1915 to 1920 by Langevin and his co-workers while he was pioneering the field of ultrasonics.[8] Söllner[7] was one of the first to observe cavitation in degassed liquids at room temperature and atmospheric pressures (1936). In the 1930s and 1940s a number of other researchers investigated various aspects of ultrasonic cavitation, among them Boyce,[7] Harvey,[7] and Kornfeld and Suvorov.[9]

The fundamental problem of cavitation was by now well formulated. Numerous measurements of the cavitation threshold (tensile strength) of water yielded results in the range of 1–25 bar (using the most careful and painstaking procedures thresholds of up to 300 bar have been obtained,[10] but they are unusual) and yet, theoretical predictions of the homogeneous (pure liquid) threshold of water are thousands of bar.[6] What accounts for this order of magnitude (best case) discrepancy? Failure in solids under tension is usually attributed to an imperfection in the solid. The same logic (i.e. imperfections in the liquid) was applied to the failure of liquids.

In the remainder of this section different types of cavitation are mentioned in an overview of the field in the last half century. The investigators of these branches of cavitation research sought these imperfections (cavitation nuclei) to explain the various aspects of the particular cavitation process. Cavitation nucleated by these inhomogeneities is referred to as heterogeneous cavitation.

In 1944, Dean[11] suggested that vapor cavities form in the center of small vortices produced by turbulent motion of water around solid objects. In the same year, Harvey[12] suggested that gas pockets stabilized at the bottom of crevices found on dirt particles accounted for the cavitation found in animals. He had concerned himself mainly with cavitation formed by compression–decompression processes in animals (the "bends"); however, he did discuss his model's application to acoustic cavitation. One of the most careful and extensive experiments of the time was performed by Briggs, Johnson, and Mason.[13] They measured cavitation thresholds for a variety of liquids as a function of dissolved gas pressure, viscosity, and pulse length. However, their theoretical explanation of their results was later shown to be incorrect.[7]

The 1950s was a decade of diversification for cavitation research. Perhaps the most important example of this breadth is the invention of the bubble chamber by Glaser. This invention gave rise to the field of radiation-induced cavitation. Seitz[14] gave an explanation for bubble formation in superheated liquids by charged particles. He concluded that the majority of bubbles were "nucleated by moderately-energetic free electrons produced by the incident particles in Coulomb encounters." The bubble chamber was developed as a tool for elementary particle physics, whereas radiation-induced cavitation research took a different avenue. Lieberman[15] measured the threshold for acoustic cavitation in pentane and acetone exposed to neutron and beta sources. He concluded that the cavitation was nucleated by the recoil of carbon ions; these results agreed well with Seitz's theory.

In the field of acoustically induced cavitation, a different type of nucleus was suggested in 1954 by Fox and Herzfeld.[16] This nucleus consisted of a gas bubble surrounded by a rigid skin of organic molecules. This organic skin model is discussed further in the next section. An extensive set of measurements was reported by Strasberg.[17] He analyzed his results in terms of the nucleation models proposed up to that time and found the crevice model offered the best explanation. Other noteworthy papers of the 1950s are by Connolly and Fox[18] and Galloway[19]; some of their results are discussed later.

Sette and Wanderlingh[20] were the first to study acoustic cavitation induced in water by cosmic rays. They concluded that the nuclei were recoiling oxygen atoms and their results agreed well with calculations based on Seitz's theory. The Soviet researcher Akulichev[21] proposed another cavitation nucleus based on his measurements of the influence of dissolved ions on cavitation thresholds. His nucleus, similar to that of Fox and Herzfeld, consisted of an ionic skin surrounding a gas bubble. Greenspan and Tschiegg[22] demonstrated that if a liquid were extensively and carefully filtered to remove the nuclei, it could momentarily sustain extremely high tensile stresses. They also made a valuable contribution to cavitation research by developing a method by which they were able to attain unusually high cavitation thresholds. This method consisted of high-Q (quality factor) resonance chambers that were driven by small tuned ceramic transducers.

In 1970, Apfel[23] developed the crevice model further to include the size of the crevice; he suggested that there existed a "critical" size for the crevice. Crevices larger than and smaller than this critical size exhibit quite different behaviors. Apfel published papers on both homogeneous and heterogeneous cavitation. In 1971,[24] he modified a theory for homogeneous nucleation of vapor cavities at the interface between a flat solid and a pure liquid to account for nucleation at the interface between two pure liquids. Apfel's crevice model was modified further by Crum,[25] who considered the variation of contact angles with surface tension. The crevice model in this latest form was successful in explaining a wide variety of data obtained from acoustic cavitation measurements. Winterton[26] also modified Apfel's model and used it to explain boiling and nonacoustic cavitation.

At about the same time Yount was developing a different type of model for the cavitation nucleus.[27] Modifying the organic skin models of Fox and Herzfeld, and Sirotyuk,[28] Yount considered his nucleus to be a bubble surrounded by polar surfactant molecules. This skin has the property of offering a varying permeability to gas, depending upon the ambient conditions. He applied his model to compression–decompression processes and was successful in explaining his data. Yount[29] has developed his model further theoretically and experimentally. Ward et al.[30,31] have done considerable work with the thermodynamics of heterogeneous nucleation in supersaturated systems of constant mass and volume. Among other things, they have predicted the rate of formation and stability of bubbles in a conical pit contained within a system of fixed volume. Their work has been applied to bubble formation in bone cells.

The crevice model has recently been reformulated by Atchley[32] and Atchley and Prosperetti.[33] This reformulation is based upon the mechanical stability of a gas pocket in a conical crevice and the predictions of this new version of the crevice model agree quite well with a broad set of measurements. The model has been applied both to acoustic cavitation and "diffusion" cavitation (i.e., bubble formation driven by gas diffusion, such as effervescence of a carbonated beverage).

Conspicuously absent from this historical review is any mention of the developments in the theory of bubble dynamics. Attention is given to this development in the theoretical portions of this chapter because it is fitting to associate the names of the founders of this field with their theoretical advances.

NUCLEATION AND TRANSIENT CAVITATION INCEPTION

In one sense, the generation of acoustic cavitation can be described as a measure of the "dynamic tensile strength" of a liquid. The "failure" of the liquid under stress is detected by the presence of a gas (or vapor) phase within the bulk of the fluid. The theoretical tensile strength of both liquids and solids is several orders of magnitude larger than the measured values, unless special circumstances occur. The discrepancy between the experimental and theoretical values is likely caused by inhomogeneities in the bulk of the material. These imperfections enable weaker molecular bonding to occur than would be present in a homogeneous material.

In a solid, the inhomogeneities take on the form of dislocations, grain boundaries, etc., and act as initial sites in crystalline plane slippage. If the imperfections are removed, as in the case of metallic whiskers, which are essentially single crystals, extremely large tensile strengths can be achieved. (In some cases, of course, imperfections in a solid actually increase the tensile strength.)

In a liquid, pockets of gas serve as the weak points that prevent a liquid from sustaining large tensile strengths. The presence of these sites greatly reduces the tensile strength of the liquid; for example, in water, the theoretical strength is nearly 1000 bar,[6] while the measured value, without protracted purification, is on the order of 1 bar.

We shall briefly discuss nucleation in a homogeneous liquid, and then consider in some detail the various models associated with cavitation inception from something foreign to the pure liquid, or heterogeneous nucleation.

Homogeneous Nucleation

In the bulk of a pure liquid at a given pressure and temperature, density fluctuations cause the rapid formation and destruction of microscopic vapor cavities.

The probability that one of these fluctuations can be made to grow to an observable size in a reasonable period of time can be calculated. The pioneering work in this area was performed by Döring[34,35]; subsequent work by Volmer,[36] Fisher[37] and Apfel[38] has redefined Döring's work and permitted us to specify the requirements for homogeneous nucleation with some confidence.

The general form of the transcendental expression that describes the pressure–temperature requirements for the onset of cavitation (here the irreversible growth of a vapor cavity) is given by Apfel.[38]

$$\frac{16\pi[\sigma(T)]^3}{3kT[P(T) - P_L]^2} = \ln C(T) \tag{1-1}$$

In this expression, T is the absolute temperature, k is Boltzmann's constant, $\sigma(T)$ is the liquid–vapor surface tension, and $[P(T) - P_L]$ is the pressure differential across the cavity interface, $P(T)$ being the pressure of the vapor on the interface and P_L the pressure in the liquid. The function $C(T)$ is discussed below.

The lefthand side of Eq. (1-1) represents the energy barrier to nucleation, W_m, divided by kT. W_m is found by calculating the reversible isothermal work to form a vapor cavity of arbitrary size, and then by calculating that value of the size that makes the work an extremum. A vapor cavity of such size, called a "critical" cavity, is in a state of unstable equilibrium and with a slight increase in temperature or decrease in pressure, will grow without bound (at least to observable size).

The righthand side of Eq. (1-1) contains the complex kinetics of the nucleation process through the expression $C(T)$. In order to express this precisely, it is necessary to know the evaporation, condensation, and accommodation coefficients, and considerable other information about the liquid state. Fortunately, approximate expressions for $\ln C(T)$ can be found that differ only by about 25% in a wide variety of liquids; thus, a reasonably good estimate for $\ln C(T)$ is 60.[38]

In order to perform measurements of homogeneous nucleation in a liquid, one must ensure that the liquid does not contain any foreign substances that may act as preferential nucleation sites. Early attempts at such measurements failed to purify the liquid sufficiently to remove such substances. Berthelot[4] was able to estimate the tensile strength of water to be in the range of 30–50 bar. Briggs[10] and Greenspan and Tschiegg[22] were able to raise that estimate to over 200 bar by systematic attempts to clean the liquid of all "motes" or solid particles. These values are still significantly less than the theoretical value in water of over 1000 bar, however.

Apfel[38] was able to design an experiment that produced a mote-free liquid by injecting small-volume droplets of one liquid into a second host liquid. Because the samples were so small ($<10^{-3}$ cm^3), the chance that a droplet could be produced containing no motes was quite large. Apfel was able to remove a second major obstacle in homogeneous nucleation experiments—the problem of producing extremely large acoustic pressure amplitudes—by raising the temperature of the host liquid to near or above the superheat limit of the sample

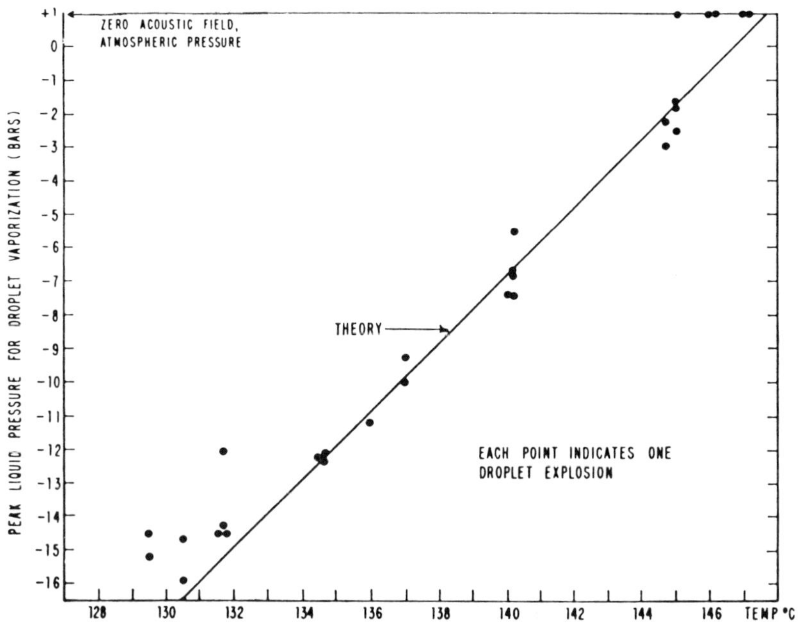

Figure 1-1. Liquid pressure at vaporization versus temperature for ether droplets in glycerine. (Reproduced from Ref. 38 with permission.)

droplet. By "trading" some temperature rise for pressure amplitude, he was able to reach the homogeneous nucleation threshold.

Shown in Fig. 1-1 are Apfel's results for diethyl ether immersed in glycerine. The symbols are experimental points indicating that homogeneous nucleation has occurred within the liquid and has triggered a droplet explosion; the theoretical curve is Eq. (1-1) with $\ln C(T) = 53$. The excellent agreement between the predicted and measured values indicate that homogeneous nucleation in liquids can occur and can be described rather well analytically.

Heterogeneous Nucleation

Since the cavitation thresholds measured for most liquids is significantly less than the homogeneous nucleation threshold, it is reasonable to assume that most liquids contain foreign substances that stabilize pockets of gas and thus weaken the liquid to stress failure. For the purposes of this communication, we shall label these weak spots "nuclei" and assume that they are associated with an inhomogeneity of some sort.

The obvious and simplest model for a nucleus is that of a free bubble. However, because of the "Laplace pressure" (i.e., an internal pressure caused by

Table 1-1. Dissolution Times for an Air Bubble Present in Water [a]

Radius (μm)	Dissolution time (s)
1000	6×10^6
100	6×10^2
10	6
1	1×10^{-2}

[a] Surface tension, 70 dyn cm^{-1}; temperature, 20°C; dissolved gas concentration equal to saturation.

surface tension and equal to $2\sigma/R$, where R is the bubble radius) these bubbles gradually dissolve as gas is forced out of the interior of the bubble into the liquid.

Epstein and Plesset[39] have examined theoretically the behavior of air bubbles present in liquids. The differential equation describing the change in equilibrium radius R_o with time for an unstirred liquid is given by:

$$\frac{dR_o}{dt} = \frac{Dd}{R_o}\left[\frac{C_i/C_o - 1}{1 + 4\sigma/3R_oP_\infty}\right]\{1 + R_o/(\pi Dt)^{1/2}\} \qquad (1\text{-}2)$$

where D is the diffusion constant, C_i/C_o is the ratio of the dissolved gas concentration to the saturation concentration, P_∞ is the ambient pressure in the liquid far from the bubble, and $d = R_gTC_o/P_\infty$, where R_g is the universal gas constant. As an example, consider Table 1-1, which describes the approximate lifetimes of air bubbles of various sizes present in water.

Thus, a liquid containing free gas bubbles that is not supersaturated will rid itself of these bubbles as the small ones dissolve and *the large ones rise* to the surface. If the liquid is supersaturated (the obvious example is that of a carbonated beverage), bubbles can grow from small stabilized gas pockets on the walls of the vessel to visible size in a short time. There is a threshold size, however, and nuclei below that size do not grow; they even dissolve if there is no mechanism for stabilization.

Figure 1-2 shows a stability curve for gas bubbles at various dissolved gas concentrations. This figure shows that bubbles of very small size (<5 μm) can be made to grow to visible size with only modest dissolved gas concentration ratios. (That's why bubbles appear out of nowhere in a champagne glass.) Note also that very small nuclei or very low supersaturations do not result in bubble growth. (That's why bubbles appear to originate from the same spot, and then stop appearing as the liquid goes flat.)

The important points to be made from the above discussion are (a) that ordinary liquids contain pockets of gas (i.e., nuclei) that can easily be activated, and (b) that these gas pockets are not free gas bubbles, but must have some mechanism to stabilize the gas pocket from dissolution. We next consider a few of the proposed stabilization mechanisms for cavitation nuclei.

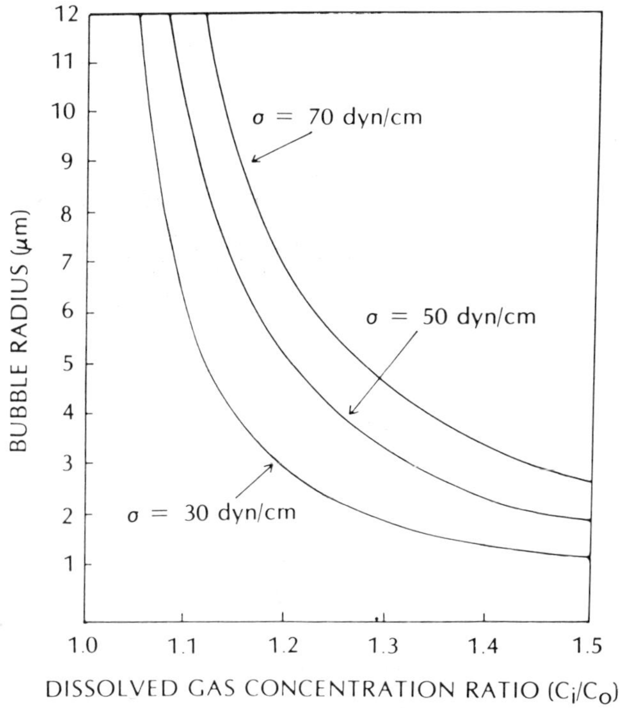

Figure 1-2. Stability curves for gas bubbles in water at various dissolved gas concentrations and surface tensions. If a gas bubble lies above the appropriate line, it will grow by diffusion at zero acoustic pressure amplitude; if it lies below, it will dissolve.

Organic Skins

Since organic molecules are abundant in our environment, and since many of these compounds are surface active, a number of investigators have proposed that these compounds form "skins" on the surfaces of gas bubbles, preventing them from dissolving.

The original suggestion that these organic molecules could stabilize cavitation nuclei was made by Fox and Herzfeld,[16] although it was suggested much earlier[40,41] that these skins influenced the hydrodynamics of gas bubbles rising in a fluid. In the model proposed by Fox and Herzfeld, the skin was rigid and stabilized the gas bubble by preventing gas diffusion. According to this model the cavitation threshold as a function of static pressure should show a discontinuity at the "skin crushing pressure." However, when Strasberg[17] made such measurements, no discontinuity was observed; instead, a gradual rise in threshold with applied static pressure was seen. Herzfeld[42] subsequently abandoned the hypothesis, but it has been taken up anew by others.

In a modified version of the organic skin hypothesis, Sirotyuk[28] examined the possible stabilization of "microbubbles" by a film of surface-active substances. Detergents and soaps are composed of a polar or charged group on one end and

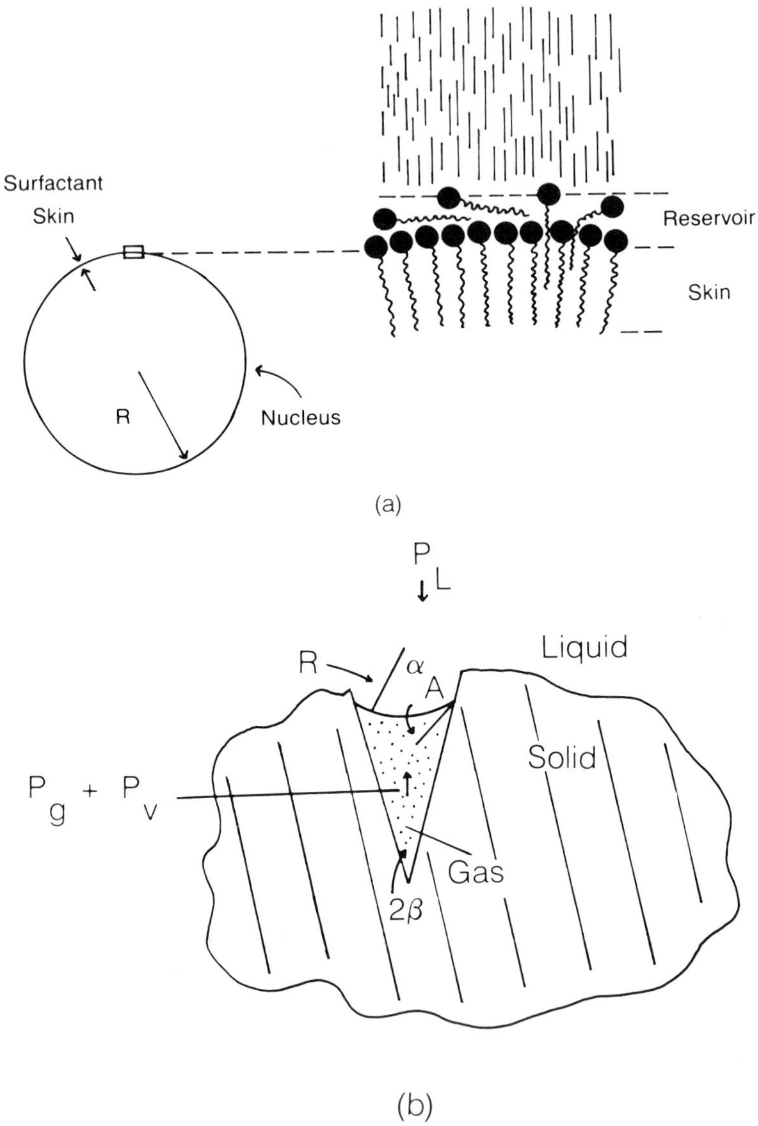

Figure 1-3. Two possible cavitation nuclei. (a) An organic skin nucleus, i.e., a bubble stabilized by a skin of polar surfactant molecules. The nucleus is surrounded by a thin layer of nonaligned molecules in a reservoir.[29] (b) The stabilization of a gas pocket in a crack in a hydrophobic solid.[25]

a long, nonpolar, hydrocarbon tail on the other (see Fig. 1-3a). The polar end of a gas bubble in water interacts strongly with the water surface and the tail extends into the gas. This "picket-fence" configuration has an inherent elasticity that may stabilize the microbubble. Sirotyuk has measured the acoustic cavitation threshold of water in a device designed to maintain a high degree of purity and contends that his experimental measurements support his claim that cavitation nuclei are microbubbles stabilized by surface-active agents. The experimental evidence is extremely weak, however, and subject to different interpretations.

A further modification of the organic skin hypothesis has been given by Yount et al.[27,29] In this model, a cavitation nucleus is formed in the following manner: a gas bubble is introduced (the actual mechanism is immaterial) into a liquid and begins to dissolve. As the dissolution is taking place, surface-active molecules accumulate on the gas–liquid interface. Eventually, the density of these molecules is sufficient to prevent the collapse of the bubble, presumably through a Coulomb interaction of some sort. Once the critical density is reached, the radius of the nucleus may make "large-scale" changes. The equilibrium condition for this nucleus is that the chemical potentials of the skin and the reservoir must be equal. The reservoir is a thin, perhaps monomolecular, layer of nonaligned surfactant molecules that surrounds the skin of aligned molecules. This equilibrium condition can be satisified by the accretion or deletion of molecules from the skin (or reservoir), resulting in large-scale changes in the radius of the nucleus. Furthermore, subsequent "small-scale" changes in radius can occur. These changes involve only the adjustment of the separation of adjacent skin molecules, not a change in the number of molecules.

This model has been applied primarily to compression–decompression studies and in that realm is assumed to work as follows: for modest values of overpressure, the skin remains permeable to gas diffusion (although one would suspect diffusion to be retarded). Thus, the gas within the bubble remains in equilibrium with that in the fluid. At some critical pressure, the skin molecules become so tightly packed that the skin becomes essentially impermeable to diffusion; now no gas can escape the bubble, and it survives compression. When decompression occurs, the bubble can expand because the surfactant molecules can withstand mechanical compression but not tension.

Yount's so called "varying-permeability" model has been used to explain bubble formation in gelatin exposed to a compression–decompression sequence. It has not yet been applied to acoustic cavitation, although Yount[29] calculates the characteristic times required for surfactant molecules to react to changes in applied pressure and bubble radius and concludes that the molecules can play a role in acoustic cavitation.

Hydrophobic Ions

It has been known for some time that air bubbles in water possess some excess electric charge,[43,44] and so it seems reasonable to assume that the repulsive nature of like electrical charges on the surface of an air bubble can stabilize it from dissolution. Akulichev[21] has examined this model in some detail and has hypothesized that certain ions, such as Cl^- and F^-, are hydrophobic and migrate to the surface of an air bubble in water. Other ions, such as OH^-, are hydrophilic and do not migrate to the surface. Accordingly, one would expect there to be a dependence of the cavitation threshold on the concentration of such salts

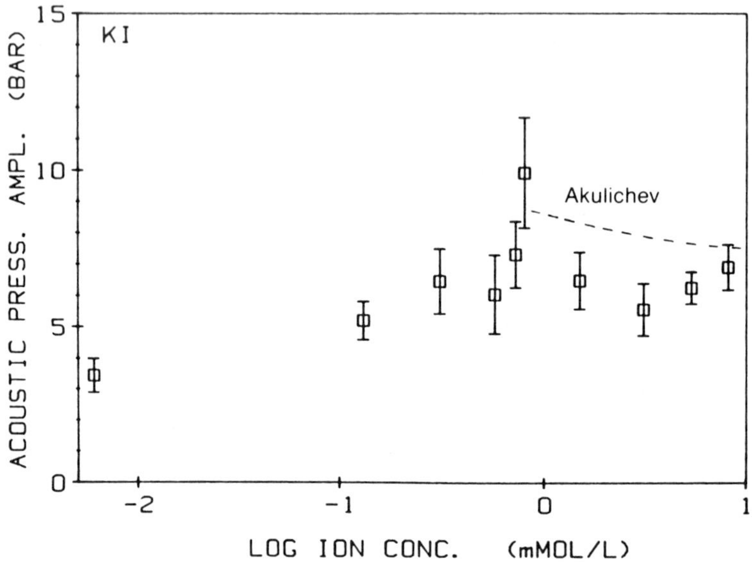

Figure 1-4. Cavitation threshold versus dissolved ion concentration of KI in water with a gas tension of 0.30 bar. The data points are Atchley's.[32] The dashed line indicates the trend of Akulichev's[21] measurements made in water with gas a tension of 0.13 bar.

as NaCl or NaF, but not of NaOH. Akulichev was able to make measurements of the cavitation threshold of water as a function of the concentration of a number of dissolved salts. His measurements indicate that if the dissolved gas concentration is reduced to below about 40% of saturation, the acoustic cavitation threshold of water is measurably reduced upon addition of the hydrophobic salts NaCl and NaF, but not affected at all by the addition of the hydrophilic salt NaOH.

Atchley[32] has repeated the measurements of Akulichev and also has extended the range of the dissolved ion concentration by two orders of magnitude. In this experiment, he first removed most of the dissolved chemical species typically present in distilled water with a series of filters that reduced the dissolved organic content to parts per million and increased the resistivity (reduced the number of dissolved ions) to 1 MΩ cm. Thus, he was able to measure the cavitation threshold as a function of dissolved ion concentration over four orders of magnitude. His measurements for one particular salt, KI, are shown in Fig. 1-4; also shown on the figure are the measurements of Akulichev (dashed line). A true comparison between Atchley's and Akulichev's measurements cannot be made because the dissolved gas concentration ratio used by Akulichev was 0.13, whereas that of Atchley was 0.30. With this caveat, however, it is seen that the two measurements are consistent.

Atchley's data, however, show an additional feature: if the water is initially cleaned of most of the dissolved ions, the cavitation threshold *increases* with addition of hydrophobic ions to a relative maximum and then subsequently decreases in a manner similar to that discovered earlier by Akulichev. Note that the relative increase in the threshold is nearly a factor of 3. These data indicate quite clearly that the dissolved ion concentration is an important aspect of the stabilization process.

Crevices in Solids

One of the earliest models proposed for microbubble stabilization was that attributed to Harvey et al.,[12] in which gas is trapped in a conical crevice in a solid inhomogeneity present in the liquid. (It appears that Gernez considerably preceded Harvey in the suggestion of crevices in solids as possible nucleation sites, but Harvey's extensive studies in this area have resulted in this model often being referred to as the Harvey model.) This model has received considerable attention because of its plausibility and its success in explaining available data, and it has been examined by many researchers, including Strasberg,[17] Apfel,[23] Winterton,[26] Crum,[25] and most recently Atchley[32] and Atchley and Prosperetti;[33] the latter analysis is used here to describe this model in some detail.

If one has ever observed the boiling of water in a pot, or the effervescence of bubbles from a carbonated beverage contained in a glass, it is apparent that there are nucleation sites on the walls from which the bubbles appear to form (and reform). Apparently, the cracks and crevices of the walls are imperfectly wetted by the liquid and so contain gas pockets that act as the sites of bubble growth.

Liquids such as water are known to contain numerous small solids (Apfel[45] estimates this number to be as high as 100,000/cm^3) that remain in suspension via Brownian motion. As mentioned above, the crevice model assumes that gas pockets are stabilized at the bottom of cracks or crevices found on hydrophobic solid impurities present in the liquid. For simplicity these crevices are assumed to be conical. The essential features of the model are depicted in Fig. 1-3b. Whereas skin-type nucleation models require that the effect of surface tension be neutralized for the stability of the nucleus, the crevice model uses the effect of surface tension for stabilization. In order to understand the stabilization mechanism of the crevice model, one must first understand the relationships of the various pressures involved. Consider a system consisting of a closed container partially filled with a liquid, as illustrated in Fig. 1-5. The space above the liquid is filled with the liquid vapor and a gas. The total pressure in this space is equal to the partial pressure of the gas plus the vapor pressure. That is, the total ambient pressure is $P_G + P_v$. Assuming that the gas is soluble in the liquid, the equilibrium concentration of the gas in the liquid is found by Henry's law:

$$C = K(T)P_G \tag{1-3}$$

where C is the concentration and $K(T)$ is a function of temperature only. Through Henry's law we can establish the concept of an effective gas tension, G, which is the gas pressure required for the liquid to have a given gas concentration at a given temperature.

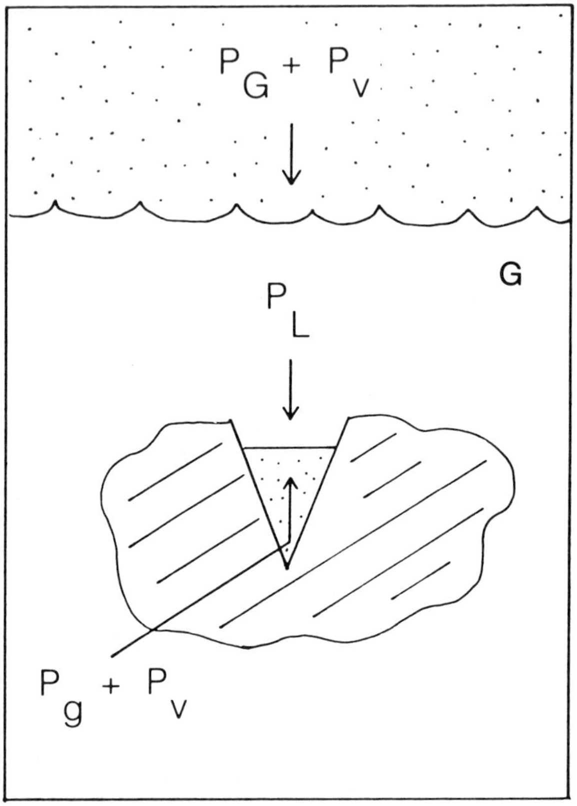

Figure 1-5. A closed system illustrating the identification of the various pressures involved in cavitation nucleation. The system consists of a gas- and vapor-filled space above a liquid in which exists a hydrophobic solid containing a gas- and vapor-filled crevice.

Within the liquid there exists a solid impurity that contains a gas- and vapor-filled crevice. We call the gas pocket a cavitation nucleus. Diffusion of gas across the interface of this nucleus maintains the partial pressure of the gas inside the nucleus, P_g, equal to the gas tension of the liquid in the neighborhood of the interface. If the entire system is in equilibrium, then $G = P_G$ and $G = P_g$, so $P_G = P_g$. The pressure within the nucleus is $P_g + P_v$. The pressure balance equation for the nucleus is:

$$P_L + \frac{2\sigma}{R} = P_g + P_v \tag{1-4}$$

where P_L is the total pressure the liquid exerts on the interface. If hydrostatic effects are neglected, then $P_L = P_G + P_v$. Because at equilibrium $P_G = P_g$, we have $P_L = P_g + P_v$. This requires that the Laplace pressure $(2\sigma/R)$ be zero or, in other words, $R = \infty$. At equilibrium, therefore, and when the liquid is saturated, the interface is flat.

Now assume that the liquid is undersaturated with respect to the "atmosphere"; that is, $G < P_G$. Since diffusion maintains the equality of the gas tension and the gas pressure inside the nucleus, we have $P_L > P_g + P_v$. In order to satisfy Eq. (1-4), the Laplace pressure must act so as to oppose P_L. This condition requires that the interface be bowed toward the apex. When equilibrium is established, the Laplace pressure exactly equals $P_L - P_g - P_v$ and the nucleus is stabilized against dissolution.

Suppose we have a degassed (undersaturated) liquid and the liquid pressure is increased. In response to this condition, the interface bows inward more, until it reaches the "advancing contact angle." At this point any subsequent motion of the interface involves the entire interface advancing toward the apex. As it advances, the radius of curvature necessarily becomes smaller, since the angle that the interface makes with the crevice wall is now fixed at the advancing contact angle. As the radius of curvature decreases, the Laplace pressure increases. Eventually equilibrium is reestablished with the gas pocket shrunk deeper into the crevice. It is readily seen that in small crevices (i.e., ones having small radii of curvature) significant levels of applied hydrostatic pressure can be withstood without destroying the nucleus. Indeed, Hayward[46] has shown that the gas pockets stabilized in solid, hydrophobic particles added to water can sustain applied pressures of 140 bar before they are deactivated.

Having established the stabilization mechanism for a gas pocket in a crevice (a Harvey nucleus), we turn our attention to the process by which such a nucleus forms a free bubble in a liquid, in other words, the nucleation process.

Assume there exists a stabilized Harvey nucleus in a degassed liquid. At this point, an acoustic field of pressure amplitude P_A is applied such that the liquid pressure exerted on the interface varies between $P_o \pm P_A$. Assume P_A is such that $P_v > P_o - P_A$. During the negative half-cycle, the interface bows outward and the nucleus attains a maximum volume. This process is isothermal if the thermal diffusion length is large compared to crevice dimensions, and this condition is normally met.

A bubble can be nucleated from this new position only if two nucleation criteria are met. The first is that the angle of contact between the interface and the crevice wall reach the receding contact angle. The second criterion is that the radius of curvature of the interface exceed a critical value R_c. This criterion, first imposed by Atchley and Prosperetti,[33] ensures that the nucleus is mechanically unstable and can grow in an unbounded manner. This point is crucial to any nucleation theory and is deserving of a brief digression.

Consider a gas-filled bubble of radius R surrounded by a liquid that exerts a pressure P_L on the bubble. Assume the bubble is in equilibrium so that the pressure inside equals the liquid pressure plus the Laplace pressure:

$$P_v + P_g = P_L + \frac{2\sigma}{R} \tag{1-5}$$

If the gas is ideal then rearranging Eq. (1-5) gives:

$$P_L + \frac{2\sigma}{R} = \frac{3nR_gT}{4\pi R^3} + P_v \tag{1-6}$$

where n is the number of moles of the gas inside the bubble, R_g is the universal gas constant, and T is the absolute temperature. To examine the stability of this system, consider the effect of a small increase in the bubble radius, R. For the bubble to be stable, the pressure that tends to collapse it, which is given by the lefthand side of Eq. (1-6), must increase more than the pressure that tends to make it grow, given by the righthand side of Eq. (1-6). This will always be true if $P_L \geq P_v$, because the gas pressure varies as R^{-3}, while the Laplace pressure varies as R^{-1}. If $P_L < P_v$, however, then the situation is different.

The stability criterion can be expressed, in general, as

$$\partial/\partial R(P_L + 2\sigma/R) > \partial/\partial R(3nR_gT/4\pi R^3 + P_v) \tag{1-7}$$

Performing the differentiation and solving for R gives:

$$R < 4\sigma/3(P_v - P_L) \tag{1-8}$$

In other words, when $P_L < P_v$, the bubble is stable against spontaneous growth only if R is less than the critical radius, given by:

$$R_c = 4\sigma/3(P_v - P_L). \tag{1-9}$$

Taking another view, a bubble can grow in a spontaneous unbounded manner when the liquid pressure is less than the vapor pressure only if $R \geq R_c$. A bubble that grows in this way will be mostly vapor filled.

It can be shown that the critical radius for any nucleus whose volume can be written as fR^3, where f is a constant, is given by Eq. (1-9). This case does not always hold for a crevice nucleus. However, the stability criterion, Eq. (1-7), can be written in more general form as

$$\frac{\partial(\text{external pressure})}{\partial V(R)} \geq \frac{\partial(\text{internal pressure})}{\partial V(R)}, \tag{1-10}$$

or

$$\frac{\partial(P_L + 2\sigma/R)}{\partial V(R)} \geq \frac{\partial(nR_gT/V(R))}{\partial V(R)}. \tag{1-11}$$

The derivative is taken with respect to the volume, rather than the radius of curvature, because instability results in a volume increase. In crevice geometry, the volume is not always directly proprotional to (some power of) the radius of curvature. It is easily verified that Eq. (1-11) yields Eq. (1-9) in the case of a spherical bubble.

Because of the complex geometry of the crevice model, threshold equations can be unwieldy and so they will not be presented in this discussion. However, the general method of determining the threshold is straightforward. Equilibrium values of P_L are determined from the pressure balance equation,

$$P_L + 2\sigma/R = P_v + nR_gT/V(R), \tag{1-12}$$

for values of the volume ranging from that in the initial, stabilized state to the volume when the interface is hemispherical at the crevice mouth. The stability of these equilibrium solutions is then checked by applying Eq. (1-11). The cavitation threshold is the lowest value of the liquid (or highest value of the acoustic) pressure which satisfies both nucleation criteria.

Apfel[23] defines crevices of larger than or smaller than critical size. A crevice is larger than critical size if the angle of contact of the interface in a degassed liquid, α_i, is equal to the advancing contact angle. A crevice is smaller than critical size if it is not. Crevices larger and smaller than critical size behave differently in terms of cavitation threshold. We return to this point later.

Concept of Threshold

The investigation of cavitation nucleation mechanisms historically is linked with measurements of the transient cavitation threshold. The process through which a nucleus produces a detectable bubble can be presumed to occur as follows: A stabilized nucleus is exposed to an acoustic field. If the amplitude of the field is sufficiently low, the nucleus undergoes no appreciable change in size. When the field reaches a sufficient amplitude the nucleus becomes unstable and rapidly (within a few acoustic cycles) grows into a mostly vapor-filled bubble. This vaporous cavity eventually reaches a size sufficiently large to be detected. (Methods of detection are discussed later.) In theory, the cavitation threshold is the value of the acoustic pressure amplitude necessary for the nucleus to become unstable. In practice, the cavitation threshold is the value of the acoustic pressure amplitude to which the liquid is exposed when an "event" is detected. Ideally the two thresholds are the same.

By measuring the cavitation threshold as a function of various liquid properties, information can be gained about the cavitation nucleus. In this section we briefly summarize the results of a number of cavitation threshold measurements on water. The summary is the compilation of the results of reported research.[16–19,21–25,32]

Dissolved Gas Tension

Liquids saturated with gas have transient cavitation thresholds near 1 or 2 bar provided that they have not been "cleaned" in any appreciable way. The type of cavitation that is detected in saturated (or nearly saturated) liquids is not the "crisp" transient cavitation expected when vaporous cavities collapse. "Crisp" cavitation is that in which a clearly distinguishable, audible snap or pop is heard, indicating the violent collapse of an individual cavity. In contrast the cavitation observed in a saturated liquid indicates that the bubbles contain an appreciable amount of gas that cushions the collapse. In this case, cavitation is associated with the presence of numerous gas bubbles that gradually appear in the bulk of the liquid. Often "streamers" or relatively long lines of bubbles are present. If any sound is heard, it will be a continual hissing or frying noise. As the gas tension is reduced from saturation to about 25% of saturation, the threshold increases almost linearly to about 7 bar and the cavitation becomes crisper. For gas tensions below about 25%, the threshold increases further.

Hydrostatic Pressure

The cavitation threshold is directly proportional to the hydrostatic pressure applied to the liquid. The threshold of a liquid can be increased permanently if the hydrostatic pressure is maintained long enough for gas to diffuse out of the nucleus.

Surface Tension

The cavitation threshold varies inversely with surface tension. This behavior is somewhat suprising because the Laplace pressure varies directly with surface tension.

Temperature

The cavitation threshold varies inversely with temperature. The decrease in threshold with increasing temperature is roughly linear except for liquids near the boiling point. Near the boiling point the threshold drops to zero.

Solid Contaminants

The cavitation threshold increases with decreasing number and size of solid contaminants (dirt).

Dissolved Ion Concentration

The behavior of the cavitation threshold as a function of dissolved ion concentration is complicated. However, in general as the concentration increases, the threshold, relative to extremely low concentrations, also increases.

Comparison with Theoretical Predictions

Of the nucleation models discussed previously, the crevice model has been the one most extensively, and most successfully, applied to acoustic cavitation. In this section we compare the predictions of the crevice model to measurements of the cavitation threshold of degassed and distilled water. We use the Atchley and Prosperetti version of the crevice model for our analysis, except where noted.

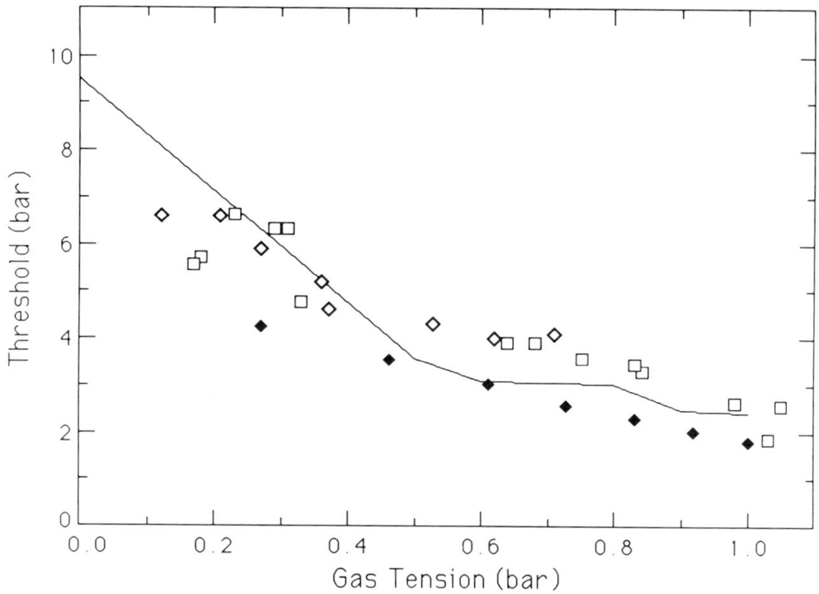

Figure 1-6. Cavitation threshold versus gas tension. The solid line is the prediction of the crevice model with $\beta = 15°$, $\alpha_H = 12.0°$, and $\sigma = 70$ dyn cm^{-1}. The data are □, Strasberg's[17]; ◇, Crum's[25]; and ◆, Connolly and Fox's.[18]

Dissolved Gas Tension

Figure 1-6 is a graph of the cavitation threshold versus gas tension. Plotted are the predictions of the crevice model as well as the experimental thresholds measured by Strasberg,[17] Connolly and Fox,[18] and Crum.[25] No attempt has been made to normalize any of the data, although the calibrations of the various experimental systems were based on different standards. Theoretical predications of the threshold as a function of gas tension for crevices smaller than critical size are presented in Fig. 1-7. The different lines correspond to values of the radius of the crevice mouth, a_1, ranging from 10 to 50 nm. It is seen that for crevices smaller than critical size, the threshold at saturation can assume large values because it is now a function of a_1. There is no threshold dependence on gas tension, which is not the case for crevices larger than critical size. In order to see this difference, we normalize the predictions presented in Figs. 1-6 and 1-7 to their values at saturation. These normalized thresholds are reproduced in Fig. 1-8. It is clear from the figure that the dependence on dissolved gas concentration is less for crevices smaller than critical size. This result is consistent with the observations of Greenspan and Tschiegg.[22] They found that after water had undergone protracted filtration through 0.2-μm filters, the cavitation threshold was about 160 bar if they waited 1 min for cavitation to occur or 210 bar if they waited only a few seconds for it to occur. In addition, the threshold was independent of gas tension; their range of gas tensions was from about 0.15 to 0.85 bar. Referring to Fig. 1-8, over this range of gas tensions the threshold for crevices

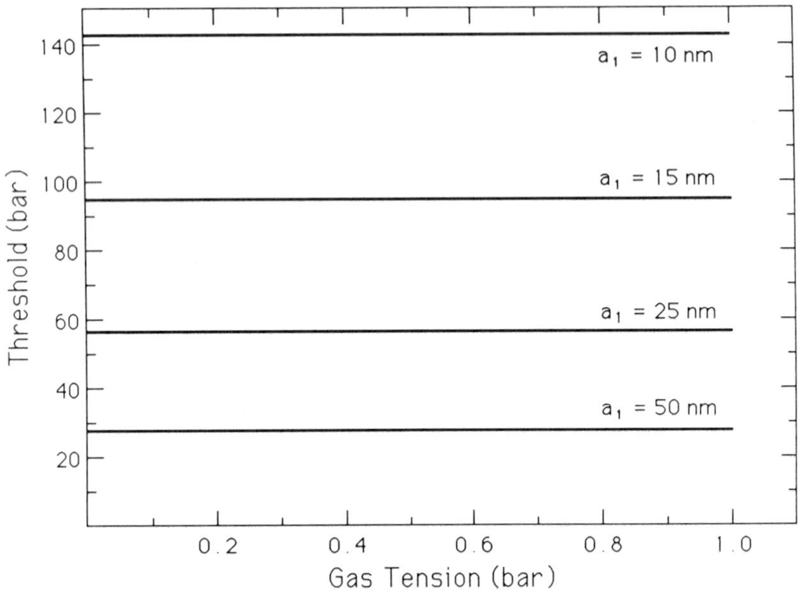

Figure 1-7. Cavitation threshold versus gas tension for crevices smaller than critical size. The lines are the predictions of the crevice model for $\beta = 15°$, $\alpha_H = 12.0°$, $\sigma = 70$ dyn cm^{-1}, and various values of a_1.

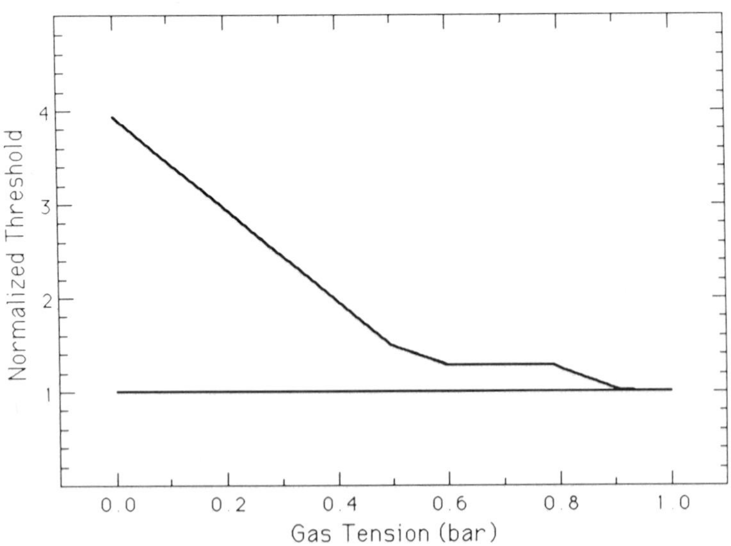

Figure 1-8. Normalized cavitation threshold versus gas tension. The upper curve is the predicted threshold from Fig. 1-6 and the lower curve is the predicted threshold from Fig. 1-7 for $a_1 = 15$ nm.

smaller than critical size is constant, while that of crevices larger than critical size varies by about a factor of 3.

Surface Tension

The functional dependence of the cavitation threshold on the liquid–gas surface tension is not as obvious as with the case of gas tension. To understand this dependence, we follow the lead of Crum,[25] who used an empirical relationship discovered by Bargeman and Van Voorst Vader[47]:

$$\cos \alpha_E = C/\sigma - 1 \tag{1-13}$$

where α_E is the equilibrium contact angle and C is a constant. For many non-polar solids C has a value of about 50 dyn cm^{-1}. This relationship is valid for surface tensions greater than about 10 dyn cm^{-1}. In the analysis given by Crum as well as the one to follow, it is assumed that $\alpha_E = \alpha_A$ and that $\alpha_R = \alpha_A - \alpha_H$, where α_H is the hysteresis contact angle. (Hysteresis is what gives rise to the asymmetrical shape of droplets sliding down windows.) Using these relationships, the crevice model becomes a model with two adjustable parameters: α_H and β. In the analysis of this model, the sum of these two parameters ($\phi = \alpha_H + \beta$) was held constant. Figure 1-9 is a graph of threshold versus surface tension. Two

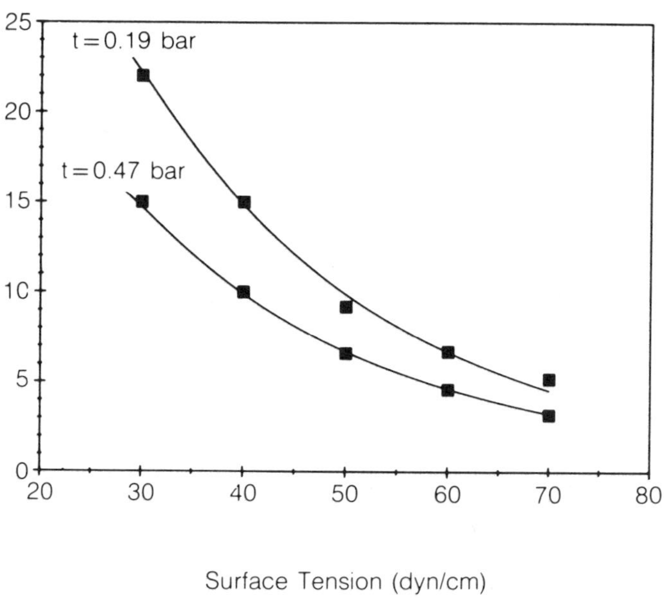

Figure 1-9. Cavitation threshold versus surface tension for water for different values of gas tension. The data are Crum's[25] and the curves are the predictions of the crevice model using $\beta = 12.85°$ and $\alpha_H = 22.15°$ for $G = 0.47$ bar, and $\beta = 10.8°$ and $\alpha_H = 24.2°$ for $G = 0.19$ bar.

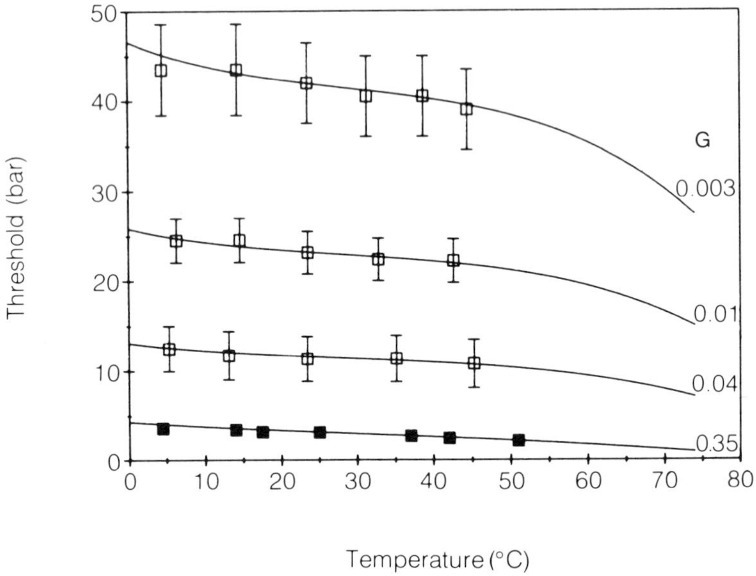

Figure 1-10. Cavitation threshold versus temperature for water. The data are □, Galloway's[19] and ■, Crum's[25] and have been normalized to fit the theory at room temperature. The curves are the predictions of the crevice model with $\beta = 12.85°$ and $\alpha_H = 22.15°$ for the indicated values of G.

sets of data are shown corresponding to two different values of gas tension. The data are Crum's.[25] It is seen that the agreement between theory and experiment is quite good.

Temperature

Temperature variations affect the cavitation threshold through three measurable parameters: gas tension, surface tension, and vapor pressure. The largest variation is associated with the temperature dependence of the dissolved gas tension; however, all of the dependencies are taken into account in the analysis, using data from standard references.[48–50]

The cavitation threshold is plotted as a function of temperature for a range of dissolved gas tensions in Fig. 1-10. The open squares represent Galloway's data[19] and the solid squares represent Crum's data.[25] The solid lines represent the threshold as predicted for $\beta = 12.85°$ and $\beta + \alpha_H = 35°$. The data have been normalized to fit the theory at room temperature (23°C). The variation with temperature is predicted well by the crevice model. Although there are no data to test the theory above 50°C, the threshold must approach the vapor pressure at the boiling point. This behavior is suggested by the "tail" at high temperatures.

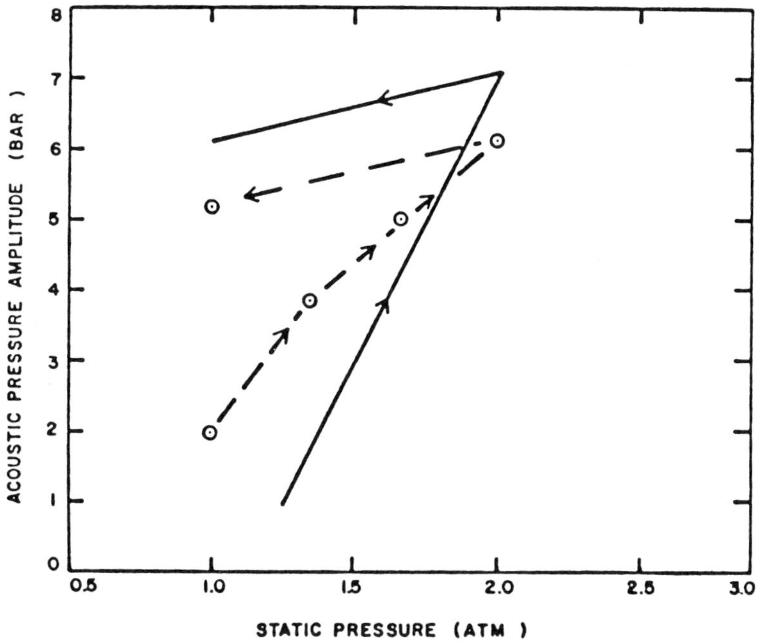

Figure 1-11. Cavitation threshold versus applied static pressure. The arrow indicates the direction of the change in static pressure. (After Crum[57].)

Hydrostatic Pressure

Crum[25] has applied Apfel's[23] version of the crevice model to Strasberg's[17] measurements of the threshold dependence on applied hydrostatic pressure. His results are shown in Fig. 1-11. It is seen the overall theoretical trend is in good agreement with the measured trend.

Dissolved Ion Concentration

As of now, no model has been able to predict the dependence of the threshold on dissolved ion concentration.

In summary, then, it is seen that for carefully controlled measurements in distilled water, the crevice model gives reasonably good predictions of the dependence of the cavitation threshold on a variety of physical parameters. It cannot explain all the dependences, however, and is thus incomplete in the current version. To have an adequate model, one must somehow introduce the obvious influences of dissolved ion concentration and surface-active materials. This problem is a difficult one but also interesting and important and awaits the ingenuity of a future investigator.

A final caveat: currently, nucleation is of considerable interest to those examining the biological effects of ultrasound, and specifically, cavitation in tissue. It is important to note that the nuclei present there may be entirely different in character, and that the successful results obtained for water may not apply to those situations.

EXPERIMENTAL TECHNIQUES FOR INVESTIGATING TRANSIENT CAVITATION

General Considerations

The origin of bubbles in liquids has been considered in the previous section. It was implied that cavitation nucleation and the inception of transient cavitation were closely linked. Measurements of the transient cavitation threshold were compared to the theoretical instability of a Harvey nucleus. It was seen that, in general, the agreement between measurement and calculation was quite good. In order to investigate nucleation processes, therefore, it is appropriate to measure the transient cavitation threshold. Measurements of the transient cavitation threshold are also made to investigate the violent side effects of transient cavitation, such as material erosion and free-radical production.

In this section we consider various aspects of the experimental investigation of transient cavitation. We begin the discussion with a brief overview of some general concerns of experimental system design and end the discussion with a description of an apparatus that we have developed in our laboratory.

In order to understand the basic aspects of any experimental system, it is useful to give a brief description of transient cavitation. Transient cavitation is produced by a high-amplitude acoustic field (usually), the threshold for its inception is strongly dependent upon the nature of the liquid, and it is short-lived (a transient cavity has a lifetime of one, or at most a few, acoustic cycles). This characterization suggests that an experimental system must consist of three components: (1) a method of generating the cavitation, i.e., a method of producing a sound field; (2) a method of preparing the liquid and controlling its properties; and (3) a method of detecting the cavitation when it occurs. Apfel[45] ascribes to each of these components a golden rule for threshold measurements: (1) know thy sound field, (2) know thy liquid, and (3) know when something happens.

"Know Thy Sound Field"

There has been a tremendous variety of experimental systems used to generate acoustic cavitation. In an early comprehensive review by Blake,[7] many of these systems are described. These early devices commonly employed a flat acoustic radiator, or transducer, to produce the sound field. These systems have one major disadvantage. Flat transducers produce high-amplitude fields in the immediate vicinity of the transducer face, as well as in the bulk of the liquid. Therefore, since solid–liquid interfaces serve as preferential sites for bubble formation, the end result of most early attempts at generating acoustic cavitation has been the formation of bubbles on the transducer face. Thus, the cavitation was as much a property of the transducer material as it was of the liquid. Cavitation generation on or near the transducer is undesirable for other reasons as well. First, the violence of the collapse of a transient cavity can cause erosion of the transducer material. In addition, since gas bubbles have an acoustical impedance vastly different than that of liquids, the formation of bubbles on the transducer creates an impedance mismatch between the transducer and the liquid, thereby diminishing the efficiency of the sound generation system. Bubbles are also good absorbers of ultrasound. If bubbles form in the region in front of the transducer, the ultrasound becomes highly attenuated as it passes through this region and the insonification of the bulk of the liquid is greatly reduced (i.e., "bubble shrouding").

In order to diminish the chances of bubble formation at the transducer, the pressure amplitude near the face must be kept to a minimum. However, high pressure amplitudes must still be produced in the bulk of the liquid in order to generate cavitation there. Both of these requirements can be met if the sound field is focused. One early method of focusing was accomplished by reflecting the acoustic field generated by a flat radiator off a curved surface. Willard[51] was one of the first to employ a curved transducer to focus a sound field so as to remove the effect of boundaries. He used a concave, spherically focused, piezoelectric transducer driven in the megahertz frequency region. Today, this focusing technique is quite common.

Perhaps it is worthwhile to digress momentarily on the subject of piezoelectric transducers. Piezoelectric materials develop an electric polarization when mechanically stressed and, conversely, they become deformed when electrically polarized. The extent of the deformation depends upon the extent of the polarization and vice versa. If the polarization alternates, first positive and then negative, the deformation alternates; that is, the thickness of the material changes at the frequency at which the polarization alternates. Many different materials exhibit piezoelectric properties: crystals, such as quartz, tourmaline, Rochelle salt (potassium sodium tartrate), and barium titanate ($BaTiO_2$ is actually electrostrictive rather than piezoelectric but can be made to behave as a piezoelectric if permanently polarized); ceramics, such as lead zirconate titanate (PZT); polar polymer films, such as polyvinyl chloride (PVC), polyvinyl flouride (PVF), and polyvinylidene flouride (PVDF); and even biological substances, such as the dried Achilles tendon of a cow.[52] Presently ceramic piezoelectrics are widely used to generate ultrasonic signals, as they are relatively inexpensive and can be cast in various shapes to fit various experimental requirements.

Most focused transducers are spherical or parabolic sections and are quite analogous to converging optical lenses. However, since piezoelectric ceramics can be manufactured in almost any configuration, they can be easily adapted to fit an experimental situation. One example of such a transducer is a hollow cylindrical tranducer. When driven in a radial mode, the acoustic field is focused inward along the cylindrical axis and, if closed on the ends, also develops standing waves along the axis. Thus, a region of intense sound can be generated in the center of a cylindrical container. This type of transducer has been used in various cavitation studies, some of which are discussed later.

For a given pressure amplitude at the transducer face, a transducer having a larger surface area produces a higher pressure amplitude at the focus than one having a smaller surface area. However, the cost of a transducer is very dependent upon surface area. In addition to higher costs, larger transducers naturally take up more space and are harder to manipulate. There is a method of effectively increasing the surface area of a transducer. This method involves resonance systems and such a system is briefly described now.

Consider a spherical flask filled with a liquid and having one or more transducers affixed to its side. When the transducers are driven, the entire flask vibrates and, if the transducer–flask system is driven at one of its resonance frequencies, the oscillation amplitude of the flask wall can be relatively large. In this manner the effective surface area of the transducer becomes that of the entire flask. As the flask oscillates radially, the acoustic field is focused at the center of the sphere. Resonance systems having different geometries have been used and, although the pattern of the acoustic field may be more complex than that produced by a spherical system, they all work on the same principle.

Resonance systems such as this have disadvantages, one being the often difficult problem of permanently mounting the transducers to the flask. The transducer is usually epoxied to the flask. During operation, however, there invariably exists relative motion between the transducer and the flask, thereby generating stress in the adhesive. Over a period of time this stress weakens the adhesive. The weakening is accelerated by larger driving amplitudes. The stress can be minimized if extreme care is taken to retune the transducer to match the resonance frequency of the flask. (The retuning of a transducer is a fine art!) However, it is virtually impossible to eliminate failure of the adhesive and in fact this failure usually dictates the upper limit to pressure amplitudes that can be achieved with these systems. A way of avoiding the adhesive-failure problem is to make the entire containment vessel from transducer material. In fact, cylindrical transducers made from piezoelectric ceramics, with thin glass windows on the ends, make excellent cavitation cells. In this case, however, the quality factor, or Q, of the system is determined by the ceramic, which typically has a larger internal friction than glass or stainless steel, and thus has a lower Q. Ceramics are also opaque and it is difficult to observe visually when cavitation occurs.

Another disadvantage of these systems is that the resonance frequency is strongly temperature dependent. Thus, when measurements extend over long periods of time, the system must be retuned continually to keep it operating at resonance. The resonance frequency is also dependent upon the nature of the liquid sample. For example, when filled with water a flask has a different resonance frequency than when it is filled with glycerine.

One advantage of resonance systems is that they can have high *Q*s'. The *Q* of an oscillator is a measure of the sharpness of the resonance or the time required for the system to dissipate energy. Generally speaking, the greater the *Q*, the higher the pressure produced at the focus. *Q*s' of 10,000 are not uncommon for resonance systems.[22] It should be pointed out, however, that this property of resonance systems makes them unacceptable for measurements involving short pulses of ultrasound. The lower limit of the duration of the pulse that can be produced by a transducer is dictated by how quickly the oscillation of the transducer damps out. The damping time is directly related to the *Q*. Therefore, for short pulse durations, the transducer must have a low *Q*.

Cavitation generated by a focused sound field offers a significant advantage over that produced by an unfocused, or uniform, sound field. This advantage is that the researcher knows more or less where the cavitation is going to occur.

Whereas cavitation generated by a uniform sound field is equally likely to occur anywhere within the sound field, cavitation generated by a focused sound field can be made to occur only in the focal region. This site-specific information can be quite useful, especially when attempting to detect the cavitation.

One aspect of experimental systems that is exceedingly important and upon which we have not touched has to do with sound field calibration. Calibration in a resonance system is most commonly accomplished with a probe hydrophone having small dimensions. The aim of calibration is to ascertain the properties of the sound field (peak pressure amplitude, field geometry, etc.) from a knowledge of the signal supplied to the transducer, so calibration usually must be performed with a sound field having a pressure amplitude lower than it is to have during actual measurements. This precaution is necessary to prevent damage to the probe by the formation of cavitation on it. We do not attempt any discussion of calibration techniques here. Standard techniques for calibration have been devised by various scientific groups (See, for example, the technique suggested by a Committee of the American Institute for Ultrasound in Medicine.[53])

"Know Thy Liquid"

The properties of the liquid dictate the outcome of cavitation threshold measurements. Therefore, in order to interpret the results of threshold measurements correctly one must know as much about the liquid as possible. This information must consist not only of the liquid properties at the time of the measurement but the history of the liquid as well. However, a simple knowledge of liquid properties is not adequate for most experimental systems. In order to test the range of validity of any theory, it is necessary to be able to vary the liquid properties as well. It is also necessary to control the properties of the liquid so that they remain unchanged during the period over which the measurements extend. Depending upon the nature of the individual experiment, this time period may be anywhere from a few minutes to several hours.

The nature of the liquid must also be considered when choosing a transducer. When ultrapurity is required or when dealing with a corrosive liquid, the transducer and the liquid should not be in contact. In such situations, a resonance system may be ideal because the resonator serves as the sample container. However, the use of a resonance system eliminates the possibility of using a short-duration pulsed ultrasound, as discussed previously. The type of detection system employed depends upon the nature of the liquid as well. For example, a system that detects cavitation optically is useless if the liquid (or container) is opaque.

"Know When Something Happens"

The measured value of the cavitation threshold depends to a great extent on how one "looks" for the cavitation event. In fact, the entire concept of a threshold is strongly dependent upon the phenomenon one is trying to detect; for example, the threshold for the onset of free-radical production may be vastly different and unrelated to the threshold for the rectified diffusion of gas within a pulsating gas bubble. Also, a detection method that is useful in one mode of operation (e.g., continuous, low-frequency ultrasound) may not be at all useful in a different mode (e.g., pulsed, high-frequency ultrasound). Likewise, if two different detection schemes are employed simultaneously, "the" threshold detected by one may be quite different than "the" threshold detected by another. Obviously, before one designs a detection system, one should know as much as possible about the phenomenon one expects to observe and the method by which this phenomenon is to be generated.

Because the result of a threshold measurement depends upon the criteria that constitute the onset of cavitation, it is useful to review some of the threshold criteria used in the past. Perhaps the most common threshold criterion is the detection of the audible "click" associated with the shockwave produced in the host liquid by a collapsing transient cavity. This criterion, although useful for cavitation produced by continuous-wave ultrasound in the kilohertz frequency region, is essentially useless for observing pulsed cavitation in the megahertz region. A second commonly used criterion, the presence of free gas bubbles in the insonified region, may only indicate growth by rectified diffusion. Such bubble growth may occur without any associated violent effects, such as free-radical production or shockwave generation.

Cavitation can be detected chemically through a variety of reactions that often involve free-radical production. This method has the advantage that it can measure the field and can be extremely sensitive; the disadvantage is that it cannot be done in real time and that some reasonably difficult assaying must be performed.

Another threshold criterion is the detection of subharmonics of the driving frequency produced by oscillating bubbles. Bubbles radiate subharmonics only when oscillating violently. Therefore bubbles that oscillate not so violently are not observed. Likewise, if one is interested in strictly transient cavitation (that is, in detecting the transient collapse of cavity) this criterion may not be appropriate. The detection of subharmonics indicates the presence of violently oscillating bubbles and not, necessarily, transient cavitation. Its use is more appropriate when violent stable cavitation is of interest.

Another method of detection that indicates that presence of bubbles is based upon the fact that bubbles are efficient absorbers of ultrasound. The cavitation threshold is marked by the excessive absorption of a reference acoustic pulse (not related to the field that generates the cavitation) that is transmitted through the fluid. The onset of absorption indicates the presence of bubbles in the liquid. Again, however, this method does not distinguish between stable and transient cavitation. This method, however, does provide a way of determining the sizes of bubbles. The absorption of ultrasound of a given frequency depends upon the radius of the bubbles. Therefore, transmitting a band of frequencies and noting which frequency components are attenuated the most can enable one to determine the size of the bubbles formed.

Bubbles can also be detected optically. The eye can be used to detect the presence of bubbles in a liquid. However, the eye has only limited usefulness as a detector because of the small sizes and short lifetimes of some cavitation events. The problem of subjectivity arises as well. Photographic methods have been employed to detect cavitation but, because of the delay associated with processing, this method is more useful for "capturing" an event than for threshold measurements. Real-time detection is usually preferable. The detection of light scattered from bubbles is also a means of detecting cavitation, but this method requires precise alignment of components and the cavitation must be generated in the path of the light beam.

The production of sonoluminescence provides another criterion for a cavitation threshold, one that we have found quite successful. Sonoluminescence is radiated light the production of which is associated with bubble activity (since sonoluminescence is the topic of a later chapter, we attempt no further discussion of its origin here). This light may be detected with a dark-adapted eye, photographic plates, or electronic light detectors (photomultiplier tubes, single-photon counters, etc.). The intensity of the sonoluminescence can be enhanced by changing the gas composition of the liquid sample or by altering the chemical properties of the liquid. The latter method usually involves the addition of light "enhancers" to the liquid. (The light produced is then a secondary result of cavitation and arises from chemiluminescence.) Optical methods of cavitation detection place a stringent requirement on the liquid and its container: they must be optically transparent. (The requirement makes optical detection of cavitation within biological systems difficult.)

In general, there are many different ways of detecting the various types of cavitation that exist. The choice of the detection scheme must be based upon the specific type of cavitation to be detected and the experimental limitations placed upon the detection system. The cardinal rule in detection is still "know when something happens."

Description of a Specific Experimental Apparatus

We have developed an apparatus that incorporates some of the ideas presented above. It is useful to describe it here as an example of a versatile, yet precise, system that can be used to measure cavitation thresholds.

Control over the liquid composition is best attained when there are few interfaces between different materials that require potentially contaminating adhesives and seals. An obvious configuration with several advantages is a spherical glass flask in which the liquid is contained inside, with the transducers mounted to the outside wall and driven so as to excite a standing wave pattern within the liquid. The spherical focusing generates a high-intensity region at the center of the flask that is well removed from the walls. Further, the system can be easily cleaned, provides high optical visibility, and is simple in construction. This configuration was suggested by Blake,[7] perhaps first utilized by Galloway,[19] and employed later with great success by Strasberg,[17] Barger,[54] and Vaughan.[55] The major disadvantages of the spherical flask configuration is that the maximum acoustic pressure attainable is about 25 bar (2.5 MPa).

We now briefly describe details of the experimental apparatus. A detailed description has been published.[56]

Fluid-Management System

The cavitation threshold depends upon the liquid "properties," such as dissolved gas content, surface tension, temperature, ionic content, particulate contamination, and chemical impurity. Accordingly, it is desirable to control these parameters as much as possible so that we can study the properties of the nuclei that give rise to cavitation. Figure 1-12 shows the fluid-management system that we have devised for this purpose.

It was determined by Greenspan and Tschiegg[22] in their careful study of cavitation inception that one needed a closed system that permitted the liquid to be continuously circulated before adequate degrees of cleanliness could be achieved. They discovered that squeezing pliable tubing or exposing the liquid briefly to the air would "reseed" the liquid with nuclei. Accordingly our system is a closed one, with the liquid exposed only to glass, teflon, and stainless steel. Although the figure is reasonably self-explanatory, a few special features are noted below.

We found it desirable to remove the dissolved air from the liquid and replace it with argon, which gives a more intense sonoluminescent response than air (because of the differences in thermal conductivities). We did this by simultaneously evacuating this reservoir, circulating the liquid, and bubbling argon through the liquid within the reservoir. By an appropriate selection of gas pres-

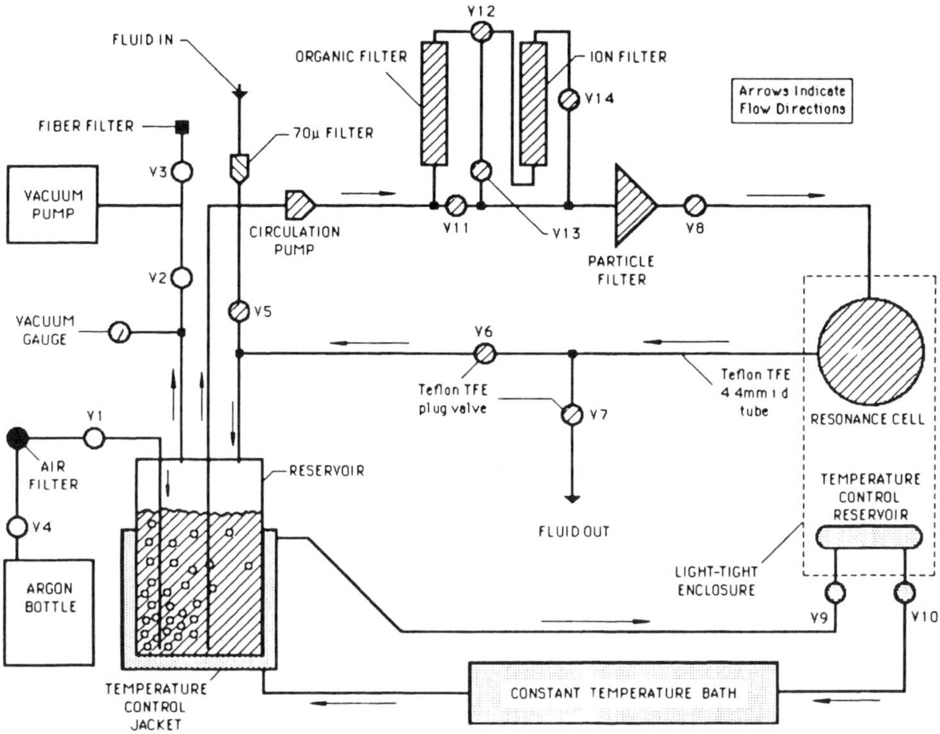

Figure 1-12. Schematic diagram of the fluid-management system of the experimental apparatus developed by Roy et al.[56]

sures and circulation rates and times, a predetermined "dissolved argon concentration ratio" (the amount of dissolved argon compared to that at saturation) could be roughly achieved. The actual value was determined by measuring a sample of the liquid in a volumetric Van Slyke apparatus. This system permits an examination of this ratio from about 0.3 to 1.0.

In order to automate the threshold measurement system fully, we found it necessary to build a resonance-tracking system to lock onto a resonant mode of the sphere. Since energy dissipation tended to raise the temperature of the sphere–transducer combination, a constant-temperature control system was required to hold the resonance frequency of the sphere to within permissible limits. Within the light-tight box containing the cavitation flask and the photomultiplier tube, the temperature change could be held to within $\pm 1°C$.

One of the motivations for constructing this system was to determine information about the nature of cavitation nuclei. In particular, we were interested in the dependence of the cavitation threshold on the size of solid particulate matter and thus, to a certain extent, on the size of the nuclei. Nucleus size control was achieved via extensive filtration; the filter size thus determined, in a sense, the nuclei size. Implicit in this statement is the assumption that the filter remains relatively clean. During the filtration process, the filters were replaced frequently and visual inspections revealed little or no observable deposits on their surfaces. The measured fluid flow rates during the filtrations were about 1 L min^{-1}. Since the total quantity of water in the system was about 2 L, we could reasonably assume the entire bulk of the fluid to pass through the filter in a few minutes. Prior to each run, the sample underwent simultaneous filtration and "argonation" for 45 min through filters that remained as clean as possible. We therefore assume that the nucleus size distribution had an upper bound that was approximately equal to, or slightly less than, the filter pore size. Filter sizes used in this study ranged from 10 to 0.5 μm in radius.

Akulichev[21] has reported that the cavitation threshold has an important dependence on the concentration of dissolved ions, particularly certain alkaline salts. We desired to reproduce and extend these measurements. In order to determine the dependence of the threshold on the ionic concentration, it was necessary to remove as many chemical impurities from the liquid as possible, so that the desired ions could be added at low concentrations. In a method similar to removing the physical impurities by filtration, the chemical impurities were removed by circulating the liquid through Sybron Barnstead organic and ionic removal columns. The manufacturer contends that the content of high molecular weight organics is lowered to a few parts per million and the resistivity of the water is raised to 1 MΩ cm.

Crum[25,57,58] and others[55] have also discovered that small additions of surface-active agents can greatly influence the threshold. These agents can be easily added to the liquid with our apparatus, but their removal is more difficult. It is possible to remove most organic surfactants from the liquid by irradiation with an intense ultraviolet source through a quartz window.

Cavitation Generation System

The generation of acoustic cavitation in a liquid such as water is relatively simple. One needs only a function generator, a power amplifier, and an appropriate transducer. To determine the threshold, one merely increases the gain of the oscillator (at a rate determined by the operator) until there is audible or visual evidence that cavitation has occurred. One then repeats the operation after waiting for a short interval for the liquid to equilibrate. It is rather obvious that this technique involves a considerable amount of subjectivity that ideally would be avoided. We have attempted to automate the system so that completely objective measurements can be made. Although we have not entirely succeeded, the following procedure is a significant advance toward operator independence.

Figure 1-13 shows a block diagram of the cavitation generation system. The heart of the system is a ramp generator (our design) that drives both the amplitude-modulation input of a function generator and a stripchart recorder. The dc voltage level of the stripchart record is linearly related to the peak-to-peak output amplitude of the function generator. This signal passes through a

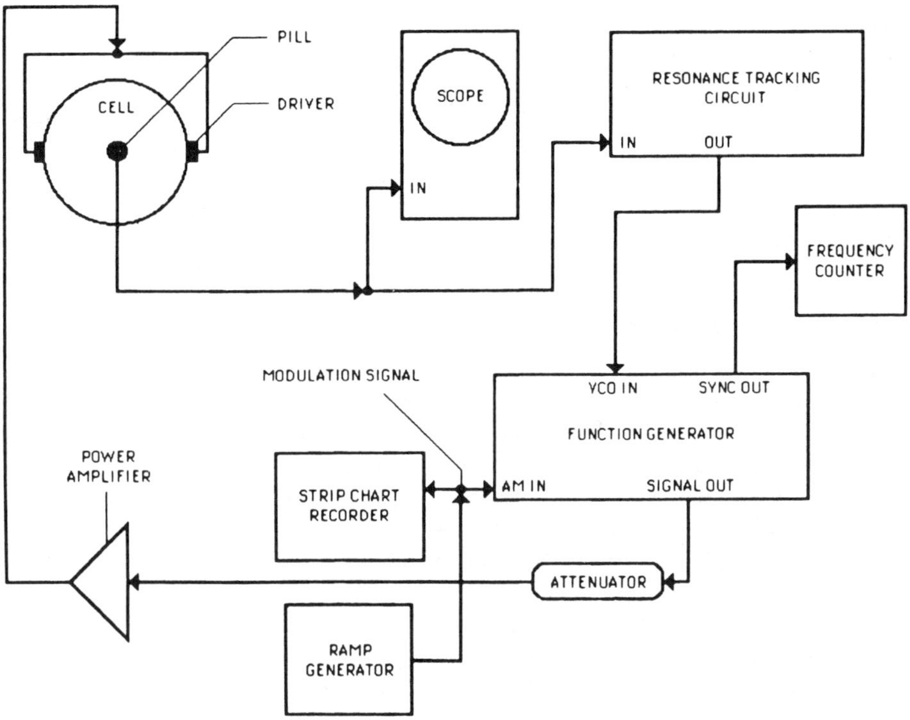

Figure 1-13. Schematic diagram of the cavitation generation system of the experimental apparatus developed by Roy et al.[56]

precision attenuator and is amplified by a power amplifier, which in turn supplies the driving voltage to two PZT-4 cylindrical transducers that are mounted on a 1 L fused-quartz resonance sphere. The sphere was custom made for us with a slightly flattened bottom (to mount the photomultiplier tube) and smaller neck (to reduce its effect on the resonant modes). The transducers are stock items, 2.54 cm long, with i.d. = 1.91 cm, o.d. = 2.54 cm, and a resonance frequency of about 50 kHz. Also epoxied to the cell are a miniature PZT-5 pill transducer (to monitor the sound field) and a linear thermistor (to monitor the temperature). The entire sphere assembly is cradled in a monofilament line harness to reduce damping. The signal generated by the pill transducer drives the vertical input of a storage oscilloscope as well as the input of a digital resonance tracking circuit (our design). This device maximizes the RMS output of the cell by adjusting the frequency of the function generator (via the voltage controlled oscillator input) to lay on resonance. A universal counter monitors the frequency.

Since the acoustic pressure amplitude is essentially determined by the dc modulation signal, a "shutting down" of the sound field is tantamount to resetting the ramp generator to its quiescent "no-ramp" state. The ramp generator exercises total control over the sound field while the stripchart recorder keeps a running record, in hard copy, of the pressure amplitude in the cell.

The essential nature of the cavitation generation system is that the sound field within the cavitation cell is increased linearly with time, at reproducible rates, until cavitation occurs. The system then resets into a predetermined quiescent state and remains there until the timer restarts the ramp (the actual sequence of events that determine a measurement is given later).

Cavitation Detection System

One of the most subjective aspects of cavitation threshold measurements is the decision as to what constitutes a threshold. The appearance of bubbles has been used by some observers[59] (sometimes called "quasi-cavitation," because what is observed is normally rectified diffusion rather than rapid growth and violent collapse), the presence of subharmonics by others[60] (primarily nonlinear oscillations of stable gas bubbles), audible clicks and pops[61] (at least a shockwave involved), as well as a variety of other chemical and physical mechanisms. Because we would like to consider only violent cavitation, for the time being, we use the presence of a significant level of radiated electromagnetic energy (sonoluminescence) as an indication of cavitation inception. Thus, with this constraint, we are considering the "transient–violent cavitation threshold."[62]

Figure 1-14 shows the cavitation detection apparatus together with the gating, timing, and triggering circuitry. Once cavitation has been initiated in the spherical resonator by the ramping voltage, the sonoluminescent flash is observed by the photomultiplier tube. The anode pulse from the photomultiplier tube base, upon amplification by three stages of a preamplifier, feeds into a pulse-height discriminator. The discriminator level, which determines the sensitivity of the apparatus, is set to 600 mV, a value that is well above the electrical background level ($\sim$150 mV) but that lies below the typical sonoluminescence pulse height ($\sim$1.0 V). If the anode pulse survives discrimination, it passes through a "noise gating" network (described below), followed by a linear gate and finally results in the "trigger" pulse that subsequently resets the system. A "system reset" refers to a reduction in the sound pressure amplitude to a minimal quiescent level for a predetermined length of time. An Ortec model 776 timer provides the necessary control signals.

As an example of how the system works in a typical run, consider the system to be in its quiescent state where the sound field is off and the timer is counting. The timer disables both the linear gate and the ramp generator via the "busy" output, which is active whenever the timer is counting. At the end of the timing sequence the busy signal goes low, which opens the linear gate and engages the ramp generator. The ramp proceeds until a transient event generates a pair of light pulses that, provided they survive amplitude discrimination and "noise" gating, reset the timer and the whole process starts over. The linear gate prevents any stray noise pulses from resetting the timer before the quiescent period is

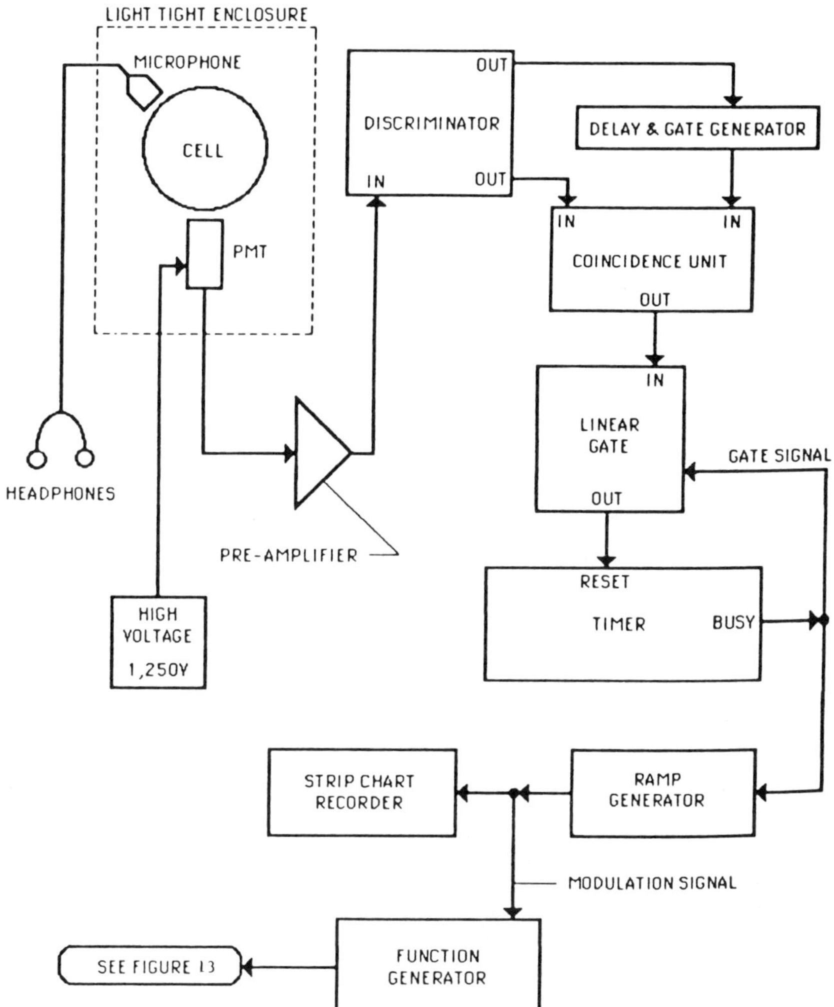

Figure 1-14. Schematic diagram of the cavitation detection system of the experimental apparatus developed by Roy et al.[56]

over. In the event that a transient collapse does not generate a sufficient quantity of light to trigger the reset, the operator, who is monitoring the situation with headphones, can manually reset the system simply by pressing a button. This system override is to prevent an unlimited ramp that could damage the cell transducers. A sketch of raw data from a typical run is shown in Fig. 1-15.

Sonoluminescence is not the only source of light one encounters in this configuration. It is found that a background "light noise" level of about five pulses per minute exists, probably resulting from Cerenkov radiation. We desire a slow

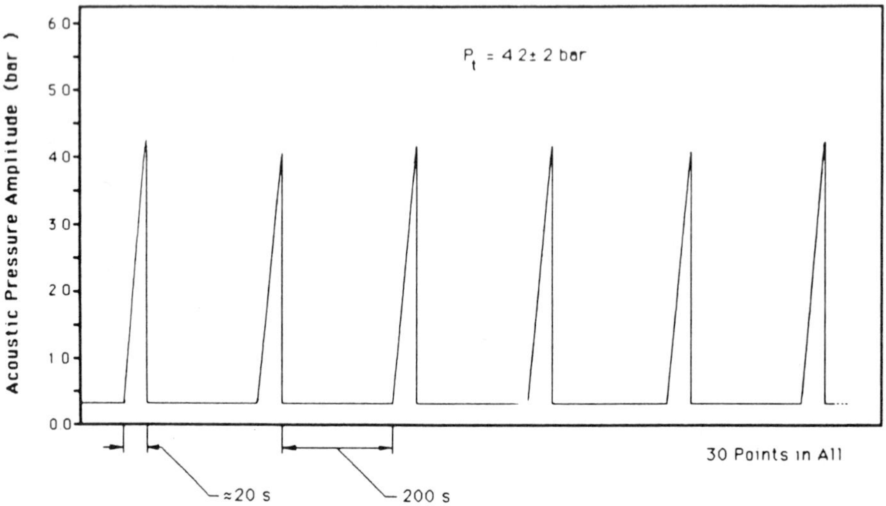

Figure 1-15. Reproduction of a stripchart record illustrating the technique for threshold measurements and the reproducibility of the data. The liquid was distilled water having a gas tension of 0.50 bar, a surface tension of 71.3 dyn cm^{-1}, and a temperature of 23.0°C and circulated through a 20-μm filter.[56]

ramp rate; however, with stray noise signals occurring every 12 s or so, one cannot wait very long for cavitation to occur. Pulse-height discrimination proves only partially effective because the brightest noise pulses are more intense than the weakest sonoluminescence events. We observe that in this system transient collapses almost always occur in bunches, some of the catastrophic events generating hundreds of pulses in a matter of milliseconds. This fact, of considerable interest in itself, inspired a "noise gating" scheme in which the criteria for the generation of a trigger pulse is the occurrence of two discriminator pulses separated by no more than 100 ms. The pulse-height discriminator has two outputs, one of which is connected directly to the "B" input of a Lecroy model 465 coincidence unit. The other undergoes a 20-ns delay (via a Tennelec model TC410A gate and delay module) and then activates the trigger of a BNC model 8010 pulser. This device produces a 100-ms gate signal that drives the "A" input of the coincidence unit. The time delay ensures that the discriminator pulse, which is about 15 ns in width, reaches the coincidence unit before the gate signal. Thus, to establish coincidence, a second discriminator pulse must follow within 100 ms. The output of the coincidence unit generates the trigger signal that resets the system. This technique allows us to halve the discriminator level while increasing the noise immunity by a factor of 7.

The system just described is a relatively complex system for generating and detecting cavitation. However, cavitation is a very complex phenomenon, subject to a variety of misinterpretations. Unless one controls the liquid and the applied sound field and knows what to detect and how to determine whether a valid detection has occurred, then the information obtained may be essentially useless.

RECTIFIED DIFFUSION

When one visually examines the effect of the direct application of a sound field to a liquid such as water, one often sees streamers of bubbles. In certain configurations, one can see gas bubbles rising from the field of acoustic activity. The process by which dissolved gas in the liquid is converted into free gas in the form of bubbles by the action of the sound field is called *rectified diffusion.* This "rectification of mass transfer" is a direct consequence of the applied sound field and can be of importance whenever the applied acoustic pressure amplitude exceeds the threshold for rectified diffusion (which is on the order of 1 bar or less for typical values of the acoustic frequency and bubble radius). For liquids that are not supersaturated with gas, a definable threshold exists arising from the Laplace pressure exerted on the gas contained within a bubble by the force of surface tension. The value of this pressure is given by $P_i = 2\sigma/R_o$, where P_i is the internal pressure and R_o is the equilibrium radius of the bubble.

For relatively large gas bubbles, on the order of 1 cm diameter, this pressure is quite small (<0.001 bar). However, for bubbles on the order of 1 μm in diameter, this pressure can be quite significant (>1.0 bar). Therefore small, free, gas bubbles in a liquid with a surface tension similar to that of water must be stabilized in some manner, as described earlier, or they dissolve rapidly. With the application of a sound field, however, a bubble can be made to oscillate about an equilibrium radius and caused to grow in size via the following mechanisms.

The first mechanism is the surface area effect. When the bubble contracts, the concentration of gas (moles per liter) in the interior of the bubble increases, and gas diffuses from the bubble. Similarly, when the bubble expands, the concentration of gas decreases, and gas diffuses into the bubble. Because the diffusion rate is proportional to the surface area, more gas enters during expansion than leaves during the contraction of the bubble; therefore, over a complete cycle, there is a net increase in the amount of gas in the bubble.

The second mechanism is the shell effect. The diffusion rate of gas in a liquid is proportional to the gradient of the concentration of dissolved gas. Consider a spherical shell of liquid surrounding the bubble. When the bubble contracts, this shell expands, and the concentration of gas near the bubble wall is reduced. Therefore, the rate of diffusion of gas away from the bubble is greater than when the bubble is at its equilibrium radius. Conversely, when the bubble expands, the shell contracts, the concentration of gas near the bubble is increased, and the rate of gas diffusion toward the bubble is greater than average. The net effect of this convection is to enhance the rectified diffusion. It has been shown that both the surface area and the shell effects are necessary for an adequate description of the phenomenon.

Apparently, rectified diffusion was first recognized by Harvey,[12] who considered the biological implications of the effect as early as 1944. Blake[63] was probably the first to attempt a theoretical explanation and his predictions were nearly an order of magnitude too small, because he considered only the surface area effect. An adequate theory was first presented by Hsieh and Plesset[64] who were the first to include the effect of convection. Strasberg,[65] who had made some experimental measurements, showed that the theory of Hsieh and Plesset[64] generally agreed with his measurements. Eller and Flynn[66] extended the analysis to include nonlinear or large-amplitude effects by treating the boundary condition of the moving wall in a slightly different way. Safar[67] showed that the Hsieh-Plesset results were essentially equivalent to those of Eller and Flynn[66] when inertial effects were added to the Hsieh-Plesset approach.

The few measurements obtained by Strasberg were extended by Eller[68] to include both the threshold and the growth rate. He found that the theory was adequate to predict the thresholds for growth by rectified diffusion but was unable to account for some large growth rates that he observed. He suggested that acoustic streaming (rotational liquid movement caused by asymmetrical bubble pulsations) may be the cause of the larger than predicted rates of growth. Gould[69] was able to observe gas bubble growth by rectified diffusion directly through a microscope and discovered that growth rates could be greatly enhanced by the onset of surface oscillations of the bubble, which in turn seemed to induce significant acoustic microstreaming. Attempts by Gould[69] to apply the acoustic streaming theories of Davidson[70] and of Kapustina and Statnikov[71] to explain his results were not successful. Extended theoretical treatments were presented by Skinner[72] and Eller[73] to account for growth through resonance.

Additional measurement of the growth of gas bubbles by rectified diffusion were later obtained by Crum,[74] who made both threshold and growth rate measurements for a variety of conditions. He extended the theory to include effects associated with the thermodynamic behavior of the bubble interior and found excellent agreement between theory and experiment for both the rectified diffusion threshold and the growth rate for air bubbles in pure water. Anomalous results were obtained, however, when a small amount of surfactant was added to the water. The rate of growth of bubbles by rectified diffusion increased by a factor of about 5 when the surface tension was lowered by a factor of about 2, with no discernible surface wave activity. Although some increase is predicted, the observed growth rates were much higher. A slight reduction in the threshold with reduced surface tension was also observed. Some explanations offered by Crum[74] for this anomalous behavior were that there was some rectification of mass transfer resulting from the surfactant on the surface of the bubble or that there was microstreaming occurring even in the absence of surface oscillations.

To bring our limited review of the literature up to date, it is noted that there have been recent applications of the equations of rectified diffusion to studies of the threshold for growth of microbubbles in biological media,[75] the threshold for cavitation inception,[62] and as a possible explanation for the appearance of gas bubbles in insonified guinea pigs.[76] Finally, a comprehensive paper on rectified diffusion was recently published by Crum[77] and can be referred to for details.

The Governing Equations

A brief outline of the theory is now given, complete with the detailed equations to be used to make calculations of the threshold and the growth rate of bubbles by rectified diffusion. A complete mathematical description of the growth of a gas bubble by rectified diffusion would require an equation of motion for the bubble, diffusion equations for the liquid and the interior of the bubble, and heat conduction equations for both the liquid and the bubble. Continuity relations at the bubble wall also are required, and the situation is further complicated by the fact that these equations are coupled. Since this general approach presents a very formidable mathematical problem, various simplifications are used to obtain a solution.

The approach used by Hsieh and Plesset[64] and by Eller and Flynn[66] is to separate the general problem into an equation for the motion of the bubble wall, and a diffusion equation for the concentration of gas dissolved in the liquid alone. We will use as the equation of motion for the gas bubble the well-known Rayleigh-Plesset[78] equation, given here as:

$$R\ddot{R} + \frac{3}{2}\dot{R}^2 + \frac{1}{\rho}\{P_o[1 - (R_o/R)^{3\eta}] - P_A \cos \omega t + \rho R_o \omega_o b \dot{R}\} = 0 \quad (1\text{-}14)$$

where R and R_o are the instantaneous and equilibrium values of the bubble radius, respectively; ρ is the liquid density; η the polytropic exponent of the gas contained within the bubble; P_A the acoustic pressure amplitude; ω the angular frequency; ω_o the small-amplitude resonance frequency; and b is a damping term applied to the bubble pulsations. It is assumed that the damping is only important near bubble resonance and so the expressions for the specific damping terms are only applicable near resonance.

Further, $P_o = P_\infty + 2\sigma/R_o$ where P_∞ is the ambient pressure. The small-amplitude (linear) resonance frequency of the bubble, ω_o, is given by:

$$\omega_o^2 = (1/\rho R_o^2)(3\eta P_o - 2\sigma/R_o) \quad (1\text{-}15)$$

The diffusion equation for the gas in the liquid is Fick's law of mass transfer:

$$dC/dt = \partial C/\partial t + \mathbf{v} \cdot \nabla C = D\nabla^2 C \quad (1\text{-}16)$$

where C is the concentration of gas in the liquid; $\mathbf{v}$ is the velocity of the liquid at a point; and D is the diffusion constant. The coupling between Eqs. (1-14) and (1-16) is through the convective term $\mathbf{v} \cdot \nabla C$ and represents a major difficulty in the solution. This term was neglected by Blake,[63] which resulted in a prediction for the threshold that was an order of magnitude too small. The problem with the moving boundary has been solved in two slightly different ways by Hsieh and Plesset[64] and by Eller and Flynn.[66] The details of their solutions can be found in their respective papers; the solution of the latter is used here.

It is shown by Eller and Flynn[66] that the time rate of change in the number of moles of gas, n, in a bubble is given by:

$$dn/dt = 4\pi D R_o C_o \cdot \left[\langle R/R_o \rangle + R_o \left(\frac{\langle (R/R_o)^4 \rangle}{\pi D t} \right)^{1/2} \right] H \qquad (1\text{-}17)$$

where C_o is the "equilibrium" or "saturation" concentration of the gas in the liquid in moles per unit volume, the angle brackets imply time average, t is the time, and H is defined by:

$$H = C_i/C_o - \langle (R/R_o)^4 (P_g/P_\infty) \rangle / \langle (R/R_o)^4 \rangle \qquad (1\text{-}18)$$

Here C_i is the concentration of dissolved gas in the liquid far from the bubble and P_g, the instantaneous pressure of the gas in the bubble, is given by:

$$P_g = P_o (R_o/R)^{3\eta} \qquad (1\text{-}19)$$

As seen by Eq. (1-19), a polytropic exponent assumption is made for the pressure of the gas in the interior of the bubble. This assumption is commonly made and has been shown to be reasonably accurate for small-amplitude pulsations. There is evidence, however, that it needs to be modified for moderate- to large-amplitude oscillations.

The values of R/R_o to be used in the equations above are obtained by an expansion solution of Eq. (1-14) in the form:

$$R/R_o = 1 + \alpha (P_A/P_\infty) \cos(\omega t + \delta) + \alpha^2 K (P_A/P_\infty)^2 + \cdots \qquad (1\text{-}20)$$

where it has been determined that:

$$\alpha^{-1} = (\rho R_o^2/P_\infty) [(\omega^2 - \omega_o^2)^2 + (\omega \omega_o b)^2]^{1/2} \qquad (1\text{-}21)$$

$$K = \frac{(3\eta + 1 - \beta^2)/4 + (\sigma/4R_o P_\infty)(6\eta + 2 - 4/3\eta)}{1 + (2\sigma/R_o P_\infty)(1 - 1/3\eta)} \qquad (1\text{-}22)$$

$$\delta = \tan^{-1} [\omega \omega_o b/(\omega^2 - \omega_o^2)] \qquad (1\text{-}23)$$

and:

$$\beta^2 = \rho \omega^2 R_o^2 / 3\eta P_\infty \qquad (1\text{-}24)$$

It is necessary to know the damping of the bubble pulsations when the bubble is driven near resonance. This damping has been calculated by Eller[79] using the expressions of Devin;[80] a similar but not identical value for the damping term has also been obtained by Prosperetti.[78]

Because the pressure in the interior of the bubble is assumed to be given through the polytropic exponent approach, an accurate expression for the thermal damping cannot be given. Prosperetti[78] has shown that an estimate of this damping can be expressed through an "effective viscosity." Since the damping is only important near resonance (or at harmonics and subharmonics of the resonance frequency), and since the rectified diffusion equations are probably not accurate near resonance anyway, this nonrigorous approach is followed for the present time.

The total damping constant, b, is assumed to be a sum of contributions of the three common damping mechanisms, viz., thermal, viscous, and radiation:

$$b = b_t + b_v + b_r \tag{1-25}$$

where:

$$b_t = 3(\gamma - 1)\left[\frac{X(\sinh X + \sin X) - 2(\cosh X - \cos X)}{X^2(\cosh X - \cos X) + 3(\gamma - 1)X(\sinh X - \sin X)}\right] \tag{1-26}$$

with:

$$X = R_o(2\omega/D_1)^{1/2} \tag{1-27}$$

and:

$$b_v = 4\omega\mu/3\eta P_\infty \tag{1-28}$$

$$b_r = \rho R_o^3\omega^3/3\eta P_\infty c \tag{1-29}$$

with:

$$\eta = \gamma(1 + b_t^2)^{-1}\left[1 + \frac{3(\gamma - 1)}{X}\left(\frac{\sinh X - \sin X}{\cosh X - \cos X}\right)\right]^{-1} \tag{1-30}$$

In these equations γ is the ratio of specific heats and $D_1 = K_1/\rho_1 C_{p1}$, where K_1 is the thermal conductivity of the gas in the bubble, ρ_1 is the density of gas, C_{p1} is the specific heat at constant pressure for the gas, and μ is the viscosity and c the speed of sound, both in the liquid. To obtain a usable expression for the growth-rate equation, the time averages $\langle R/R_o\rangle$, $\langle (R/R_o)^4\rangle$, and $\langle (R/R_o)^4(P_g/P_\infty)\rangle$ are required. It is found that:

$$\langle R/R_o\rangle = 1 + K\alpha^2(P_A/P_\infty)^2 \tag{1-31}$$

$$\langle (R/R_o)^4\rangle = 1 + (3 + 4K)\alpha^2(P_A/P_\infty)^2 \tag{1-32}$$

and:

$$\langle (R/R_o)^4(P_g/P_\infty)\rangle = \left[1 + \frac{3(\eta - 1)(3\eta - 4)}{4}\alpha^2(P_A/P_\infty)^2 + (4 - 3\eta)K\alpha^2(P_A/P_\infty)^2\right]\left(1 + \frac{2\sigma}{R_o P_\infty}\right) \tag{1-33}$$

By assuming an ideal gas behavior, we can relate the equilibrium radius to the number of moles of gas:

$$P_o = 3\eta R_g T/(4\pi R_o^3) \tag{1-34}$$

where R_g is the universal gas constant and T is the absolute temperature. This equation should be a good approximation for small acoustic pressure amplitudes, so that the density fluctuations do not become excessive, and for low host-liquid temperatures, so that the ratio of vapor to gas within the bubble remains low. Such conditions are often met in common experimental situations.

By combining Eqs. (1-17), (1-18), and (1-34), an expression is obtained for the rate of change of the equilibrium bubble radius with time:

$$\frac{dR_o}{dt} = \frac{Dd}{R_o}\left[\langle R/R_o\rangle + R_o\left(\frac{\langle (R/R_o)^4\rangle}{\pi Dt}\right)^{1/2}\right]$$
$$\left(1 + \frac{4\sigma}{3R_oP_\infty}\right)^{-1}\left(\frac{C_i}{C_o} - \frac{\langle (R/R_o)^4(P_g/P_\infty)\rangle}{\langle (R/R_o)^4\rangle}\right) \qquad (1\text{-}35)$$

where $d = R_g T C_o / P_\infty$.

In Eq. (1-35), the time averages $\langle R/R_o\rangle$, $\langle (R/R_o)^4\rangle$, and $\langle (R/R_o)^4(P_g/P_\infty)\rangle$ are given by Eqs. (1-31), (1-32), and (1-33), respectively. The threshold acoustic pressure for growth of a gas bubble is obtained by setting $dR_o/dt = 0$, and results in:

$$P_A^2 = \frac{(\rho R_o^2\omega_o^2)[(1-\omega^2/\omega_o^2)^2 + b^2(\omega^2/\omega_o^2)](1 + 2\sigma/R_oP_\infty - C_i/C_o)}{(3+4K)(C_i/C_o) - \{[3(\eta-1)(3\eta-4)/4] + (4-3\eta)K\}(1 + 2\sigma/R_oP_\infty)} \qquad (1\text{-}36)$$

The expressions in Eqs. (1-35) and (1-36) are the governing equations for rectified diffusion and enable one to obtain the threshold for inception of growth and the rate of this growth. The growth rate equation (1-35) also applies below the threshold and, with $P_A = 0$, describes the case for diffusion in the absence of an applied sound field. Crum[77] has shown that these equations can be considerably reduced in complexity when one examines restricted ranges of bubble size, acoustic frequency, etc. The reader is referred to that reference if Eqs. (1-35) and (1-36) appear too formidable. However, the authors have found that the equations need to be evaluated with a computer anyway, and one may just as well put in the general forms that are applicable over a wide range of parameters.

Experimental Technique for Examining Rectified Diffusion

In this section is presented a brief description of the principal experimental technique used to make measurements of growth of bubbles by rectified diffusion. This technique was first utilized by Strasberg[65] and later with modifications added by Eller[68] and Gould.[69]

The main feature of the technique is the isolation of a single gas bubble in an acoustic stationary wave, with the ability to determine the bubble radius accurately and quickly. The specific description of the technique used by Crum is now given; Fig. 1-16 is a diagram of the experimental apparatus.

The stationary wave system was constructed by cementing a hollow glass cylinder between two matched hollow cylindrical transducers, fitted with a flexible pressure-release diaphragm on one end and open at the other. The composite system was approximately 7.5 cm in diameter by 10 cm in height, with the width of the glass in the middle about 2.5 cm. This system was driven at its $(r, \theta, z) = (2, 0, 2)$ mode at a frequency of 22.1 kHz.

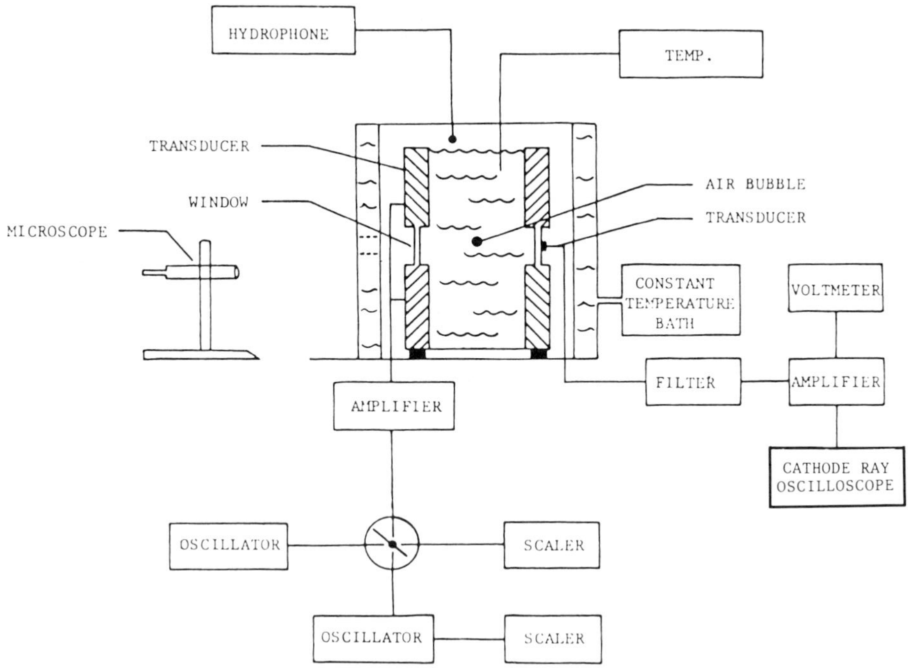

Figure 1-16. Experimental apparatus for use in rectified diffusion measurements.

When a standing wave exists in the liquid, an acoustic radiation pressure force, commonly called the primary Bjerknes force,[81] is exerted on air bubbles; this draws them toward a pressure antinode if they are less than resonance size and forces them toward the nodes if they are larger than resonance size. Because of gravity, there is also an essentially constant force of buoyancy on a gas bubble in a liquid, and this force is always directed upward. Since the primary Bjerknes force[82] is proportional to the acoustic pressure amplitude, which varies with position, there exists a position of stable equilibrium for the bubble in the "levitation chamber" somewhat above the position of the pressure antinode. With a little practice, one can easily position a bubble within the field and monitor this position with a cathetometer.

The normal approach is then to establish a fixed acoustic pressure amplitude within the chamber, which essentially fixes the position of the bubble (the position is independent of the radius if the bubble is sufficiently smaller than resonance size). The radius is then measured with some technique (such as permitting the bubble to rise at its terminal velocity, measuring this velocity, and then determining the radius from an applicable drag law) as a function of time. For threshold measurements, the acoustic pressure amplitude is constantly adjusted to ascertain the value for which a bubble of a particular measured size neither grows nor dissolves.

With this technique it appears that rectified diffusion threshold measurements can be made to within an accuracy of approximately 5%. The radii measurements depend upon the applicability of the drag law, and its specific accuracy, but are probably good to within a few percent also.

Some Experimental Results

In this section a graphical representation of the equations is presented as a function of the various physical parameters that are applicable. Some measurements of thresholds and bubble growth rates are also given for comparison. In these figures, unless specific mention is made otherwise, it is assumed that we are considering an air bubble in water with the following set of experimental conditions: $P_\infty = 1.01 \times 10^6$ dyn cm^{-2}, $\sigma = 68$ dyn cm^{-1}, $\rho = 1.0$ g cm^{-3}, $D_1 = 0.20$ cm^2 s^{-1}, $\mu = 1.0 \times 10^{-2}$ g cm^{-1} s^{-1}, $c = 1.48 \times 10^5$ cm s^{-1}, $\gamma = 1.4$, $T = 293$K, $D = 2.4 \times 10^{-5}$ cm^2 s^{-1}, and $d = 2.0 \times 10^{-2}$.

Central to the study of rectified diffusion are measurements of the acoustic pressure amplitude required for bubble growth and the agreement between these measurements and the applicable theory. Show in Fig. 1-17 are measurements of the threshold for a frequency of 22.1 kHz along with the theoretical calcula-

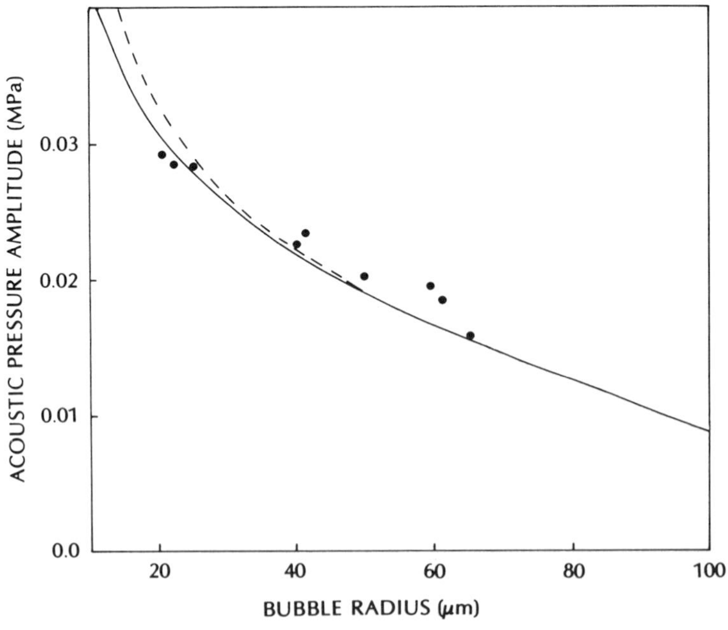

Figure 1-17. Variation of the threshold for growth by rectified diffusion of an air bubble in water as a function of the bubble radius for a frequency of 22.1 kHz, a surface tension of 68 dyn cm^{-1}, and a gas tension of 1 bar. The data are experimental measurements; the solid line is calculated from Eq. (1-36); the dashed line is from a less complicated expression from Ref. 77.

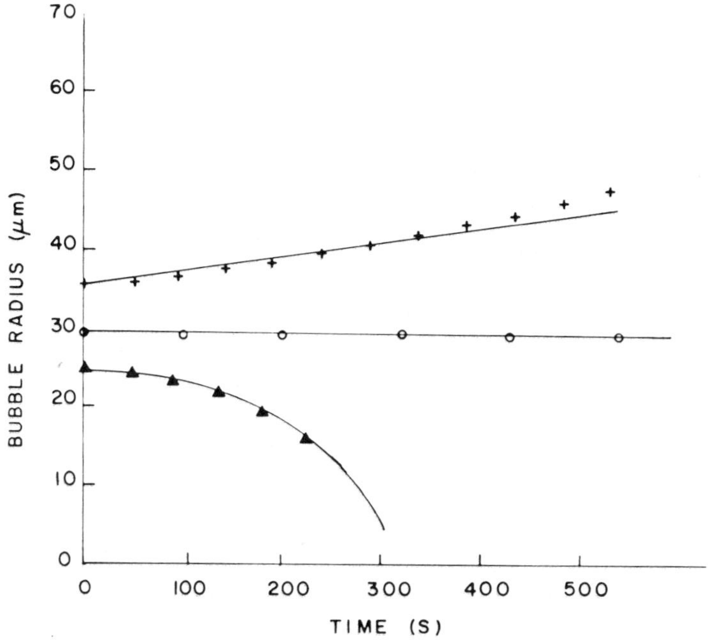

Figure 1-18. Variation of bubble radius with time for a liquid surface tension of 68 dyn cm^{-1} and an acoustic pressure of 0.27 bar. The liquid has a gas tension of 1.0 bar. The solid lines are calculated by numerical integration of Eq. (1-35).

tions. The solid curve is the prediction of Eq. (1-36); the dashed line is a simplified version of the threshold.[76]

The agreement between theory and experiment (for this limited range of bubble radii) is seen to be rather good and within the range of experimental error. This bubble range is particularly important at this frequency because it partially bridges the gap between isothermal behavior (for $R_o = 20\ \mu m$, $\eta = 1.01$) and adiabatic behavior (for $R_o = 80\ \mu m$, $\eta = 1.23$) of the bubble pulsations. The gradual reduction in the threshold as the bubble radius increases occurs mostly because as the bubble grows toward resonance size its pulsation amplitude increases, which results in more rectified mass transfer per cycle.

So far, our analysis in this section has been limited to the rectified diffusion threshold. It is also proper to examine the rate of growth by rectified diffusion when the acoustic pressure amplitude exceeds the threshold. An examination of the growth (or decay) of a bubble is shown in Fig. 1-18. The symbols are experimental measurements; the curves were obtained by numerical integration of Eq. (1-35). It was observed by Eller[68] that his measurements of the growth rate were considerably larger than predicted by theory. Gould[69] later confirmed the excessive growth rates but also noticed that surface oscillations, with their associated acoustic streaming, were probably the explanation for the increased growth rates that he observed.

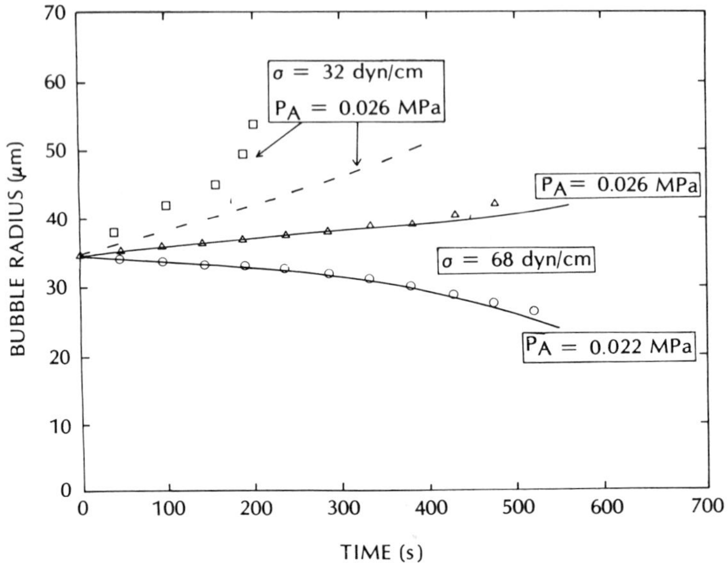

Figure 1-19. Dependence of bubble radius on time for an air bubble undergoing rectified diffusion in water. The curves are the result of numerical integration of Eq. (1-35). The lower two curves are for distilled water having a surface tension of 68 dyn cm^{-1} and different values of the acoustic pressure amplitude. The upper curve is for water containing surfactants.

Shown in Fig. 1-19 are some observations of growth rates made by Crum.[74] He found an agreement between the measured and predicted rates provided surface oscillations were avoided, and provided that the water was relatively pure and free of surface-active containants. When surface-active agents were added to the water, growth rates much higher than predicted were observed. Among the possible explanations offered for this effect were that (1) the surface active agents were behaving as a rectifying agent, permitting more diffusion in than out, and (2) acoustic streaming was occurring in the absence of surface oscillations, induced somehow by the surface contamination. The origin of this interesting anomaly is not yet determined.

In general, then, one can say that except for regions of low surface tension, the concept of rectified diffusion is rather well understood. If the bubble is driven in its linear range of pulsation amplitudes, then rectified diffusion can be described with considerable accuracy by a straightforward analytical theory.

BUBBLE DYNAMICS

In order to examine the effect of an acoustic field on a liquid it is desirable first to understand the response of a single bubble to an oscillating pressure field. In this section we present the general equations of bubble dynamics and examine, in a restricted way, the theoretical and experimental behavior of a pulsating bubble in a liquid.

Stable Cavitation

We consider a spherical bubble containing gas and vapor present in a liquid that extends to infinity. To account adequately for the motion of the bubble interface, it is necessary to solve the equations of conservation of mass, momentum, and energy for both the gas and the liquid phases and, as a boundary condition, to match these equations at the interface between the gas and the liquid. These equations are:

Conservation of mass

$$\frac{1}{\rho}\left[\frac{\partial p}{\partial t} + u\frac{\partial p}{\partial r}\right] = -\left[\frac{\partial u}{\partial r} + \frac{2u}{r}\right] \tag{1-37}$$

Conservation of momentum

$$\frac{\partial u}{\partial t} + u\frac{\partial u}{\partial r} = -\frac{1}{\rho}\frac{\partial \rho}{\partial r} \tag{1-38}$$

Conservation of energy

$$\rho c_v\left[\frac{\partial T}{\partial t} + u\frac{\partial T}{\partial r}\right] = K\nabla^2 T - p\nabla \cdot u \tag{1-39}$$

In these equations, ρ is the density, u is the radial velocity, r is the radial distance, p is the presure, T is the temperature, c_v is the specific heat at constant volume, and K is the thermal conductivity of the fluid.

It would be a formidable task indeed to solve these coupled equations for the variables of interest within the bubble: the temporally and spatially dependent pressure and temperature, the radial displacement, and the velocity of the bubble interface. A great deal of effort has been expended to approximate these equations in such a way as to obtain results that can be compared with the limited experimental results presently available. It is beyond the scope of this chapter to describe these approaches in any detail; instead, we refer the reader to the classic papers of Rayleigh,[6] Plesset,[83] Noltingk and Neppiras,[84] and Poritsky.[85] Further, some recent reviews by Neppiras,[86] Apfel,[45] Flynn,[87] and Prosperetti[88] provide a wealth of information on this topic.

In order to describe some results applicable to this study, we shall examine perhaps the most widely used equation, the Rayleigh-Plesset equation, given earlier as Eq. (1-14) and repeated below in a somewhat different form:

$$\rho(R\ddot{R} + 3\dot{R}^2/2) = [P_i(R, t) - P_A(t) - 2\sigma/R - 4\mu\dot{R}/R] \tag{1-40}$$

where in this case we have not made the polytropic approximation:

$$P_i = P_\infty(R_o/R)^{3\eta} \tag{1-41}$$

although in some cases we shall wish to do so. There is a problem with applying Eq. (1-41) to realistic situations. If the polytropic approximation is made, then the viscous damping term does not adequately account for the damping unless very small bubbles or very high frequencies are considered. For regions of common interest, the principle damping mechanism is thermal in origin; however, introducing the internal pressure via the polytropic exponent removes the phase information that accounts for this form of damping. On the other hand, retaining the internal pressure as a spatially and temporally varying quantity introduces a nearly intractable complication. The damping problem was circumvented somewhat by Prosperetti, who introduced an "effective viscosity" that could account for the additional contributions of thermal and radiation damping. Thus, one can obtain a linearized solution to the Rayleigh-Plesset equation that gives a first-order approximation to the behavior of a pulsating gas bubble in a liquid.

The linear or first-order approach to this problem is to assume the internal pressure is given by the polytropic approximation and the radius of the bubble can be written as:

$$R = R_o(1 + x) \tag{1-42}$$

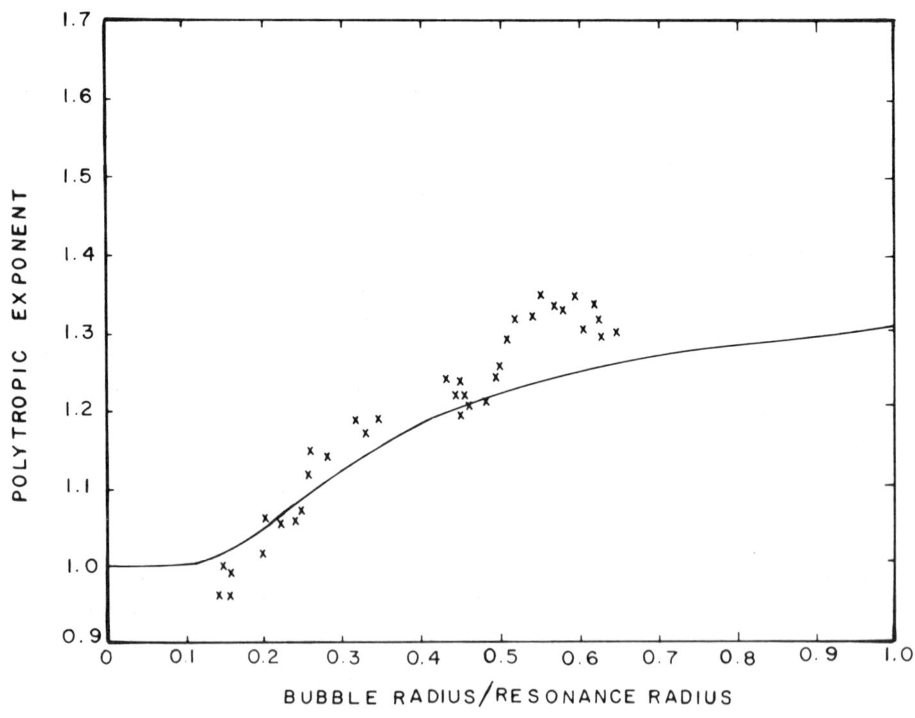

Figure 1-20. Variation of the polytropic exponent with bubble radius for air bubbles in water. The symbols are experimental measurements and the solid line is the theoretical prediction.

where x is given by Eqs. (1-20) through (1-24). This approach enables one to obtain an analytical expression that closely approximates the threshold and growth rate for rectified diffusion, as shown earlier in Figs. 1-17 and 1-18, and enables one to measure indirectly the variation of the polytropic exponent with bubble radius, as shown in Fig. 1-20. This figure shows that the linear approximation described above gives a general description of the variation of the polytropic exponent with bubble radius until the ratio of bubble size to resonance size approaches 0.5. At this point, the second harmonic resonance of the bubble is excited and the agreement between the linear theory and the experimental measurements fails.

This failure demonstrates an important caveat that must be acknowledged when dealing with the oscillations of a gas bubble in a liquid: a pulsating gas bubble is a highly nonlinear system, and its nonlinear behavior can be activated at very low values of the acoustic pressure amplitude. For example, the average acoustic pressure amplitude required to introduce these nonlinear effects was on the order of 0.25 bar (0.025 MPa).

To follow up on this point consider Fig. 1-21, which is a plot of the relative pulsation amplitude, $X_m = (R_{max} - R_o)/R_o$, versus the ratio of the driving frequency to the resonance frequency for a 1.0-μm bubble in water. The figure

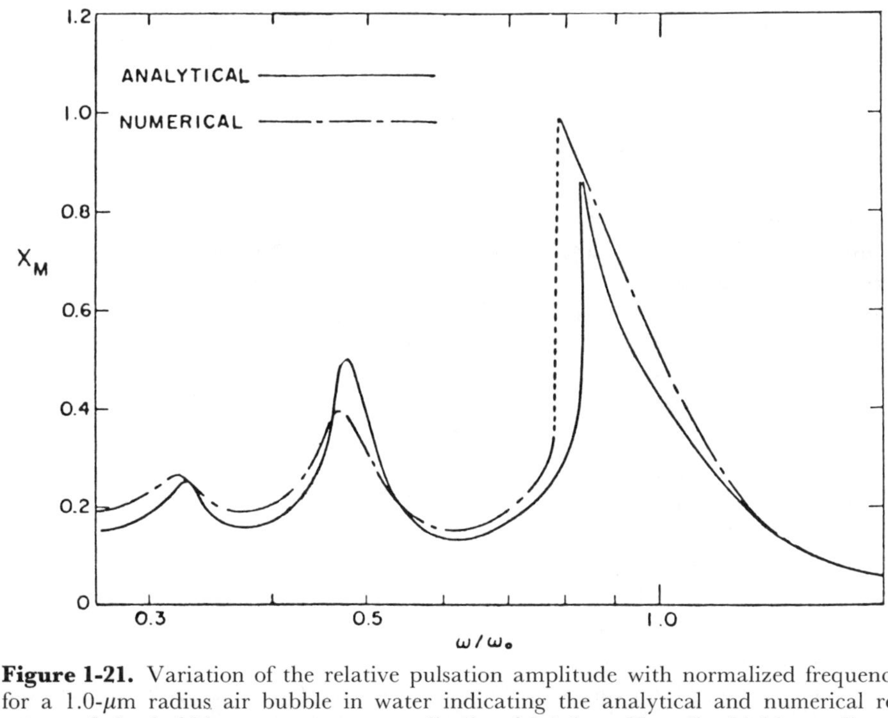

Figure 1-21. Variation of the relative pulsation amplitude with normalized frequency for a 1.0-μm radius air bubble in water indicating the analytical and numerical response of the bubble to a pressure amplitude of 0.9 bar. Note the highly nonlinear behavior.

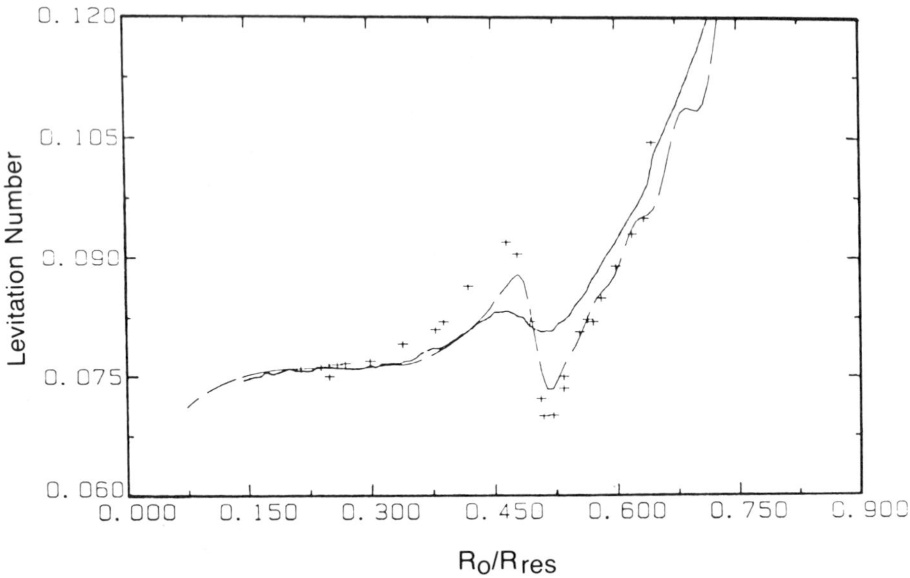

Figure 1-22. Variation of the levitation number with normalized bubble radius for an air bubble in a glycerine–water mixture. The solid line represents a numerical solution of Eq. (1-14) using a polytropic approximation and an "effective viscosity" used to account for thermal damping. The broken line is a result from Ref. 89, which calculates thermal damping directly from internal pressure. The driving frequency was 22.2 kHz and the resonance radius of the bubble was 138 μm.

shows the results of a numerical calculation by Lauterborn and an approximate analytical solution by Prosperetti[89] of the Rayleigh-Plesset equation. Note that a smooth linear resonance curve is not observed; there is, in fact, considerable nonlinear structure in the form of higher harmonics. The amplitude of these higher harmonics can be quite considerable. Some experimental confirmation of these results is shown in Fig. 1-22. This plot shows the variation of the levitation number, a useful nondimensional number that is quite similar to the X_m of Lauterborn and Prosperetti, plotted versus the ratio of the measured radius to the resonance radius (approximately equal to ω/ω_o) for an air bubble in a glycerine–water mixture. The experimental points show that this nonlinear behavior is easily observed; the solid curves are theoretical calculations indicating that there is not good agreement between the calculated and measured values. The solid curve shows the analytical result expected if one assumes a polytropic approximation; the dashed curve is a numerical result in which the internal pressure is calculated directly, and the thermal damping obtained from this result. It is seen that the damping is still overestimated and further work needs to be done in this area.

When oscillations of a stable gas bubble are driven at higher and higher amplitudes, it is seen that not only is a wide spectrum of harmonics generated by

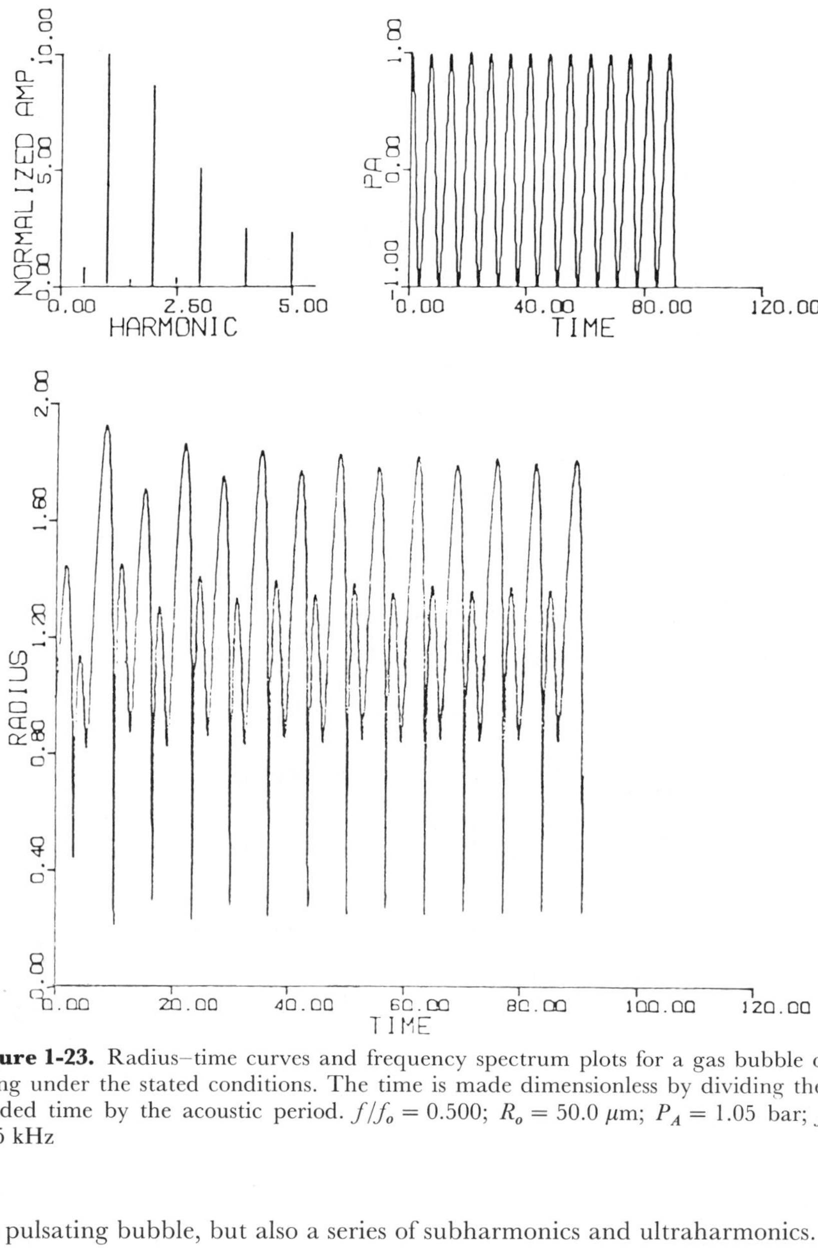

Figure 1-23. Radius–time curves and frequency spectrum plots for a gas bubble oscillating under the stated conditions. The time is made dimensionless by dividing the expended time by the acoustic period. $f/f_o = 0.500$; $R_o = 50.0\ \mu$m; $P_A = 1.05$ bar; $f_o =$ 60.5 kHz

the pulsating bubble, but also a series of subharmonics and ultraharmonics. We have examined numerically the large-amplitude oscillations of a gas bubble to observe the spectral content of these pulsations. Figure 1-23 shows that when a 50-μm bubble is driven at an acoustic pressure amplitude of 1.05 bar, the bubble radius repeats itself with the same period as the driving frequency. This is called period-one behavior and the Fourier spectrum shows only harmonics of the

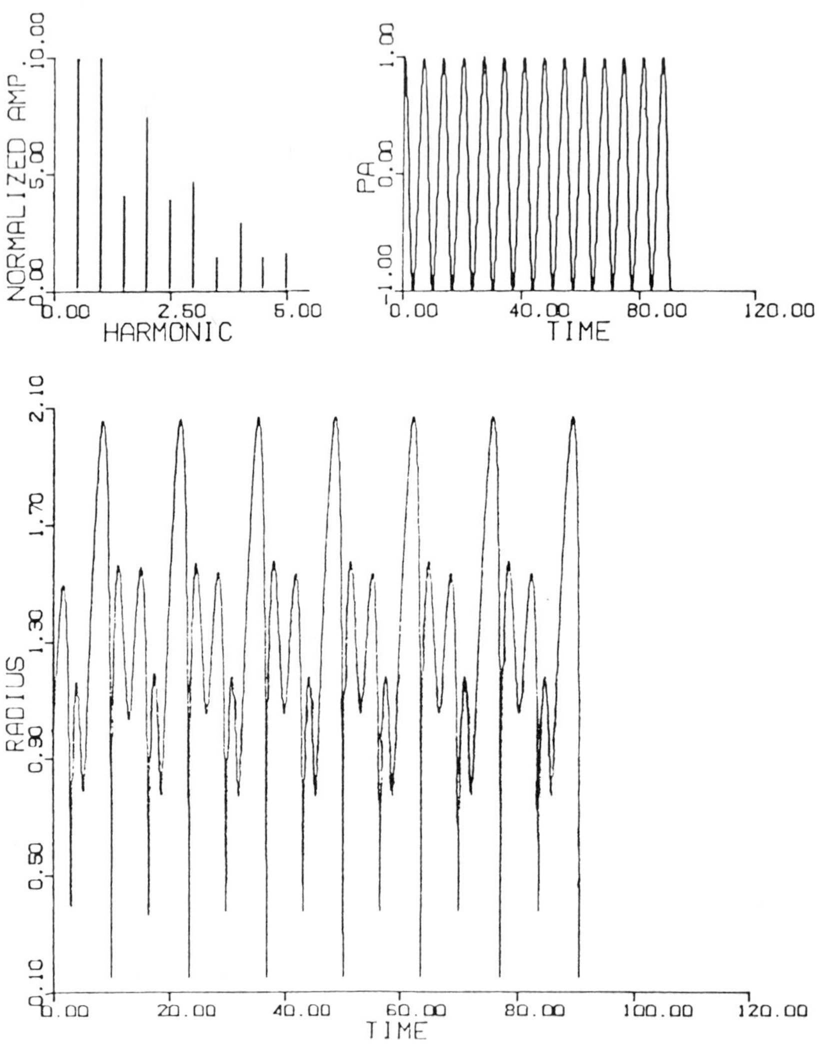

Figure 1-24. Radius–time curves and frequency spectrum plots for a gas bubble oscillating under the stated conditions. The time is made dimensionless by dividing the expended time by the acoustic period. $f/f_o = 0.500$; $R_o = 50.0$ μm; $P_A = 1.15$ bar; $f_o =$ 60.5 kHz

driving frequency. Figure 1-24 shows that when the amplitude is increased slightly to 1.15 bar, the radius now repeats itself after two cycles of the driving frequency. This is called period-two behavior and the spectrum now indicates the presence of a strong subharmonic. Because the presence of the subharmonic is easily detected and because it is somewhat indicative of violent bubble

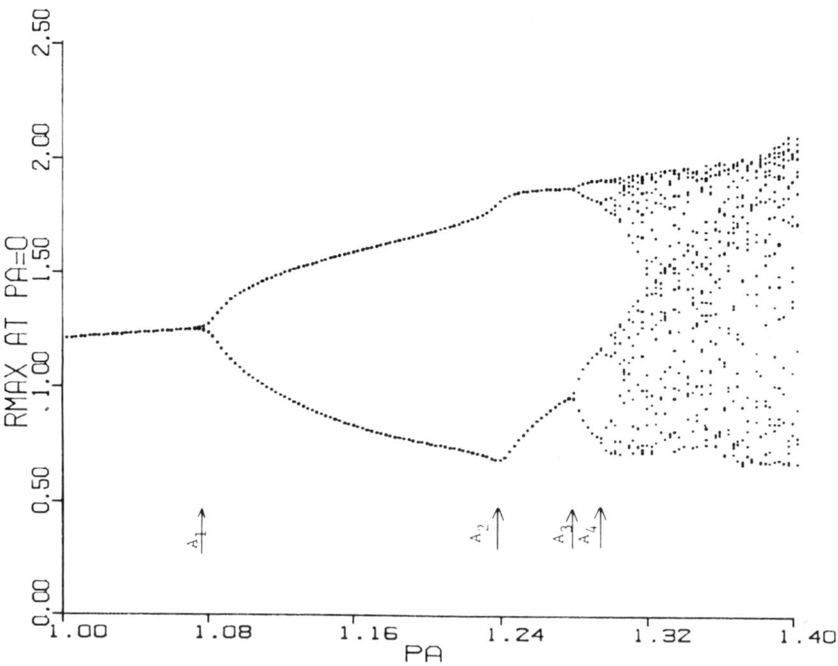

Figure 1-25. Maximum bubble radius at zero pressure amplitude. The pattern, or Feigenbaum "tree," shows the systematic sequence of bifurcations leading to chaos for a gas bubble of 50 μm driven at approximately half its resonance frequency. The acoustic pressure amplitude is given in bars.

behavior, it is often used as a criterion for cavitation inception. These figures indicate to us that such a definition of cavitation inception is arbitrary, at best.

As an interesting sidelight of bubble oscillations, one can follow this bifurcation sequence from period one to period two, from two to four, from four to eight, etc., until the bubble exhibits chaotic behavior that has some inherent determinism. Figure 1-25 shows a plot of a systematic progression to chaos following the bifurcation scheme described above. This new area of nonlinear mechanics is of considerable current interest and activity.

Finally, Prosperetti et al.[90] have recently obtained a more precise formulation of the bubble-oscillation problem enabling one to estimate numerically the temporal and spatial dependence of the temperature and pressure within a "stable" oscillating bubble. Figure 1-26, for example, shows that a 50-μm bubble driven at a relatively modest acoustic pressure amplitude of 0.5 bar can reach an internal temperature of nearly 3000K when driven at a frequency that is near its second harmonic resonance. Such a cavity could quite conceivably maintain its

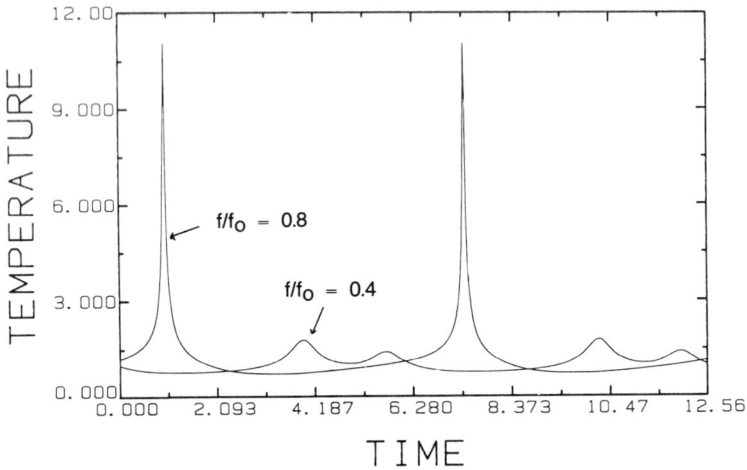

Figure 1-26. Variation of the temperature inside a pulsating gas bubble as a function of time for different values of the driving frequency. The bubble has a radius of 50 μm and the acoustic pressure amplitude is 0.5 bar. The time is normalized by the acoustic period.

integrity (not break up) and would probably also radiate light by sonoluminescence. Figure 1-27 describes sonoluminscence observed by an image intensifier camera from a cavitation field that in the view of the investigator contained stable cavities rather than transient ones.[96]

Some important points can be made from these observations of the violent behavior exhibited by stable bubbles: if the bubbles oscillate with pulsation amplitudes sufficient to produce sonoluminescence, then free radicals can be produced. If the free radicals are produced each cycle, then an enormous number of these highly reactive species may be produced as a result of the ultrasound. In fact, a larger number would be produced by "stable" bubbles that create free radicals each cycle than by transient events that destroy themselves after a few cycles. The authors believe that these results have important implications in the biological effects of therapeutic ultrasound in the light of the recent discovery that individual gas bubbles have been observed in the insonified region of living tissue.[59,91]

Transient Cavitation

A small gas bubble subject to an acoustic pressure amplitude of more than a few bars experiences such violent radial pulsations that the wall velocity of the collapsing bubble approaches the velocity of sound. At this point, the likelihood is small that the bubble can maintain its physical integrity and it will most probably shatter upon collapse. Apfel[92,93] and Flynn[94] have recently examined this problem in some detail and have published theoretical threshold curves for

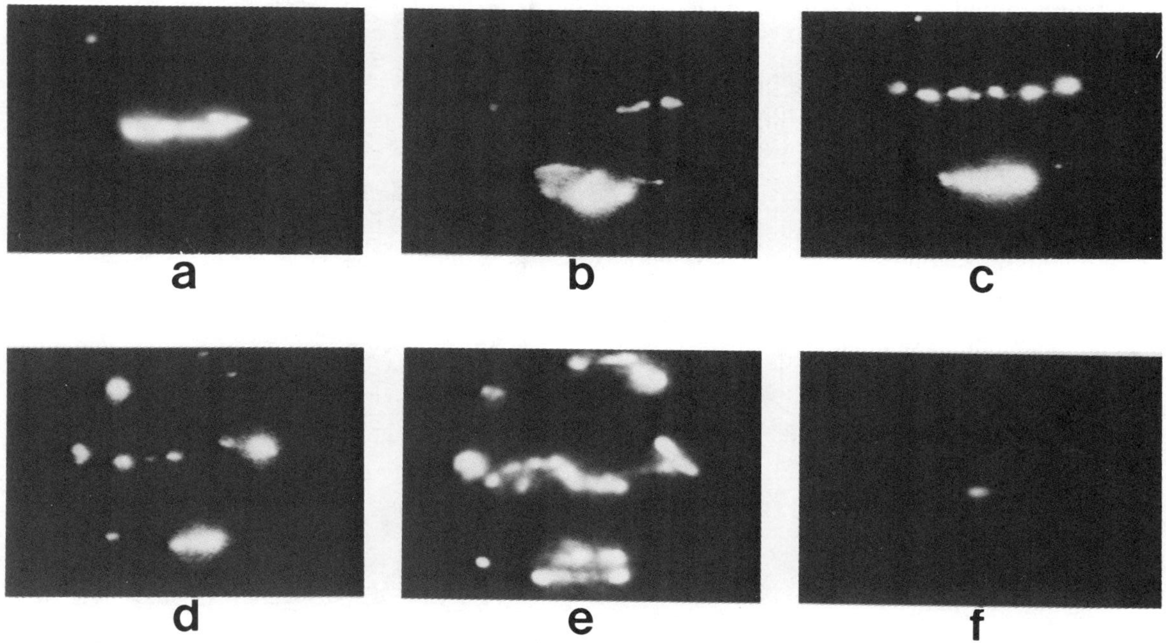

Figure 1-27. Photographs taken from a video monitor of sonoluminescence activity viewed by an image intensifier tube for increasing levels of acoustic pressure amplitude. Cavitation was produced by an acoustic horn radiating into a flash of water. The horn was driven at its lowest level in (a) and at its highest level in (f), where the standing wave pattern is disrupted by violent cavitation on the horn tip. The "layers" of activity arise from bubbles trapped near antinodal planes separated by approximately a half wavelength.

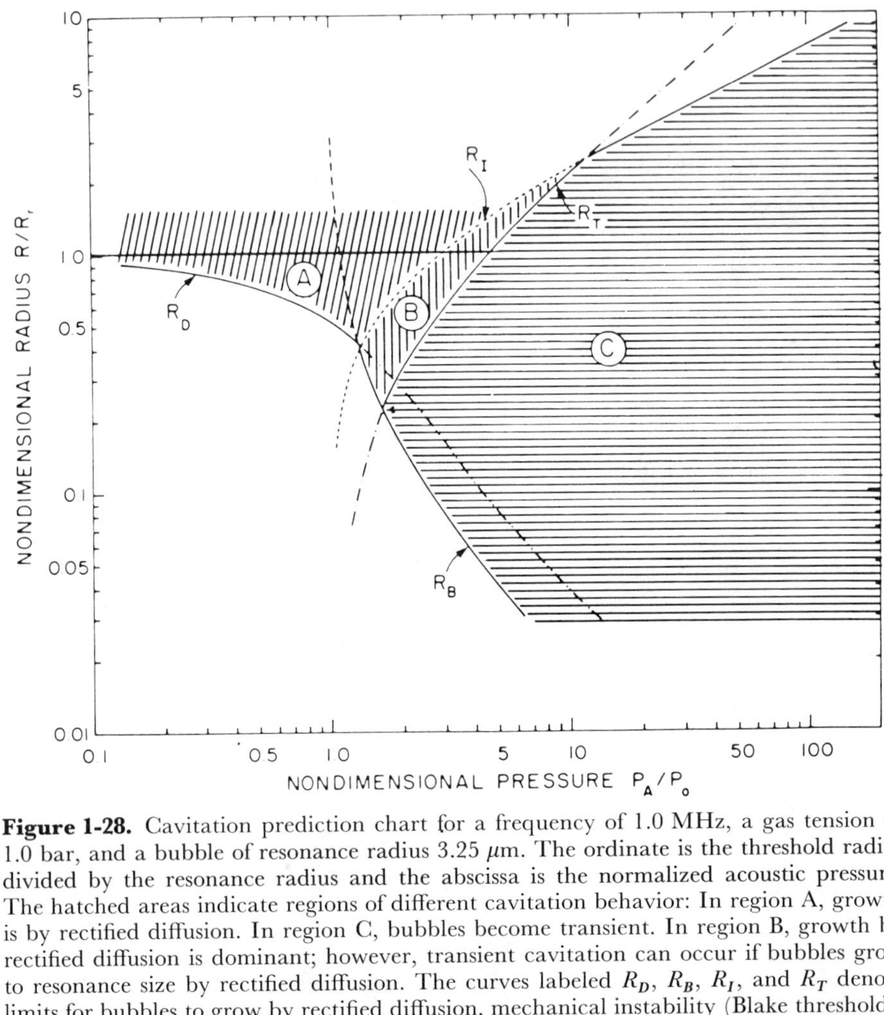

Figure 1-28. Cavitation prediction chart for a frequency of 1.0 MHz, a gas tension of 1.0 bar, and a bubble of resonance radius 3.25 μm. The ordinate is the threshold radius divided by the resonance radius and the abscissa is the normalized acoustic pressure. The hatched areas indicate regions of different cavitation behavior: In region A, growth is by rectified diffusion. In region C, bubbles become transient. In region B, growth by rectified diffusion is dominant; however, transient cavitation can occur if bubbles grow to resonance size by rectified diffusion. The curves labeled R_D, R_B, R_I, and R_T denote limits for bubbles to grow by rectified diffusion, mechanical instability (Blake threshold), inertial processes, or by becoming transient, respectively. (Reproduced from Ref. 93 by permission.)

transient cavitation inception. Figure 1-28 shows regions where transient cavitation is to be expected for insonification of water with high-intensity ultrasound. Apfel's numerical study enables one to obtain an estimate of the thresholds for cavitation inception and what type of cavitation is to be expected. Flynn has examined the case of short-pulse excitation also. In Flynn's study, the response of a small microbubble to a Gaussian acoustic pulse is examined. The wholly expected, but vitally important, result is that violent transient cavitation can be expected from acoustic pulses similar to those used in clinical diagnostic ultrasound devices. Figure 1-29 shows some typical results obtained by Flynn.

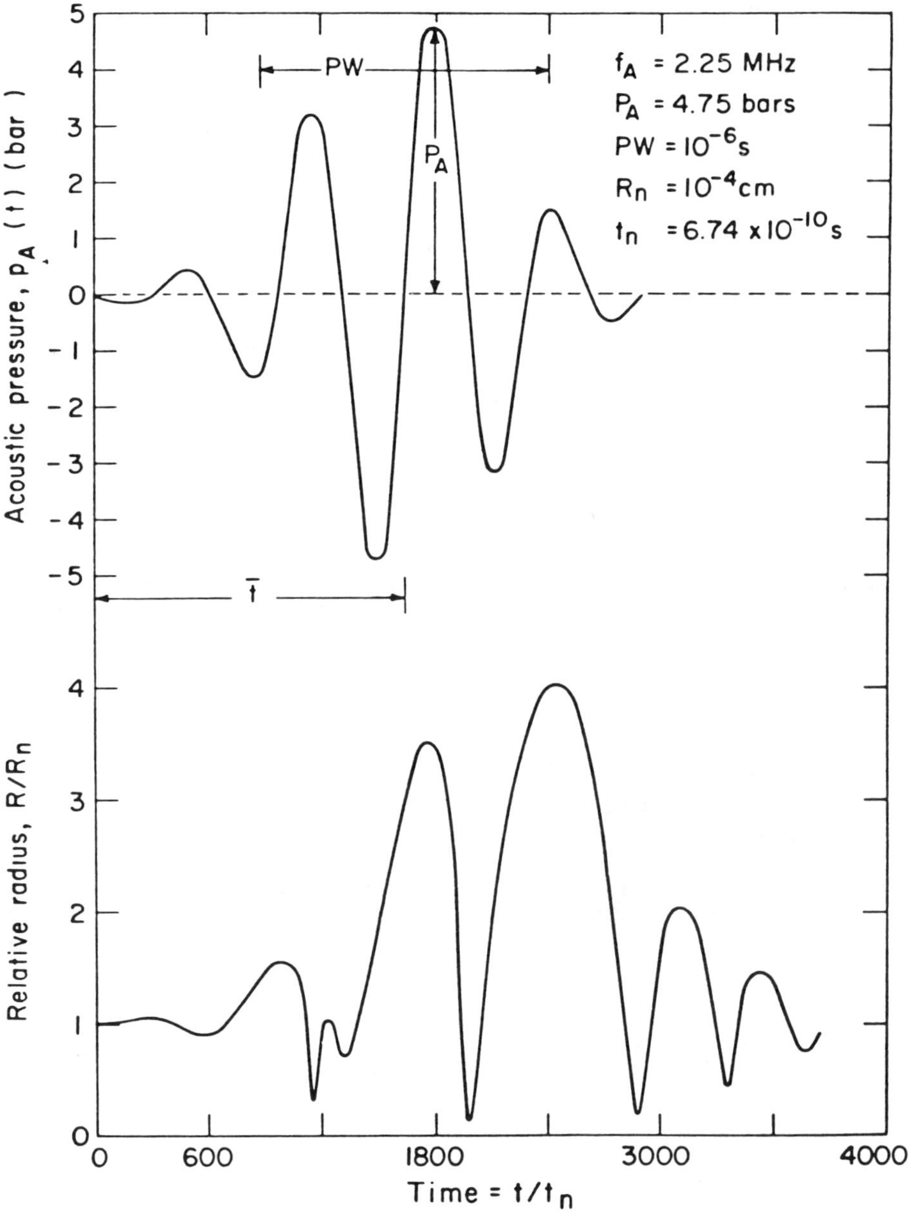

Figure 1-29. Acoustic pressure amplitude (upper curve) and relative bubble radius (lower curve) versus nondimensional time. The acoustic pulse has a frequency of 2.25 MHz, a maximum pressure amplitude of 4.75 bar, and a width of 1 μs. The bubble has an equilibrium radius of 1 μm. (Reproduced from Ref. 94 with permission.)

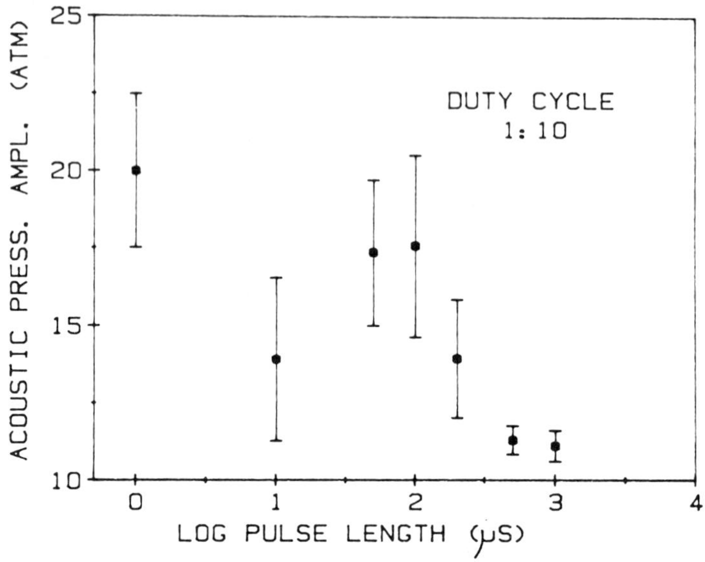

Figure 1-31. Variation of the cavitation threshold of water containing luminol with duty cycle for a pulse length of approximately 1 μs. The acoustic frequency was 1.0 MHz and the data are from Ref. 95.

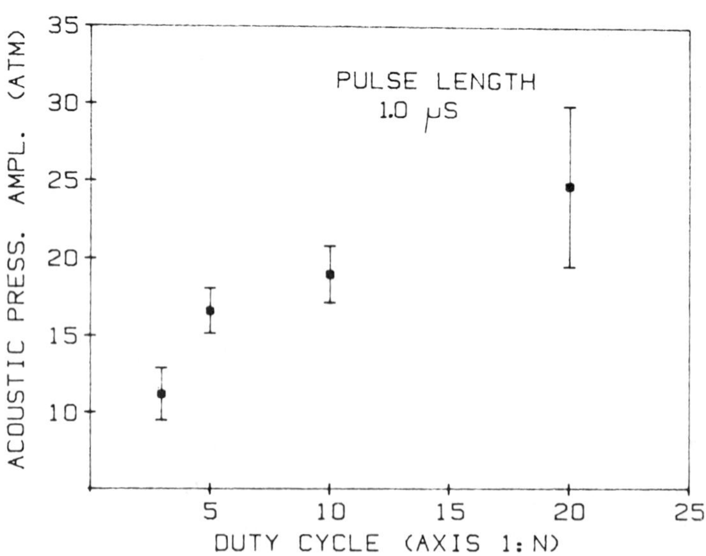

Figure 1-30. Cavitation threshold versus pulse length for a duty cycle of 1 : 10. The points are from Ref. 95; the error bars indicate ±1 SD. The acoustic frequency was 1.0 MHz and the liquid was water containing luminol and saturated with argon.

Recently, some experimental confirmation of these predictions has been obtained by Crum and Fowlkes.[95] In their study, they irradiated a liquid containing luminol, a chemiluminescent material, and observed light emission indicative of chemiluminescence for acoustic pulses as short as 1.0 μs. Shown in Fig. 1-30 are measurements of the cavitation threshold, as determined by sufficient levels of emitted light, as a function of pulse width for approximately 1.0 μs pulses of 1.0-MHz ultrasound. For this case the duty cycle was 1:10. Here duty cycle refers to the ratio of the "on" time to the total time expended. Since there was some concern that the cavitation nuclei might be responding to the pulse repetition frequency rather than the individual pulses, an examination of the threshold as a function of duty cycle was made. Shown in Fig. 1-31 is the variation of the threshold with duty cycle and there are indications that a limiting threshold is expected.

These data give strong evidence that an extensive risk assessment of diagnostic ultrasound should be carried out in the immediate future. Cavitation is a violent phenomenon, and at high frequencies, hard to detect. Further research in this area is greatly needed and promises exciting rewards.

CONCLUDING REMARKS

The authors wish to take this opportunity to point out some of the phenomena of acoustic cavitation that they believe will be worthy of future investigation and that have promise of meaningful contributions to the advancement of science.

Cavitation Produced in Biological Tissue

High-intensity ultrasound was once thought to be perfectly safe for widespread clinical use (and still is by the overwhelming majority of investigators). However, there is an increasingly large body of information that suggests that cavitation is produced by both therapeutic and diagnostic ultrasound and that there may be risks associated with its use. Some areas that need future study are:

1. *Microbubble nucleation and stabilization.* What is the nature of stabilized microbubbles in tissue? What is the stabilization mechanism? In what ways can they be activated? What is their anatomical distribution?
2. *Biological effects.* What are the damage mechanisms for cavitation in tissue: shockwaves, high-velocity jets, free-radical production? Are the effects observable on a short-term macroscopic level, or only on a cellular or long-term level?
3. *Continuous-wave studies.* Can the cavitation nuclei present in tissue be made to grow by rectified diffusion? Will the stable bubbles generate sufficiently high temperatures to produce free radicals? What are the requirements for growth in tissue?

4. *Pulsed studies.* Can one make the translation from cavitation in water to cavitation in tissue? What are the effects of nucleus stabilization mechanisms on cavitation inception by pulsed ultrasound? Are there frequency bandwidths to be avoided?

Nonlinear Bubble Dynamics

Nonlinear bubble dynamics may seem esoteric and mathematically complex, but most sound fields of interest in sonochemistry are larger than 0.50 bar, a pressure amplitude above which nonlinear effects in bubble dynamics must be taken into account. Some areas that are as yet unexplained are:

1. *Damping mechanisms.* What evidence is there that the damping at relatively large bubble pulsation amplitudes is properly accounted for by linear theories? How important is the discrepancy?
2. *Noise production.* What correlation exists between spectral components of the radius–time curve of a bubble and the radiated acoustic signals? Can the noise radiated by a cavitation field be related to the pulsations of individual gas bubbles?
3. *Acoustic chaos.* What relationships exist between the concept of deterministic chaos in nonlinear mechanics and the oscillation of a single gas bubble as it is driven at higher and higher amplitudes? Can this systematic approach to chaos be observed experimentally with a single bubble?

Cavitation Fields

Cavitation fields have not been discussed in this chapter. This term refers to the fact that most cavitation utilized in sonochemistry arises not from isolated single bubbles but from a collection of clouds of bubbles. These bubbles interact with each other and behave as a group, rather than individually. Some questions of interest are:

1. *Bubble–bubble interactions.* What is the nature of the interactions between bubbles? Is it governed by simple Bjerknes forces or is it more complex? Can a cloud of bubbles give rise to stronger effects than individual bubbles?
2. *Propagation of sound through bubble mixtures.* How is the propagation of sound affected by a bubbly mixture? At what concentration of bubbles does a saturation effect set in? What theoretical approaches are necessary to examine these phenomena?

ACKNOWLEDGMENTS

The authors wish to gratefully acknowledge the financial support of the Office of Naval Research and the National Institutes of Health.

REFERENCES

1. Tomlinson, C. *Phil. Mag.* **1867**, *34*, 136, 220.
2. Gernez, M. *Phil. Mag.* **1867**, *33*, 379.
3. Bernoulli, D. "Hydrodynamica, sive de viribus et motibus fluidorum comentarii"; Strasbourg, 1738.
4. Freed D.; Walker, W. F. In "Cavitation amd Multiphase Flow Forum—1984; Hoyt, J. W., Ed.; American Society for Mechanical Engineers, New York, **1984**, p. 1.
5. Thornycroft, J.; Barnaby, S. W. *Inst. Civ. Eng.* **1895**, *122*, 51.
6. Lord Rayleigh. *Phil. Mag.* **1917**, *34*, 94.
7. Blake, F. G. "The tensile strength of liquids: A review of the literature"; Tech. Memo. No. 9, Harvard Acoustics Research Laboratory, 1949.
8. Klein, E. *J. Acoust. Soc. Am.* **1948**, *20*, 601.
9. Kornfeld, M.; Suvorov, L. *J. Appl. Phys.* **1944**, *15*, 495.
10. Briggs, L. J. *J. Appl. Phys.* **1950**, *21*, 721.
11. Dean, R. B. *J. Appl. Phys.* **1944**, *15*, 446.
12. Harvey, E. N.; Barnes, K. K; McElroy, W. D; Whitely, A. H; Pease, D. C.; Cooper, K. W.; *J. Cell. Comp. Physiol.* **1944**, *24*, 1.
13. Briggs, H. B.; Johnson, J. B.; Mason, W. P. *J. Acoust. Soc. Am.* **1947**, *19*, 664.
14. Seitz, F. *Phys. Fluids.* **1958**, *1*, 2.
15. Lieberman, D. *Phys. Fluids.* **1959**, *2*, 466.
16. Fox, F.; Herzfeld, K. *J. Acoust. Soc. Am.* **1954**, *26*, 984.
17. Strasberg, M. *J. Acoust. Soc. Am.* **1959**, *31*, 163.
18. Connolly, W.; Fox, F. *J. Acoust. Soc. Am.* **1954**, *26*, 843.
19. Galloway, W. *J. Acoust. Soc. Am.* **1954**, *26k*, 849.
20. Sette, D.; Wanderlingh, F. *Phys. Rev.* **1962**, *125*, 409.
21. Akulichev, V. *Sov. Phys. Acoust.* **1966**, *12*, 144.
22. Greenspan, M.; Tschiegg, C. *J. Res. Natl. Bur. Stds.* **1967**, *71C*, 299.
23. Apfel, R. *J. Acoust. Soc. Am.* **1970**, *48*, 1179.
24. Apfel, R. *J. Chem. Phys.* **1971**, *54*, 62.
25. Crum, L. *Nature (London)* **1979**, *278*, 148.
26. Winterton, R. *J. Phys. D: Appl. Phys.* **1977**, *10*, 2041.
27. Yount, D. *J. Acoust. Soc. Am.* **1979**, *65*, 1429.
28. Sirotyuk, M. *Sov. Phys. Acoust.* **1970**, *16*, 237.
29. Yount, D. *J. Acoust. Soc. Am.* **1982**, *71*, 1473.
30. Ward, C. A.; Balakrishnav, A.; Hooper, F. C. *J. Basic Eng.* **1970**, *92*, 695.
31. Tikuisis, P; Ward, C. A.; Venter, R. D. *J. Appl. Phys.* **1983**, *54*, 1833.
32. Atchley, A. A. Ph.D. Dissertation, The University of Mississippi, Oxford, MS 1984.
33. Atchley, A. A.; Prosperetti, A. Submitted for publication, *J. Acoust. Soc. Am.*, **1987**.
34. Döring, W. *Z. Phys. Chem.* **1937**, *B36*, 371.
35. Döring, W. *Z. Phys. Chem.* **1936**, *B38*, 292.
36. Volmer, M. In "De Chemische Reaktion, Band IV"; Verlag von Theodor Steinkopff: Dresden, Leipzig, 1939.
37. Fisher, J. *J. Appl. Phys.* **1948**, *19*, 1062.
38. Apfel, R. *J. Acoust. Soc. Am.* **1971**, *49*, 145.
39. Epstein, P.; Plesset, M. *J. Chem. Phys.* **1950**, *18*, 1505.
40. Arnold, H. D. *Phil. Mag.* **1911**, *22*, 755.
41. Langmuir, I. *J. Am. Chem. Soc.* **1917**, *39*, 1848.
42. Herzfeld, K. "Proc. First Sympos. Naval Hydrodynamics"; F. S. Sherman, Ed.; 1957, p. 319.
43. Alty, T. *Proc. R. Soc.* **1924**, *A106*, 315.
44. Alty, T. *Proc. R. Soc.* **1926**, *A112*, 235.
45. Apfel, R. In "Ultrasonics"; Edmonds, P. Ed.; Academic Press: New York, 1981, Vol. 19, Ch. 7, pp. 355–441.
46. Hayward, A. *J. Phys. D: Appl. Phys.* **1970**, *3*, 574.
47. Bargeman, D.; Van Voorst Vader, F. *J. Colloid Interf. Sci.* **1973**, *42*, 467.
48. Washburn, E. W., Ed. "International Critical Tables"; McGraw-Hill: New York, 1928, Vol. 3, p. 257.

49. Weast, R. C., Ed. "Handbook of Chemistry and Physics," 56th ed.; CRC Press: Boca Raton, 1975, F-43.
50. Eisenberg, D.; Kauzmann, W.; "The Structure and Properties of Water"; Oxford University Press: Oxford, 1969, pp. 61–62.
51. Willard, G. W. *J. Acoust. Soc. Am.* **1953**, *25*, 669.
52. Meyer, E.; Newmann, E. "Physical and Applied Acoustics—An Introduction"; Academic Press: New York, 1972.
53. Committee of the American Institute for Ultrasound in Medicine. *J. Ultrasound. Med.* **1983**, *2*(4), S7.
54. Barger, J. "Thresholds of Acoustic Cavitation in Water"; Tech. Memo. No. 57, Harvard Acoustics Research Laboratory, 1964.
55. Vaughn, P. W.; Graham, E.; Leeman, S. *Appl. Sci. Res.* **1982**, *38*, 45.
56. Roy, R. A.; Atchley, A. A.; Crum, L. A.; Fowlkes, J. B.; Reidy, J. J. *J. Acoust. Soc. Am.* **1985**, *98*, 1799.
57. Crum, L. A. In "Cavitation and Inhomogeneities in Underwater Acoustics"; Lauterborn, W., Ed.; Springer-Verlag: New York, 1980, p. 84.
58. Crum, L. A.; Brosey, J. E. *J. Fluid Eng.* **1984**, *106*, 99.
59. ter Haar, G.; Daniels, S. *Phys. Med. Biol.* **1981**, *26*, 1145.
60. Sommer, F. G.; Pounds, D. *Med. Phys.* **1982**, *9*, 1.
61. Neppiras, E. A.; Coakley, W. T. *J. Sound Vib.* **1970**, *45*, 341.
62. Apfel, R. E. *J. Acoust. Soc. Am.* **1981**, *69*, 1624.
63. Blake, F. G. "The Onset of Cavitation in Liquids"; Tech. Memo. No. 12, Harvard Acoustics Research Laboratory, 1949.
64. Hsieh, D. Y.; Plesset, M. D. *J. Acoust. Soc. Am.* **1961**, *33*, 206.
65. Strasberg, M. *J. Acoust. Soc. Am.* **1961**, *33*, 359.
66. Eller, A. I.; Flynn, H. G. *J. Acoust. Soc. Am.* **1965**, *37*, 493.
67. Safar, M. H. *J. Acoust. Soc. Am.* **1968**, *43*, 1188.
68. Eller, A. I. *J. Acoust. Soc. Am.* **1969**, *46*, 1246.
69. Gould, R. K. *J. Acoust. Soc. Am.* **1974**, *56*, 1740.
70. Davison, B. J. *J.. Sound Vib.* **1971**, *17*, 261.
71. Kapustina, O. A.; Statnikov, V. G. *Sov. Phys. Acoust.* **1968**, *13*, 327.
72. Skinner, L. A. *J. Acoust. Soc. Am.* **1972**, *51*, 378.
73. Eller, A. I. *J. Acoust. Soc. Am.* **1975**, *57*, 1374.
74. Crum, L. A. *J. Acoust. Soc. Am.* **1980**, *68*, 203.
75. Lewin, P. A.; Bjorno, L. *J. Acoust. Soc. Am.* **1981**, *69*, 1624.
76. Crum, L. A.; Hansen, G. M. *Phys. Med. Biol.* **1982**, *27*, 413.
77. Crum, L. A. *Ultrasonics* **1984**, *22*, 215.
78. Prosperetti, A. *J. Acoust. Soc. Am.* **1977**, *61*, 17.
79. Eller, A. I. *J. Acoust. Soc. Am.* **1970**, *47*, 1469.
80. Devin, C. *J. Acoust. Soc. Am.* **1959**, *31*, 1654.
81. Bjerknes, V. F. K. "Die Kraftfelder"; Vieweg and Sohn: Braunschweig, 1909.
82. Crum, L. A. *J. Acoust. Soc. Am.* **1975**, *57*, 1363.
83. Plesset, M. In "Cavitation in Real Liquids"; Davies, R., Ed.; American Elsevier: New York, 1964, p. 1.
84. Noltingk, B. E.; Neppiras, E. A.; *Proc. Phys. Soc. London* **1951**, *B64*, 1032.
85. Poritsky, H. In "Proc. First. U.S. Naval Congress. Appl. Mechj." Sternberg, E., Ed.; Am. Soc. Mech. Eng.: New York, 1952, p. 813.
86. Neppiras, E. A. *Phys. Rep.* **1980**, *61*, 159.
87. Flynn, H. G. In "Physical Acoustics"; Mason, W. P., Ed.; Academic Press; New York, 1964, Vol. IB, Ch. 9, pp. 57–172.
88. Prosperetti, A. *Ultrasonics* **1984**, *22*, 77.
89. Prosperetti, A. *J. Acoust. Soc. Am.* **1974**, *56*, 878.
90. Prosperetti, A.; Commander, K.; Crum, L. A. To be published.
91. ter Haar, G.; Daniels, S.; Eastaugh, K. C.; Hill, C. R.; *Br. J. Cancer* **1983**, *45*, Suppl. 5, 151.
92. Apfel, R., *Br. J. Cancer.* **1982**, *45. Suppl. 5*, 140.
93. Apfel, R. In "Cavitation and Inhomogeneities in Underwater Acoustics"; Lauterborn, W., Ed.; Springer-Verlag: New York, 1980, p. 79.
94. Flynn, H. G., *J. Acoust. Soc. Am.* **1982**, *72*, 1926.
95. Crum, L. A.; Fowlkes, J. B. *Nature* **1986**, *319*, 52.
96. Crum, L. A.; Reynolds, G. T. *J. Acoust. Soc. Am.* **1985**, *78*, 137.

2

Other Nonlinear Acoustic Phenomena

JAMES A. ROONEY

Jet Propulsion Laboratory,
California Institute of Technology

INTRODUCTION

Many of the phenomena in acoustics can be described and understood using theory that is linear in acoustic amplitude or uses the first terms of nonlinear equations. These phenomena typically occur in situations in which the displacement amplitude of the sound wave is small. For other phenomena, particularly those relating to or causing chemical, physical, and biological effects, the displacement amplitudes are sufficiently large that nonlinear or higher order terms in the equations must be retained to describe them accurately.

These nonlinear aspects of acoustics are particularly important for the topics and effects discussed in this volume. Their impact ranges from effects on the propagation of sound to mechanisms for the effects described and calibration of the sound fields used. In addition, the nonlinear phenomena are of interest in themselves and provide a unique opportunity to gain insight into both the interaction of sound with matter and the structure of matter itself.

NONLINEAR ASPECTS OF THE PROPAGATION OF SOUND

The pressure amplitudes of ultrasound used in medical imaging, as well as for producing biological or chemical effects, can be such that linear theory may no longer be accurately applied. These high-pressure (finite amplitude) waves distort as they propagate through the medium, which leads to harmonic generation and greater loss of energy. These nonlinear effects can therefore influence the acoustic field distribution and energy deposition, and they may possibly be used for tissue characterization. The basic theory for nonlinear propagation is briefly summarized here to provide a basis for discussion.

An extensive literature[1–5] has been developed concerning the theory describing the nonlinear effects on propagation of sound. Only certain aspects are outlined here in order to place the experimental methods used to study the associated phenomena in perspective. Since the propagation of sound can be described as an isoentropic process, the equation of state can be expanded in a Taylor series in the following manner:

$$P = P_0 + \left(\frac{\partial P}{\partial \rho}\right)_{s,\rho=\rho_0} (\rho - \rho_0) + \frac{1}{2}\left(\frac{\partial^2 P}{\partial \rho^2}\right)_{s,\rho=\rho_0} (\rho - \rho_0)^2 + \cdots \quad (2\text{-}1)$$

where P is pressure, P_0 is static pressure, ρ is density and ρ_0 is unperturbed density. It has become customary to keep only the terms shown and express Eq. (2-1) as:

$$P - P_0 = A\left(\frac{\rho - \rho_0}{\rho_0}\right) + \frac{B}{2}\left(\frac{\rho - \rho_0}{\rho_0}\right)^2 \quad (2\text{-}2)$$

Here A and B are temperature-dependent coefficients with $A = \rho_0 c_0^2$ and the quantity B/A, which is often cited in the literature, has the form:

$$B/A = 2\,\rho_0 C_0 (\partial c/\partial P)_{s,p=p_0} \quad (2\text{-}3)$$

Here C_0 is the velocity of sound in the unperturbed medium.

This expansion leads to a nonlinear equation of motion that has, for the nondissipative case and waves propagating in the $+x$ direction, a solution of the form:[6–9]

$$u(x, t) = U_0 \sin\left[\omega t - \frac{\omega x}{c_0}\left(1 + \frac{B}{2A}\frac{u}{c_0}\right)^{-2(A/B)-1}\right] \quad (2\text{-}4)$$

Here u is the particle velocity, c the velocity of sound, and $\omega = 2\pi f$, the angular frequency. This equation shows that large values of u are propagated more rapidly than small ones and the wave becomes distorted as it travels through the medium. If viscosity is taken into account in the theory, the predicted distortion of the waveform will include that which results from the more rapid attenuation of the higher harmonics, since the absorption coefficient is proportional to the square of the frequency. The theoretical analysis of propagation of finite but moderate amplitudes was performed by Airy.[10] An approximate solution for the case of larger amplitudes has been given in Fay.[11] Other approximate solutions have been developed by Mendousse[12] and Rudnick.[13] Numerical solutions to the problem have been given by Fox and Wallace[14] as well as by Cook.[15]

Other approaches to the solution of this problem have included attempts to modify the form of the equations for finite-amplitude sound propagation and consider a boundary value problem[16] or to limit regions of solution.[17] In addition, Blackstock[18] has been able to connect the earlier solutions of Fay[11] and Fubini.[7] Whereas only plane waves are discussed here, the cases of cylindrical and spherical waves of finite amplitude are described in detail by Soluyan and Khokhlov[17] and Blackstock.[18]

Experimenters have long been aware that the use of high intensities leads to falsely high values of the attenuation coefficient. Therefore, early observers tended to measure the absorption coefficient as a function of intensity. Unfortunately, they did not also measure the coefficient as a function of distance from the source. In addition, in this earlier work, a ceramic transducer, sensitive principally to the odd harmonics of the incident wave, was used as a detector.[14]

Krasilnikov et al.[19] made improvements in the experimental methods such that the harmonic content of a sound beam with a frequency of 1.5 MHz could be measured. To filter out unwanted harmonics, glass or metal plates were oriented with respect to the beam. The same group has also used thermal detectors to measure the total incident acoustic radiation and determine effective absorption coefficients.

Barnes and Beyer[20] have developed a modification of a radiation pressure microphone to determine the effective absorption coefficient at finite amplitudes. In their experimental arrangement, the carrier wave applied to the transducer is totally modulated by a square wave. As a result, the radiation pressure oscillates at the audio frequency. The signal, whose amplitude is proportional to the intensity, can be easily measured at various distances from the source. Good agreement is obtained between theory and experiment.

As shown in Eq. (2-3), the parameter B/A plays a significant role in nonlinear acoustics. Besides being a basic parameter of the medium, B/A is of interest because it quantifies the distortion process. Once B/A is known, the distortion of the propagating acoustic wave can be predicted. The importance of B/A in biological media has been understood only recently. Measurements of B/A for some media have been made by Law et al.[21] and new methodologies developed by Cobb,[22] Apfel,[23] and Zhu et al.[24] In addition, the effects on absorption coefficient and lesion production have been studied and preliminary tests of the use of B/A for tissue characterization have been made.

Measurements of B/A for biological media have only begun to be made. Some of the first measurements were made by Law et al.[21] They measured the amplitude of the fundamental frequency at the sound source and that of the second harmonic, twice the frequency, as a function of propagation distance from the source. They found values of B/A for Blood Serum Albumin, hemoglobin, and whole blood to be in the range of 5.2–7.5, with the values increasing as the concentration of the samples was increased. Cobb[22] noted that techniques such as those used by Law et al.[21] are limited to media that are not highly attenuative. He overcame these difficulties by measuring both the fundamental and second harmonic amplitudes as a function of distance from the source and applying theory discussed earlier. Zhu et al.[24] have developed methods to determine B/A from phase measurements, and Apfel[23] has developed a relationship for the effective nonlinear parameter of a system of immiscible liquids.

The fact that nonlinear propagation could influence beam shape or absorption in biological media was suggested by Muir and Carstensen[25] in 1980. Some aspects of these predictions were demonstrated by Carstensen et al.[26] in that same year. Further studies of the distortion of finite-amplitude ultrasound in lossy media have been made by Haran and Cook.[27] Goss and Fry[28] have demonstrated the importance of finite-amplitude effects in focused ultrasonic fields for spatial peak–temporal peak intensities greater than 100 W cm^{-2}. They measured intensity-dependent ultrasonic absorption in several biological tissues, including an order of magnitude increased absorption in an exteriorized mouse fetus. Carstensen et al.[29] have measured the thresholds for lesion production in tissues by unfocused ultrasound. They found that inclusion of nonlinear effects in the effective absorption coefficient were necessary in order to model the thresholds for lesion in excised bovine liver.

Investigations have now begun of the possibility of using the parameter B/A for imaging or characterizing biological tissues. Ichida et al.[30] have developed a system in which an impulsive high-power pumping wave is applied perpendicular to a continuous low-intensity, high-frequency probing wave. The phase of probe wave depends on the parameter B/A and the pressure of the pumping wave. Processing of this phase modulation permits determination of B/A and the formation of images based on this parameter. A second approach to imaging based on B/A has been developed by Nakagawa et al.[31] They use a parametric array in which the primary waves at 5.8 and 5.2 MHz so that the resultant secondary wave is at 600 kHz. The sound pressure in the secondary wave depends upon attenuation and B/A. Using this approach, it is possible to reconstruct both nonlinear parameter and conventional attenuation tomograms.

ACOUSTIC STREAMING

Theory

Several phenomena associated with nonlinear acoustics, including enhanced transfer of heat and mass, changes in reaction rates, emulsification, depolymerization, and sonically produced biological effects, are directly related to acoustic streaming. By "acoustic streaming" we mean the time-independent flow of fluid induced by a sound field. Its origins lie in the conservation of momentum dissipated by the absorption and propagation of sound. To introduce the theory for acoustic streaming, we shall outline arguments from the thorough discussion by Nyborg.[32]

For a linear, homogeneous, isotropic fluid the dynamical equation is:

$$f = \rho\left[\frac{\partial \vec{u}}{\partial t} + (\vec{u} \cdot \nabla)\vec{u}\right] \tag{2-5}$$

where:

$$f = -\nabla P + \left[\mu' + \frac{4}{3}\mu\right]\nabla\nabla\vec{u} - \mu\nabla X\nabla X\vec{u} \tag{2-6}$$

and the quantities P, ρ, and u are, respectively, the pressure, density, and particle velocity, and μ and μ' are the shear and bulk viscosity coefficients, respectively. In addition, we shall make use of the equation of continuity:

$$\partial\rho/\partial t + \nabla \cdot \rho\vec{u} = 0 \tag{2-7}$$

Also, we must consider a nonlinear equation of state relating P and ρ.

The usual approach to solution of these equations uses the method of successive approximations. Thus, we expand each variable as a series of terms of decreasing magnitude as follows:

$$\begin{aligned} P &= P_0 + P_1 + P_2 + \cdots, \\ \rho &= \rho_0 + \rho_1 + \rho_2 + \cdots, \\ \vec{u} &= \vec{u}_1 + \vec{u}_2 + \cdots \end{aligned} \tag{2-8}$$

Here the zero-order quantities P_0 and ρ_0 give the static pressure and density. First-order quantities P_1, ρ_1 and u_1 are the usual solutions of the linear wave equations. Thus, a quantity such as P_1 will vary sinusoidally in time with an amplitude proportional to the source amplitude A. There exist two types of contributions to the second-order quantities P_2, ρ_2, and u_2: the first are second harmonic contributions, which vary sinusoidally in time with a frequency $2f$, whereas the second are time-independent contributions. The magnitudes of both types of contributions are proportional to A^2.

Here we are interested only in the time-independent contributions to u_2, which represent the flow generated by the sound. Proceeding with the derivation using approximations appropriate for our applications we obtain

$$\mu\nabla X\nabla X\vec{u}_2 = \nabla P_2 - \vec{F} \tag{2-9}$$

where u_2, and P_2 are now only the time-independent contributions to the second-order velocity and pressure, while $\vec{F}$ is given by

$$\vec{F} = -\rho_0\langle(\vec{u}_1 \cdot \nabla)\vec{u}_1 + \vec{u}_1(\nabla \cdot \vec{u}_1)\rangle, \tag{2-10}$$

where the angle brackets indicate a time-averaged quantity. We note that $\vec{F}$ depends only on the first-order velocity field and has the units of force per unit volume. For sound fields where the first-order term u_1 represents a perfectly spherical cylindrical or plane wave in an unlimited medium, $\vec{F}$ is irrotational and there is no streaming. However, when there are boundaries present on which the wave impinges, boundary layers form, and $\vec{F}$ is no longer irrotational; then streaming exists. Numerical techniques have been applied to the problem and further discussion of nonlinear terms that may be included in the theory has been given by Kukarkin and Rudenko[33] and Ostrovskii and Popilova.[34]

Experimental Techniques

Experimentally, acoustic streaming can be divided into two general types. The first is associated with plane acoustic waves in the volume of an attenuating medium. This type of streaming is often called "quartz wind" for historical reasons. The second type is associated with either inhomogeneities in a sound field or interactions of a sound field with a boundary or surfaces, including objects within the sound field.

The typical situation for volume streaming is an attenuated plane wave traveling in the positive x direction in an unbounded medium. Experimentally the configuration can be achieved by mounting an ultrasonic disk transducer in one end of a tank whose dimensions are much larger than the transducer element. Lieberman[35] has used this configuration with a diaphragm on the wall opposite the sound source to provide a nonreflecting boundary. He found that streaming existed in the volume of the fluid flowing away from the transducer. The region of outward streaming corresponded closely to the cross-sectional area of the transducer and the return flow was outside that region. Thus, in this case the effective force per unit volume $\vec{F}$ from Eq. (2-10) has a value of $\rho_0 \alpha A^2$ and acts in the direction away from the transducer surface. Here, α is the absorption coefficient.

Boundary configurations for production of streaming can involve both plane waves and other acoustic field patterns. Historically, acoustic streaming was first studied by Rayleigh.[36,37] The experimental conditions were those associated with a standing wave set up in a rectangular channel. He found that acoustic streaming in the form of a regular array of eddy flows, each separated by a quarter wavelength, occurred along the channel. Typically, changes in the streaming velocities are greatest near the walls and smaller in the central regions of the channel. This type of streaming is not only of historic interest but is pertinent here because the motion is largely responsible for the aggregations of dust in the Kundt's tube experiments performed in elementary physics laboratories. This type of streaming can have significant effects on reaction rates and mass transfer occurring in fluid-filled channels,[38,39] although for the case of liquid-filled channels the situation may be complicated by the presence of cavitation.

Many experimental arrangements exist that can be used to produce acoustic streaming in the absence of cavitation. As one can see from Eq. (2-10) one important aspect of the "driving force" for acoustic streaming is the spatial variation of the first-order velocity, $\vec{u}_1$. Such gradients can be created in a variety of ways. Probably the streaming situation which has been most extensively studied theoretically is that of an infinitely long circular cylinder driven transversely with respect to its own axis. Theory for this situation has been developed by Schlichting,[40,41] Holtsmark et al.,[42] Skavlem and Tjotta,[43] Westervelt et al.,[44,45] and Raney et al.[46] All investigators have found that streaming occurs in four symmetrical counterrotating circulation cells around the cylinder. Theory for the related case of an oscillating rigid sphere was derived by Lane.[47] The streaming motions are similar to those seen near the transversely vibrating cylinder. Some aspects of the streaming patterns near a sphere were investigated by Rowe and Nyborg,[48] while Sorokodum and Timoshenko[49] have considered the elliptical cylinder.

Another important experimental situation for producing acoustic streaming is a vibrating gas bubble. The streaming is generated when a gas bubble is set into oscillation near a surface or if surface wave activity is present on a bubble oscillating in a volume of liquid. Experimental arrangements for studying such streaming include those of Kolb and Nyborg,[50] in which a small bubble is driven by a metal cone, and Elder,[51] who observed a bubble positioned on a surface that was being driven acoustically. In general, the types of patterns set up near the bubble are highly dependent on the viscosity of the fluid, surface condition of the bubble, and driving amplitude of the sound source.

A variety of other experimental arrangements involving solid acoustic-impedance-transforming horns of different configurations can be used to study features of acoustic streaming. Such horns are widely used and have been described by Mason and Wick.[52] These include bringing a vibrating bar near a plane boundary of a rigid solid, as demonstrated by Jackson,[53] and placing a vibrating horn near a membrane, as described by Jackson and Nyborg.[54] In both cases streaming velocities are particularly high near the edge of vibrating horns. Acoustic streaming patterns near exponential horns have been studied by several investigators, including Hughes and Nyborg[55] and Fill.[56]

A particularly useful arrangement for studies of acoustic streaming and its effects on a variety of systems where only a small volume of sample is available consists of a tungsten wire driven transversely by a transducer with a horn and wire. The vibration pattern for such a driven wire is composed of standing waves, as described by Williams and Nyborg[57] for the case of a transversely driven tube. To achieve maximum displacement amplitude at the tip, the length of the wire is adjusted to an odd multiple of a quarter wavelength. Results have been obtained using a resonant length of 0.025-cm diameter tungsten wire clamped at one end to the tip of a velocity transformer driven at 20 kHz by a PZT-4 ceramic transducer. In studying the details of streaming patterns that occur near the wire, two important precautions should be taken. The velocity of streaming and, in fact, the type of streaming pattern observed is highly dependent on the configuration of the tip of the wire. For well-ordered streaming patterns that permit direct comparison and use of theory, it is necessary to have a hemispherical tip. If the wire tip is merely the cut or jagged end, then the streaming will be controlled by local irregularities of this tip and no comparison or use of theory is possible. Second, orderly, predictable streaming patterns depend on driving the wire only transversely. This is usually accomplished by using only small amplitudes. If other vibrational modes are induced in the wire at high amplitudes, then drastic changes can be seen in the streaming pattern. Evidence for such changes in related systems has been described elsewhere.[58]

Since stable acoustic streaming is a time-independent flow of fluid, methods of studying the patterns produced included all of those typically used for flow visualization. In addition, because of the unique nature of acoustic streaming and the effects produced by it, unusual methods can be used to determine some aspects of the patterns. The most commonly used techniques for visualizing streaming patterns involve the use of indicators. These include smoke,[59] ink-dyed liquids,[60] aluminum, latex, or similar particles in suspension.[61] Semenova[62] has used different color particles to highlight various aspects of the streaming. Techniques based on flow birefringence for observing acoustically induced flow have been developed by Hargrove and Thurston[63] and used by Durelli and Clark.[64] In their technique, a solution of milling yellow dye that is optically isotropic when at rest is used. The region of interest can be viewed through crossed polarizing plates, and the fluid becomes anisotropic only when there is motion causing velocity gradients and internal shearing stresses to exist. The flow pattern produced can be photographed or studied visually. A related technique has been described by Nagai et al.[65] In their experiment they measured the amount of birefringence and opacity induced by acoustic streaming in oriented layers of a nematic liquid crystal. The sensitivity of these acoustooptical cells to ultrasound is described in terms of configuration of the liquid crystal and its thickness.

The effects of acoustic streaming can sometimes be used to provide information about the streaming patterns themselves. Thus, Jackson and Nyborg[54] placed vibrating horns of various geometries in contact with a membrane that was coated on the other side with an opaque film of viscous paint. Outlines in the paint obtained with different sizes and shapes of the horns corresponded to regions in the streaming fields where the highest values of shearing stress existed near the surface and removed the paint from the film. In a similar type of experiment, Nyborg et al.[66] demonstrated that selective darkening of photographic plates occurred in streaming fields.

Applications of Acoustic Streaming

The phenomenon of acoustic streaming is of interest not only because it is an unusual type of fluid flow but because it has associated with it velocity gradients, shearing stresses, and related characteristics that prove useful in specialized experimental procedures.

Consideration of the theory of acoustic streaming for the case of plane waves indicates that the ultrasonic absorption coefficient can be determined by a measurement of the streaming velocity and the energy density of the sound field. An example of this method is the work of Bhadra and Roy[67] on a variety of liquids. They measured the streaming velocity using tracer particles and the energy density using radiation force techniques. In comparison to the requirements of other techniques, larger energy densities for the ultrasonic field are needed. The results by streaming yield higher values than those predicted classically but lower than those obtained by pulse and optical methods.

Experimental methods for producing acoustic streaming have been used to demonstrate that biological effects can be produced by flow and shearing stresses associated with the streaming. Also, they have been used to determine some physical properties of biomaterials and to gain insight into function of biosystems. Although he did not describe it as such, the fact that acoustic streaming could be induced in cellular systems was demonstrated as early as 1929 by Schmitt.[68]

One of the more widely used and successful experimental arrangements for the study of details of interactions of streaming consists of a small exponential or stepped horn connected to a ceramic transducer driven by an oscillator and power amplifier. Typically, these experiments are conducted in the frequency range from 50 to 100 kHz. The tip of the horn is brought into contact with the biological system of interest and motions are observed through a microscope. Induced changes are evaluated using a variety of techniques. Examples of application of this method of investigation include work by Hughes and Nyborg[55] on the hemolysis of erythrocytes, Dyer and Nyborg[69] and Gershoy et al.[70] on the motions induced in plant cells, El'piner et al.[71] and Wilson et al.[72] on deformation of cellular structures, and Schnitzler[73] on the changes induced in skeletal muscle.

A vibrating wire has also been used for studies of the interaction of acoustic streaming with biological systems. Specifically, Williams[74–76] has placed such a wire in contact with intact blood vessels and produced a variety of effects including aggregation and rupture of platelets and formation of thrombi. Data on hemolysis obtained by Williams et al.[77] using the wire system and related data obtained by Rooney[78] using a stable bubbles as a source of streaming provided information concerning the mechanical response of the erythrocyte to shearing stress. Results of hydrodynamic experiments are in reasonable agreement with the ultrasonically obtained results.[79]

Bundy[80] has made a theoretical study of the interesting case in which acoustic streaming occurs near objects such as cells or organelles with sizes on the order of the boundary layer thickness. His expressions for the streaming near small spheres are in agreement with Lane's.[47] The threshold amplitudes for stresses that might cause lysis of cells from this type of streaming are at least an order of magnitude greater than those measured for cells in streaming flows produced by other sources.

The phenomena of acoustic streaming has been studied in many liquids. Perhaps the most unusual liquid studied has been superfluid helium. Time-independent flow such as acoustic streaming should not occur in an ideal nonviscous fluid. Therefore, one would expect streaming in the normal fluid component of superfluid helium but no streaming in the superfluid component, provided there are no vortex lines present that can couple the two fluid components via mutual friction. Studies of streaming in helium II by Rooney et al.[81–83] have shown that the Khalatnikov two-fluid model[84] can be adopted to describe the flow. The normal fluid component behaves as a conventional liquid and the temperature dependence of the streaming velocity is determined by the temperature dependence of the ratio of the coefficient of second viscosity to the coefficient of first viscosity.

In addition, the possibility that streaming may enhance heat transfer in liquid helium has been studied by Smith et al.[85] They found that ultrasound had no observable effect on the heat transfer in the superfluid phase. However, in the normal fluid phase the thermal conductance was found to increase linearly with sound amplitude which is in agreement with other results for normal liquids.[86]

EMULSIFICATION AND AGGREGATE DISPERSAL

Acoustic Methods of Emulsification

A variety of sonic and ultrasonic techniques have been used for the formation of stable emulsions of chemical or biological systems as well as dispersal or aggregates. These techniques typically involve nonlinear acoustic mechanisms in their modes of action. For use with biological systems, techniques have been developed that can make use of these mechanisms without causing cell damage.

The two principal ultrasonic techniques used to form emulsions are flowthrough sonifying systems operating in the low-ultrasonic frequency range and liquid whistle arrangements. Sonification systems used for emulsification typically consist of a velocity transformer or horn coupled to a transducer capable of oscillating in a longitudinal mode. The horn is immersed in the solutions of interest and activated so that collapse cavitation is produced in the liquids. The intensity of the cavitation, which depends on the power delivered to the horn, must be correctly chosen for efficient treatment of the liquids. Special horns, horn tips, or vessels are available for specific processing applications. These include flowthrough horns, continuous-flow attachments, and sealed-atmosphere treatment chambers. Typical descriptions of such apparatus, related experimental methods, and research results may be found in articles by Weinstein et al.[86] and Genta et al.[87]

Hislop[88] and Last[89] have developed sonifying systems for emulsification that use a different technique to concentrate the ultrasonic energy. This systems use solid focusing horns that concentrate longitudinal ultrasonic waves internally by using a parabolic horn tip. The liquids to be emulsified are passed with their host liquids through a channel that encloses the focus of the parabola.

The other major ultrasonic technique used for emulsification and also dispersal consists of various types of liquid-driven whistles. Whistles are distinguished from other sonification systems by the fact that production of the sound associated with the system is by mechanically induced oscillations instead of the standard electromechanical transduction. One typical whistle consists of a resonant chamber of variable depth upon which a jet of liquid formed by a collimation tube impinges. All components of the whistle in contact with the high-pressure jet must be made of hardened steel. An example of this type of whistle was described by Crawford.[90] In it, two jets of the driving liquid cross to impinge on a blade positioned at the crossover point. This blade is mechanically resonant at a desired frequency and is mounted at its nodal points to minimize damping.

Modifications of such whistles include enclosure in chambers designed to confine the emulsification process to the region near the vibrating blade. These whistles typically operate at frequencies in the 30-kHz range and can produce acoustic intensities of the order of 2 W cm^{-2} in the liquid in the resonant chamber. Such levels are sufficient to produce cavitation in the liquids, thus increasing the efficiency of emulsification.

Other whistle configurations have also been developed; these include the vortex whistles described by Hislop[88] and Last.[89] In these devices the driving liquid enters a large swirl chamber tangentially. As this liquid moves from the large radius of a swirl chamber to the smaller radius of an exit pipe, the liquid is accelerated. The liquid phase to be dispersed or emulsified flows through an annular tube mounted along the axis of the whistle. The tube ends at the exit pipe of the vortex so that the two liquids interact. The geometry and location of the exit orifice and tube are found to be critical for greatest efficiency of emulsification.

A somewhat similar device has been developed by Greguss.[91] In this device the driving liquid enters a cavity tangentially by one or more inlets. The liquid is then forced to move from a larger to a smaller radius because the outlet tube is smaller in diameter than the central cavity but larger than the inlet tube. Because of the angular acceleration of the driving liquid, a lower pressure exists in the central cavity and the liquid is forced to the center to be ejected through the outlet into a larger outer region. If the geometric form and dimensions of the inlet, central cavity, and outlet tubes are in proper relation, a sustained vibration develops and acoustic waves whose intensity is sufficient to cause emulsification are produced. The frequency of the waves depends on both the geometric dimensions and the liquid driving pressure.

Mechanisms and Efficiency of Emulsification

The mechanisms for the ultrasonic emulsification process are similar to those of other emulsifiers, namely, action of shearing stresses and deformation of the larger droplets in the liquid. El'piner et al.[71] have demonstrated that acoustic streaming with its associated shearing stresses can begin the emulsification process. The more subtle details of the emulsification process have been examined both experimentally and theoretically by Hinze[92] and are summarized by Goldsmith and Mason.[93] The important parameters that determine the velocity gradients necessary to cause deformation of larger droplets, leading to their disruption and efficient emulsification, include the liquid viscosities and interfacial tension.

Specifically, Rumscheidt and Mason[94] have shown the relationship developed by Taylor[95] to be valid. Taylor stated that a drop will burst when the shearing stresses become greater than the restoring forces associated with the interfacial tension. Details of the condition for droplet disruption and therefore emulsification are given as:

$$\eta_0 G = 8T(P+1)/b(19P+16) \tag{2-11}$$

where P is the ratio of the viscosity η of the disperse medium to that of the continuous medium η_0, G the velocity gradient, T the interfacial tension, and b the radius of the undistorted drop.

Studies of particle size distributions and efficiency of emulsification processes have been conducted by Hislop,[88] Last,[89] and Rajagopol.[96] These studies involve microscopic examination of emulsions after formation by the various techniques described above. Particular care was taken to sample a sufficient number of emulsions to achieve statistically significant results. The droplet sizes in the emulsions were found to be on the order of microns, ranging from 1 to 40 μm in diameter, depending on the procedure and duration of the emulsification process. Neduzhii[97–99] has found similar results and also demonstrated that the emulsification rate was proportional to the intensity of the sound used above the threshold intensity for the onset of emulsification. It should be noted that prolonged sonification led to an increase in the mean size of the particles and a broadening of the size distribution.

Dispersal of Biological Cells

A problem related to emulsification and dispersal of liquid systems is that of dispersal of aggregates of living cells without morphological damage or decrease in their viability. Sonic and ultrasonic techniques have been developed that have many advantages for such an application. Direct adaptation of some experimental methods used for emulsification already discussed was attempted by Mullany et al.[100] Using leukocytes and Chinese hamster cells they conducted tests of the effectiveness of such sonification procedures. After preparation, cell clumps were dispersed with an ultrasonic immersion horn operated at 20 kHz. However, this procedure typically led to an unacceptable increase in the amount of fluorescent debris present in their automated analysis system.

In efforts to develop more controlled ultrasonic methods for cell aggregate dispersal, Williams et al.[101] chose to take advantage of the well-characterized shearing stresses associated with acoustic streaming. Two methods were developed. In the first, using a vibrating wire method, they dispersed bacterial cell aggregates and flocculated sludge. They were able to demonstrate the effectiveness of the technique and showed that the number of single viable cells increased with both sonic displacement amplitude and duration of treatment. Specifically, for an exposure time of 10 min using a displacement amplitude of 15 μm, they found a 260% increase in the number of viable colonies. Of course, such dramatic increases do not continue indefinitely with increased sonification but are limited by saturation of the cellular system. Parry et al.[102] have demonstrated that cell aggregates obtained from cervical scrapings could be dispersed effectively using this method. Williams and Slade[103] modified the arrangement by using a long stretched wire driven at 18.5 kHz and demonstrated the successful dispersal of octets of *Sarcina lutea*. While these vibrating wire methods have proven to be effective, they do have the disadvantages that they treat only small samples to cells at one time and that unstable oscillations may occur in the long stretched wire, which can cause undesirably large shearing stresses near the wire, as demonstrated by Rooney.[58]

Brakey and Rooney[104] have developed experimental methods for cell aggregate dispersal that use a type of vortex whistle. Liquid containing the cell aggregates to be dispersed is forced by a hydraulically driven syringe to flow through a needle; then it tangentially enters a large swirl chamber, passes through a region of smaller diameter, and finally exits through a small orifice. The acceleration of the liquid in the whistle causes sufficiently high velocities and associated gradients to disperse the cells. The dispersal efficiency was found to be greatly increased by the use of an insert within the swirl chamber, as well as by machining a thread on its surface. These measures limit the trajectories of the cells within the whistle to regions of high shearing stress.

In tests of the effectiveness of aggregate dispersal, the vortex whistle was used successfully to disperse 95% of the samples of the aggregates of the algae *Scenedesmus* and monkey kidney cells grown in tissue culture after a single pass through the whistle. Light microscopy and cell growth studies indicated that the cells were morphologically normal and completely viable. Further experiments on monkey kidney cells demonstrated that cells could be dispersed efficiently without structural changes being detectable when using electron microscopic techniques.

ATOMIZATION AND DROPLET FORMATION

That it is possible to form a fog or mist ultrasonically was first discovered by Wood and Loomis[105] in 1927. Since then there have been extensive studies and debate about the mechanisms causing the process as well as development of various types of instrumentation for atomization. The principal advantages of the ultrasonic methods for atomization are that the sprays produced have drops that are relatively uniform in size and that the drops have low inertia; therefore they can be easily entrained in a gas flow and transferred from the region of atomization. The methods for ultrasonic atomization, as well as discussion of techniques to investigate the mechanisms, have been summarized by Topp and Eisenklam[106] and by Eknadiosyants.[107]

Mechanisms for Ultrasonic Atomization

After the observation of ultrasonic atomization by Wood and Loomis,[105] Sollner[108] was the first to attempt to explain the phenomenon. He believed that cavitation occurred directly below the surface of the liquid that resulted in the formation of the fog or mist. However, work by Bisa et al.,[109] Sorokin,[110] and Eisenmenger[111] indicated that instabilities in surface waves were the mechanism involved. Thus, for many years there were two rival theories. The capillary wave theory is based on a Taylor instability of a liquid–gas interface exposed to a force perpendicular to it, as described by Peskin and Raco.[112] In this theory the capillary waves produce drops at their crests, whose size is proportional to wavelength. The droplets are essentially monodispersed and have little momentum. In contrast, the cavitation theory developed by Eknadiosyants[113] postulates that the hydraulic shocks associated with collapse cavitation cause the atomization. These two theories have now been related by Boguslavskii and Eknadiosyants[114] by the suggestion that shocks from cavitation events can interact with the finite-amplitude capillary waves, thereby forming drops in the same way. In particular, cavitation is important in so-called "fountain atomization" such as used in medical inhalents, but not in "liquid-layer atomization" as used in oil-fired burners. In either case the drop size can be estimated from the equation derived by Lang:[115]

$$D = 0.34(8\pi\gamma/\rho f^2)^{1/3} \tag{2-12}$$

where D is the droplet diameter, γ the surface tension, ρ the fluid density, and f the driving frequency.

Experimental tests of the various theories of atomization have generally used two methods. The first uses high-speed photomicrography to examine details of the droplet formation. Experiments using this technique to support the capillary wave theory include those conducted by Lang,[115] as well as by Fogler and Timmerhaus.[116] Related experiments have been performed by Topp,[117] Bassett and Bright,[118] and Chiba.[119] These latter investigators present photographic evidence that both surface waves and cavitation play roles in atomization.

The second experimental method involved the examination of droplets formed. For this purpose, Lang[115] atomized molten wax. The wax droplets solidified quickly in flight and were collected and sized. Investigations by Lierke and Greisshamer[120] have used a variety of molten metals. The particles obtained were characterized by sieving and both optical and electron microscopy. Particle sizes measured were in agreement with theory.

It should be noted that details of the atomization mechanism occurring in whistles are not totally understood. While they are probably similar to those associated with the Taylor instability in capillary waves, Wilcox and Tate[121] have demonstrated that for certain whistle atomizers operating under specified conditions the presence of the sound field plays little or no part in the atomization process. In contrast, for fuel oil atomization the presence of the sound field greatly enhances the process.[122]

Droplet Production

For a variety of experiments, including studies of the interactions of droplets and of the sorting of various types of materials, it is desirable to have a controlled method for the production of uniform, single, large droplets rather than a cloud or mist. Ultrasonic methods similar to those discussed in the topic of atomization are capable of such droplet production. The essential idea is that if periodic oscillations are applied to a nozzle, pressure and velocity fluctuations are produced at the orifice that cause minute perturbations in the surface of the liquid cylinder leaving the orifice. For a velocity of the jet v and period of vibration T, the wavelength of these disturbances λ is described by the usual equation $\lambda = vT$. If λ exceeds the circumference of the jet, then Rayleigh[37,123] has shown that the disturbance grows exponentially until the column divides into detached droplets separated by an interval and passing a fixed point with velocity v and frequency $1/T$. Considering the energy relationships, he showed that the instability that grows most rapidly has a wavelength λ_m such that:

$$\lambda_m = 9.016b \tag{2-13}$$

where b is the radius of the jet. The mass of fluid M associated with this disturbance can be expressed as:

$$M = \pi b^2 e \lambda \tag{2-14}$$

where ρ is the density of the fluid in the jet. Therefore, the radius r of the droplet formed is given by:

$$r = (3b^2\lambda/4)^{1/3} \tag{2-15}$$

Upon substitution of λ_m, we find that the radius of the droplets formed is just:

$$r = 0.946A \tag{2-16}$$

where A is the diameter of the fluid jet. Equation (2-16) has been found to be accurate to within 0.33% by Lindblad and Schneider.[124]

A variety of experimental methods for droplet production have been developed. These include the method of Lindblad and Schneider,[124] who used a capillary tube orifice driven by a PZT bimorph transducer at audio frequencies to produce drops of various sizes and corresponding rates. In the higher frequency range Fulwyler[125] ultrasonically drove a nozzle at 40 kHz to produce 40,000 uniform droplets per second. A new pinhole-plate ultrasonic atomizer has been developed by Machara et al.[126] An additional option available to the experimenter is to charge the droplets formed and selectively deflect them at some later time using electric fields. Such systems have provided the basis for biological cell sorting methods,[127] as well as for the jet ink printer.[128]

When considering the use of such methods, the experimenter should be aware of the influence of other parameters on the systems involved. For example, work by Stovel[129] has demonstrated the influence that particles in the liquid have on jet breakoff. Using a 50-μm diameter nozzle driven at 40 kHz, he found that particles in the liquid whose diameters were about 2 μm had no effect on droplet formation, those 20 μm in diameter totally disrupted normal breakoff, and those in the range from 7 to 10 μm in diameter caused erratic droplet formation.

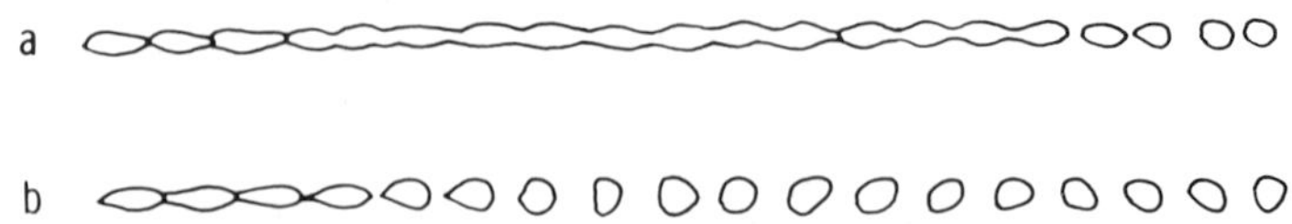

Figure 2-1. Droplet formation using ultrasound. (a) Onset of Rayleigh instability leading to drop formation in (b) (see Ref. 130).

Klopovsky and Fridman[130] have shown that the variation of the driving frequency from that predicted by the Rayleigh relation can also affect the efficiency of drop formation. At the optimal frequency jet disintegration is periodic and monodispersed drops are formed at a distance of from 2 to 3 cm away from the nozzle. An increase in frequency of forced vibrations by 15–25% will result in the formation of drops of uniform size. However, they begin to move in pairs and may coalesce, becoming nonuniform. A decrease in the driving frequency by the same amount will result in the formation of small "satellite" droplets between the larger ones. The formation of these satellites causes the major drops to differ from spherical form and to change their form in flight. The amount of deviation from optimal driving frequency permissible will depend on the particular liquid used. Results from these experiments can be seen in Fig. 2-1.

RADIATION FORCE

Introduction to the Theory

Any object irradiated with an acoustic field will experience a force whose magnitude and direction depend on the intensity of the source and field parameters, as well as the size, shape, and construction materials in the irradiated object. It is this force that has often been misnamed "radiation pressure" for historical reasons. A history of early theory and experiments has been compiled by Post.[131] The observation of the deflection of a target by acoustic radiation force was probably made by Dvorak[132] in 1876, although Eckart[133] believed the phenomena observed by Dvorak to be acoustic streaming. Preliminary discussion of the theory was completed by Rayleigh.[134] It should be noted that indeed two types of radiation force exist, which differ according to the physical situation and boundary conditions applied to the problem. Detailed discussions of the appropriate theory are now available[4,135–137] and are summarized here.

Historically the first situation considered was that of the Rayleigh radiation force resulting from the nonlinearity of the equation of state. It is applicable to situations that have several restrictions: (1) sound waves are propagating in a closed vessel with a fixed mass of fluid, and (2) regions of the fluid not directly in the sound field can only interact with the field provided that the density and pressure changes in the fluid are such as to maintain equilibrium. For this experimental arrangement and considering plane adiabatic waves, Rayleigh[138] found that the radiation force F could be given by:

$$F = \frac{1}{2}(1 + \gamma)ES \tag{2-17}$$

where E is the energy density in the sound field, S the area of the target perpendicular to the sound field, and γ the ratio of specific heats. If second-harmonic terms are also considered, Beyer[4] found that the first coefficient of Eq. (2-17) should be 1/4 instead of 1/2.

However, experiments either for investigating the phenomenon of radiation pressure or using it as a technique for dosimetry usually are conducted in open vessels. These free boundary conditions correspond to the Langevin situation. In this case the radiation force is caused by a nonlinearity of the basic Euler equation from which the sound field equations are derived. In the equations for a free sound field, inclusion of the nonlinear terms leads to a lower mean pressure in the fluid as seen by a stationary observer. If the sound beam is surrounded by a medium at rest, then this lowered pressure is compensated by the external medium. For an observer who is moving with the sound beam, this compensation has the same effect as an increased mean pressure in the fluid in the region occupied by the sound field. The difference between the two types of radiation force is discussed in detail in Beyer[4] and Rooney and Nyborg.[135]

Several early investigators[139–141] showed that for the Langevin case the radiation force is simply:

$$F = ES \tag{2-18}$$

Questions have been raised concerning the effect that nonlinear properties of the propagation medium may have on this expression for the radiation force and the appropriateness of the boundary conditions used in the early theory.[3,142] Beyer[4,137] and Rooney and Nyborg[135] have reconsidered the theory and shown again that Eq. (2-18) is indeed valid for the Langevin case. For the Rayleigh case, they obtained:[135]

$$F = (1 + B/2A)ES \tag{2-19}$$

where B/A is a parameter associated with nonlinearities of the medium.[3] However, Beyer[4,137] showed that if higher order terms of the propagating waveform have to be included, then the magnitude of the force must be reduced by a factor of 2. Thus, the correct expressions for evaluating the radiation force may depend both on the nonlinear properties of the medium and the waveform of the propagating sound.

The questions of appropriate boundary conditions and the influence of nonlinearity of the medium have been reexamined recently by Chu and Apfel.[143,144] While there is agreement and many tests of the much used Langevin radiation force, definition of exact boundary conditions for the Rayleigh case remain difficult.[145] Fortunately, the former radiation force is the one of most interest in biological and chemical applications of ultrasound.

Experimental Techniques

In the past, researchers have measured radiation force by noting the deflection of a target in a force field, such as a gravitational field, or by measuring the force required to maintain the target in a given position. Experimental arrangements have included the use of magnetic forces,[146,147] counterweights or mechanical torque,[148] buoyant forces,[149,150] as well as the deflection of fluid surfaces.[151]

A particularly simple experimental setup has been developed by Henry[149] and modified by Kossoff.[150] The arrangement consists of a float in an equilibrium position with its stem projecting vertically through the surface of the water. When the sound is turned on, the float is deflected upward, its effective weight is increased, and a new equilibrium is reached such that the radiation forces and the decrease in buoyant forces cancel. For calibration, one needs only to place a small known mass on top of the stem and note the displacement of the float. A small hole in the center of the float permits escape of air bubbles. If the target surface of the float is an absorber, there is no horizontal component to the radiation force. This fact has been experimentally demonstrated by Herrey,[152] who measured the component of the force parallel to an ultrasonic beam. He found that the force on an absorber in the direction of propagation is independent of the angle of incidence θ, whereas for the case of a reflector, the component varies as $\cos^2 \theta$.

The float system just described provides a simple inexpensive technique for determining total acoustic power output using radiation pressure. Kossoff[150] improved the sensitivity of the system by using two liquids whose densities were only slightly different, instead of water and air as shown in the figure. At present, the limit of the sensitivity appears to be associated with the surface forces on the stem.

The float concept has been optimized by Cornhill.[153] This system is shown schematically in Fig. 2-2. It uses a float that is magnetically suspended and a light-emitting diode/phototransistor position-sensing system. The system is portable and permits simple calibration and easy testing of transducers whose output is less than 50 mW.

One typical method for measuring radiation force uses a balance from which an appropriate target whose dimensions are much greater than the wavelength of sound is suspended. Course balances[154,155] have adjustments for zeroing and standard weights that permit calibration of the radiation force. For sensitive measurements of radiation force, electromagnetic servomechanisms have been used.[146,147,156]

For investigations of the phenomenon of radiation force and the calibration of ultrasonic medical units, Rooney[147] has developed a sensitive radiometer. The apparatus consists of a Cahn RG electrobalance from one arm of which a cylindrical absorbing target is suspended; appropriate tare weights are suspended from the other arm. The target is hung in the medium of interest within an inner vessel. This small interior vessel is closed at the lower end with a stretched membrane to shield the target from acoustic streaming.

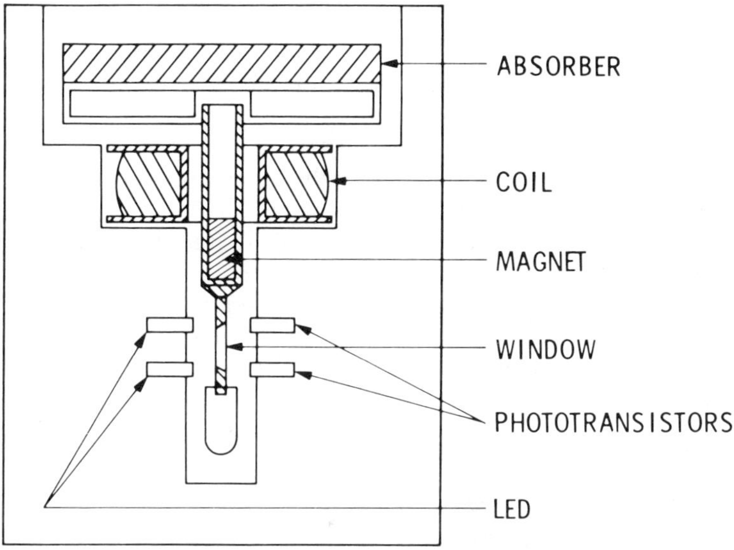

Figure 2-2. Radiation force balance using optical sensing system. (Modified from Ref. 153.)

The electrobalance is calibrated by noting the deflections on the recorder which correspond to forces resulting from placing a series of calibrated weights on the balance. The transducer element is brought into contact with a second stretched membrane at the base of the large vessel and a coupling gel is used between the transducer and the membrane. The outer vessel containing water has a large volume so that the thermal fluctuations in the inner vessel are reduced. When the transducer is activated, the target is deflected momentarily from equilibrium. This deflection changes the amount of light reaching a photocell which initiates the response of a magnetic feedback system within the Cahn balance, thus returning the target to its original position. A portion of the feedback voltage to the magnet is recorded on a chart recorder. We note that the use of the feedback system is advantageous because it eliminates errors that result from changes in surface tension forces on the suspending wire and minimizes those that are the result of other "surface" effects. The air gap between the balance and the large vessel is enclosed to prevent disturbing effects of air currents on the suspending wire.

The inner vessel proved to be important for the stability of the system. The vessel should have a diameter only slightly larger than the target in order to minimize drift in the signal, which results from convection currents. The response of the system is such that transients decay within 6 s. The size of the target also proved to be an important consideration for accuracy of the measurements. In considering the appropriate diameter of the target, a compromise had to be made. The diameter of the target must be larger than the diameter of the beam to insure that the sound is completely absorbed, but the volume of the target should be kept to a minimum to reduce excessive noise in the system. Sound-absorbing rubber materials have proved useful; no appreciable standing waves could be detected in the frequency range from 1 to 10 MHz. In addition, the experimenter should realize that building vibrations can set up pressure gradients in the liquid. In the presence of these gradients, the target will be subjected to a force proportional to its volume. Therefore, reduction of noise was achieved by using a target of minimum volume to meet the other requirements and isolating the experimental arrangement from the building.

When an absorbing target for the Langevin situation is used, the radiation force equals the total acoustic intensity divided by the velocity of sound even when nonlinearities of the propagation medium and distortion of the waveform are taken into account. Rooney[157] has tested this relationship using the apparatus just described. The radiation force was found to be independent of the parameter B/A as predicted.

The functional dependence of radiation force and acoustic power has been measured for self-consistency and comparisons have been made of the absolute magnitude of the radiation force with other absolute calibration techniques. Tarnocy[158] has completed a comparison of the calculated output of a transducer, the radiation force measured on a balance and the acoustic power measured with a calorimeter. In the intensity range between 0.5 and 2.5 W cm^{-2} he found that the measured radiation force was 20% smaller than that predicted from the theoretical power output of the transducer. However, the radiation force technique produced results closer to the theoretically expected value than calorimetry, which gave results 35% smaller than the predicted value. Better agreement has been obtained in other studies. In a comparison between a radiation force measurement and a calorimeter, Wells et al.[159] found agreement between the two techniques to within 6% in the power range from 1 to 2.2 W. The study by Haran et al.[160] of the power output from a transducer showed no statistically significant difference between determinations by radiation force and by an optical system. Based on these comparisons, it appears that indeed there is a good agreement between theory and experiment.

In discussions of radiation force, the possibility of a time-varying component usually does not arise. However, a method of measuring radiation force that uses modulation has been developed by Greenspan et al.[161] Earlier methods by Belin and Beyer[162] also used modulation. In the Greenspan method, the input to the transducer is modulated at a low frequency and the output acoustic power is intercepted by a target that experiences a component of radiation force at the modulation frequency. The target is mounted on the armature of an electromagnetic receiver, provided with an independent coil that is fed with a current at the modulation frequency; the frequency is adjusted in amplitude and phase, either manually or automatically by feedback, to arrest the motion of the armature. When the armature is stationary, the force is proportional to the current; the apparatus can be calibrated using weights. This technique avoids the drifts present in dc measurements. Narrowband ac signal-processing techniques can be used to improve the sensitivity of the system which is quoted to be 10 μW of acoustic power.

Holographic Applications of Radiation Force

While the radiation force has been widely used to determine the power output of transducers, it is of interest in other biological applications including holographic medical imaging.[163]

The Langevin radiation force can be exerted on the interface between two liquids. This effect can most easily be demonstrated following a classical experimental arrangement developed by Hertz and Mende,[151] which demonstrates an important aspect of radiation force. In their experiment, layers of different liquids, carbon tetrachloride, water, and aniline, were used because of their different densities and velocities of sound. They irradiated the fluids from above using a 5-MHz transducer, reflected the sound from a plate oriented with respect to the transducer, and displayed the image of the transmitted and reflected beams. The liquid interfaces no longer remain planar but are deformed by the radiation forces of the sound beams. In particular, the boundary between the water and aniline is deflected in a direction antiparallel to that of the sound propagation for the transmitted beam and parallel to the direction of sound propagation for the reflected beam.

This phenomenon can be explained using Langevin radiation force. Because the acoustic characteristics of the liquids are well matched, the acoustic powers and intensities, I, are uniform throughout the vessel. In contrast, the energy density $E = I/c$ is different in the three liquids because of the different velocities of sound. The energy density is less and thus the radiation force is less in water than in aniline because of its larger sound velocity.

In liquid surface acoustical holography, the levitation of the surface is used to record the image. The object to be imaged is insonified by a transducer operating in a tone burst mode. The sound emerging from the object is focused by an acoustic lens system onto the liquid surface. The sound field from a second (reference) transducer operating in the same mode reaches the liquid surface at the same time as that from the object. The radiation pressure of the combined fields causes deformation of the surface. The hologram is formed by scanning a laser across this modulated surface. Details of this surface deformation have been studied by Clark.[164]

Interactions with Small Objects

A nonlinear effect of the interaction of a sound field with an object within an acoustic field provides the experimenter with the unique possibility of manipulating an object in a host medium using only the acoustic field. This is a very special application of radiation force, and, in fact, much theoretical and experimental work has been completed concerning the time-averaged force exerted by a sound field on a spherical inclusion.

The first such study was completed by Bjerknes,[165] who investigated the effects on a spherical bubble placed in an oscillating fluid using analogies to electrical and magnetic interactions. A more detailed study of the force on a sphere in a plane wave field was published by King,[166] but the effects of compressibility were neglected. These results were extended by Yosioka and Kawasima[167] to include compressibility. Measurements on a variety of materials have been made by Chivers and Anson.[168] In addition, Gor'kov[169] has used a fluid dynamics approach to obtain similar results for small objects in more general sound fields. Westervelt[170] has derived the more general expression for the radiation force on an object with arbitrary shape; Embleton[171] has extended the work of King to include spherical waves; and Nyborg[172] has calculated the force exerted by a sound field on a small rigid sphere. Also, Dystke,[173] with corrections as noted by Crum,[174] has given a comprehensive treatment of the force exerted on solid, liquid, and gas inclusions in a stationary sound field. Hasegawa has derived theory[175] and conducted experiments[176] on the radiation force on a sphere in a quasi-stationary field.

It is found that two general terms exist in the equation for the force on an object in a sound field: the first results from the radiation pressure exerted on a finite-sized inclusion in the second field, while the second arises from the finite compressibility of the object. Thus, for hard spheres, the force is obviously dominated by the first term, while in the case of air bubbles in water, the compressibility term dominates the equation. A liquid droplet lies midway between the two extremes and, for a droplet trapped by a stationary sound field, the terms are of equal magnitude and both must be considered in the theory.

Following the suggestions of Crum,[174] we can obtain the expression for the force on a fluid droplet in a sound field by combining two independent approaches. The result is in agreement with results obtained by Yosioka[167,177] as well as Gor'kov.[169] Nyborg[172] has developed a theory that can be used to calculate the force F_r exerted on a noncompressible sphere by a stationary sound field, where F_r is given by:

$$F_r = V_0[B(\partial\bar{T}/\partial z) - (\partial\bar{U}/\partial z)] \tag{2-20}$$

and $\bar{T}$ and $\bar{U}$ are, respectively, the second-order approximations to the time-averaged kinetic and potential energies, and V_0 the equilibrium volume of the sphere. Here $B = 3(\delta - 1)/(2\delta + 1)$ and $\delta = \rho/\rho_0$, where ρ and ρ_0 are, respectively, the densities of the sphere and host medium, Eller[178] has shown that the time-averaged radiation force exerted on a sphere at a position z in the stationary sound field from its compressibility alone is given by:

$$F_c = -\langle V(t)\, \partial P(z, t)/\partial z \rangle \tag{2-21}$$

where $V(t)$ is the instantaneous volume and $P(z, t)$ the instantaneous pressure along the axis. Adding the two terms and evaluating the quantities for a standing wave acoustic field, one obtains the expression for the net force on the droplet:

$$F = \frac{V_0 P^2 k \sin(2kz)}{4\rho c^2}\left[\frac{1}{\delta\sigma^2} - \left(\frac{5\delta - 2}{2\delta + 1}\right)\right] \tag{2-22}$$

where $k = 2\pi/\lambda$ and σ is the ratio of the velocity of sound c in the droplet to that in the host liquid c_0. This equation describes the cases of a liquid droplet and a rigid sphere but it must be modified for the case of a bubble, as described in Yosioka and Kawasima,[167] Gor'kov,[169] or Eller.[178]

Experimental testing of the theory for radiation force on solid spheres has been conducted by Embleton[179] and Hasegawa et al.,[180] as well as by Breazeale and Dunn.[181] For their studies, Breazeale and Dunn compared three optical techniques for measurement of the intensities of a sound field with the radiation force on a small sphere and a transient thermoelectric method. All experimental curves were within $\pm 10\%$ of the mean values for all techniques. However, this mean was approximately 27% less than the acoustic output calculated from the voltage.

Testing of the theory for the case of a liquid droplet and an air bubble has been conducted by Crum[174] and Crum and Eller,[182] respectively. While qualitative observations of the trapping of bubbles in standing wave acoustic fields had been made by several people, it was only in 1968 that Eller[178] developed techniques for quantitatively testing the theory. The experiment took place in a cylindrical standing wave excited in a vertical tube filled with water and driven by a piston at the bottom. The frequency used was 27 kHz and the wavelength in the vertical direction was 23 cm. The procedure used was to trap the bubble in the sound field above the pressure maximum located about 5.75 cm below the surface. The pressure amplitude necessary to trap the bubble was measured and the bubble size determined from measurement of rise times of the free bubble and Stoke's law. Results from the experiment demonstrated that the average force on a bubble could be described by Eq. (2-22) and that these results were consistent with the qualitative results by earlier workers.

For his experiments on levitation of a fluid droplet in a host liquid, Crum[174] used a thin-walled cylindrical vessel coupled to a PZT-4 ceramic transducer. He found that the $(r, \theta, z) = (1, 0, 3)$ resonant cylindrical mode was useful for the experiments. In this case, the resonant frequency was 41.7 kHz and the wavelength along the vertical axis was 7.8 cm. The acoustic pressure amplitude was measured by a small calibrated ceramic transducer mounted externally to the system. The test droplet was injected into the system with a syringe and its radius was measured with a calibrated microscope. Crum then measured the magnitude of the smallest acoustic pressure amplitude at the nearest antinode of the stationary wave system required to trap a given droplet and compared the result with the theoretical prediction. Based on his results we can conclude that it is possible to levitate droplets of liquid in an acoustic stationary wave and that the theory correctly predicts the acoustic pressure amplitudes required to trap these droplets in the sound field.

The basic problems associated with the levitation apparatus include choosing and optimizing the appropriate resonance and maintaining the resonance despite changes that occur in the system. A more sophisticated experimental arrangement for conducting levitation experiments has been developed by Baxter et al.[183] The design described has the advantageous features that any manifestation of resonance can be used as input to the device and that resonances can be scanned manually, viewed, and locked onto by the circuitry.

Acoustic levitation of one drop of liquid immiscible in a host liquid has been used by Apfel[184–187] to study several properties of fluids. This levitation technique has the advantages that only small samples of liquid need to be used and that the possibility of heterogeneous nucleation is greatly reduced. Similar techniques have been used by Apfel to measure the adiabatic compressibility, density, and velocity of sound in submicroliter liquid samples. The interparticle forces on erythrocytes in standing waves have also been investigated by Weiser and Apfel.[188] The levitation technique for measuring these parameters has the advantages that since it is a comparison technique the results are independent of droplet size and that the field parameters need not be known; only the voltage applied to the transducers to levitate the droplet to the same position in the field as a reference droplet must be determined.

NONLINEAR OSCILLATIONS OF BUBBLES

The acoustically excited oscillation of bubbles plays an important role in the chemical and biological effects of ultrasound. These oscillations range from low-amplitude volume oscillations to collapse cavitation. Important aspects and effects of these oscillations, particularly collapse cavitation, are discussed throughout this book. Subharmonics and harmonics resulting from bubble oscillation will be topics of this section. Minnaert,[189] in a classic paper entitled "On Musical Air-bubbles and the Sounds of Running Water," developed theory for the volume oscillations of bubbles in water. While he was considering the mechanisms for production of ". . . the murmur of the brook, the roar of the cataract, and the humming of the sea," his results apply equally well to the case of an ultrasonically driven bubble. Minnaert recognized that a pulsating bubble was analogous to a mass on a spring with air within the bubble providing the elasticity of the spring and the surrounding water being the mass. Using conservation of energy, he was able to derive an equation for the frequency of oscillation of a spherical bubble:

$$\omega_0 = (3\gamma P_0/\rho R_0^2)^{1/2} \tag{2-23}$$

Here $\omega_0 = 2\pi f$, γ is the ratio of specific heats; P_0, the hydrostatic pressure; ρ, the density of the surrounding liquid; and R_0, the mean radius of the bubble. Using constants suitable for an air bubble in water, Eq. (2-23) can be simplified to:

$$R_0\ (\text{mm}) = \frac{3.3}{f\ (\text{kHz})} \tag{2-24}$$

These equations neglect the effects of surface tension but are accurate up into the low-megahertz frequency range. Numerical studies of details of bubble oscillation have been conducted by Eatoch et al.[190] They were able to show that nonlinear scattering was important for bubbles that were smaller than or close to resonant size. In addition, Strasberg[191] has demonstrated that the resonant frequency of the bubble is relatively independent of the shape of the oscillating bubble.

Subharmonic Response

The acoustic response of liquids to ultrasound has been the subject of many investigations. A principal focus of attention has been cavitation phenomena and subharmonic responses. Esche[192] was the first to report that the frequency spectrum of cavitation noise in water contains white noise superimposed upon harmonics and subharmonics of the fundamental frequency. Mellen[193] and Negishi[194] observed similar behavior. In 1968, Neppiras[195] showed that, as the ultrasound amplitude is increased, harmonics, then subharmonics, and finally white noise and visible bubbles appear. The harmonics are explained in terms of nonlinear oscillations of cavitated bubbles. White noise apparently arises from strong bubble oscillation and bubble collapse.

Four possible sources for the subharmonic response have been suggested: (1) noise associated with shockwaves generated by the collapse of transient cavities; (2) noise associated with waves along bubble surfaces; (3) parametric amplification[196,197] of a low-level signal already present in the liquid; (4) large-amplitude radial oscillations of cavitated bubbles that are unstable to subharmonic response when the free resonant frequency of the bubble is one-half the applied frequency. It is unlikely that mechanisms (1) and (2) are important, however, since in related hydrodynamic studies of shockwaves, strong subharmonic signals are not observed and since there is only a weak coupling between surface waves and the bulk liquid is low-viscosity liquids.

Detailed theoretical and experimental studies have been conducted to further understand subharmonic phenomena. Theoretical studies by Houghton[198] and Eller and Flynn[199] have used perturbation techniques to show the presence of higher harmonics and unstable subharmonics. The threshold for the subharmonic of order one-half has been analyzed by Safar[200] and Storm.[201] The threshold for the subharmonic was found to be lowest when the driving frequency was twice the resonance frequency of the target bubbles. In a series of careful experiments, Neppiras[202] measured the acoustic spectrum in a viscous liquid seeded with a single bubble or discrete bubbles of known size. He detected acoustic emissions as harmonic and subharmonic frequencies as well as fractional harmonics. His determination of the driving pressure for the subharmonics agrees well with theoretical predictions.

In addition to the use of the subharmonic in studies of cavitation, as discussed elsewhere in this volume, it has proved to be a particularly useful tool for the study of phenomena in superfluid helium. Studies by Carey et al.[203] and Smith et al.[204] compared experimental results obtained in superfluid helium with the bubble model of Eller and Flynn[199] and to predictions based upon parametric amplification. They showed that the subharmonic harmonic response in superfluid helium was directly associated with bulk turbulence. More recent studies by Smith et al.[205,206] have shown that below the superfluid transition, the generation of subharmonics quantitatively follows the Feigenbaum universal convergence for period doubling. In addition, the physical nature of the onset of the first bifurcation was identified as the threshold for the generation of quantum vortex line. Additional studies[207] have used light diffraction techniques to visualize the onset of the subharmonic.

Second-Harmonic Response

In addition to subharmonic activity, pulsating bubbles produce harmonics of the fundamental (first harmonic) frequency. An example of this type of response is shown in Fig. 2-3. One may note that the second harmonic is sharply peaked at

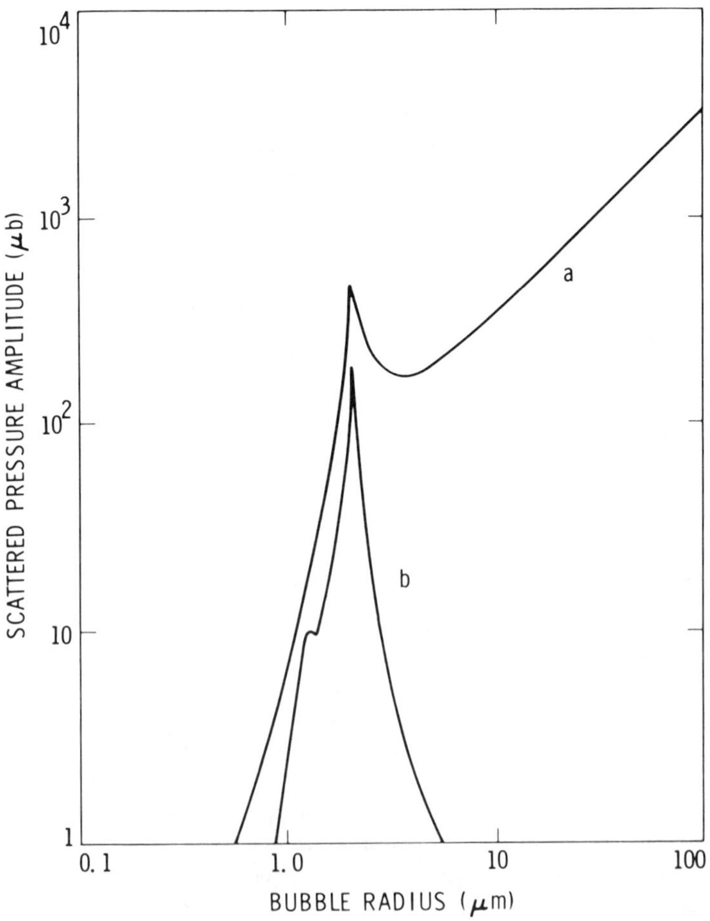

Figure 2-3. Pressure amplitude of fundamental (a) and second-harmonic (b) backscattered signals from bubble insonified at 1.64 MHz (see Ref. 223).

the resonance frequency of the pulsating bubble. The second-harmonic response of ultrasonically driven bubbles has been studied by several investigators.[208–211] The case of bubbles trapped in a membrane has been discussed theoretically by Miller and Nyborg.[212] Results from experiments by Neppiras et al.[213] and Miller and Neppiras[214] are in agreement with the theory. The second harmonic response is of particular interest because it is a characteristic of bubbles and can be used to distinguish them from other scatters of ultrasound. The capability of characterizing bubbles is important in the detection of decompression sickness, as well as for research and processing of materials.

Bubbles have been detected both by their direct backscattering and by the Doppler effect associated with their motion. Pulse-echo techniques have been used to image bubbles in cardiac chambers during echocardiographic examination[215] as well as in soft tissues during studies of decompression sickness.[216] Medical Doppler instrumentation has been used to detect bubbles through tubes and blood vessels.[217–222] The problems of Doppler detection result primarily from a poor signal to noise ratio since the methodology relies on a higher scattering of ultrasound by the bubbles than from the flowing erythrocytes. Thus, the presence of a bubble in the blood stream results only in a momentary increase in the amplitude of a Doppler frequency component that is already present in the normal blood flow Doppler signal.

In order to overcome these signal to noise problems, approaches using the nonlinear emission of harmonics with emphasis on the second harmonic have been developed. Miller,[223] following suggestions of Tucker and Welsby,[224] developed a system for monitoring the second-harmonic responses of a 1.64-MHz fundamental ultrasonic field. He found that 4.2-μm diameter resonant bubbles produced signals 40 times larger than bubbles of 500-μm diameter. A disadvantage of that system was that it only operated at a single frequency and therefore was optimized only for bubbles of a single size.

In order to cover a broader bubble size distribution, a system has been developed that uses a unique swept-frequency signal processing system called time delay spectrometry (TDS) developed by Heyser.[225] The TDS system[226] uses a linear frequency sweep from 0 to 10 MHz. The signal is transmitted into the test medium and the signal from the region of interest is received and coherently mixed with the input signal. This signal can be low-pass filtered and a Fourier transform used to covert the frequency differences into the range to the target bubbles, or the signal may be spectrum analyzed.

In order to test the feasibility of applying the JPL system to bubble detection, experiments were performed using target bubbles in various media. Specifically, the backscattered and attenuated signals were obtained from air entrapped in micropores of hydrophobic membranes similar to those described by Miller.[212–214] These membranes have a pore radius of 2.0 ± 0.14 μm and a thickness of 12 μm. The frequency was swept from 0 to 10 MHz in 20 ms. The resonance nature of the backscattering and absorption were easily detected in distilled water, saline, whole blood, and synovial fluid. The sizes of the bubbles can be determined by their resonance frequency distribution and the amplitude of the signal is related to the number of bubbles present.

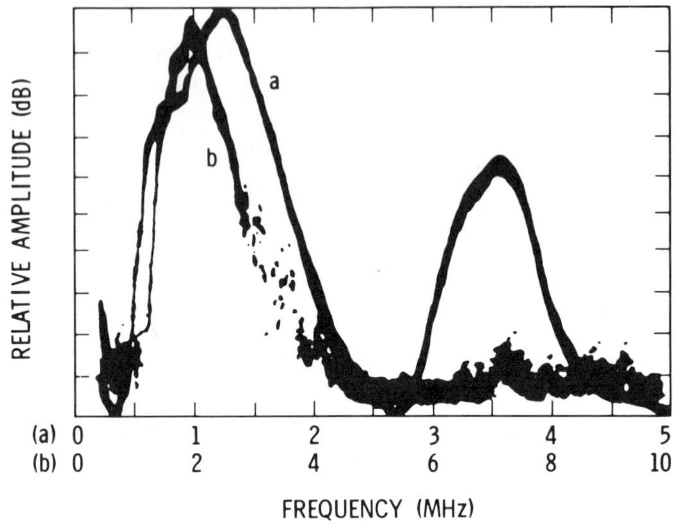

Figure 2-4. Measured fundamental (a) and second-harmonic (b) signals from bubbles of approximately 30-μm radius. The fundamental backscattering includes the transducer response as well as bubble resonance.

In addition, measurements have been made of the second-harmonic emissions from the target bubbles. For these measurements, the sweep program of the tracking receiver was driven synchronously with the transmitting program but at a rate of 1 GHz s^{-1} and a delay offset corresponding to the arrival time of the backscattered sound from the membrane. The measured frequency spectrum of the second harmonic backscatter from the bubbles corresponds closely to the predicted spectrum shape, showing a dominant peak at twice the bubble resonance frequency at which a strong absorption dip occurs for sound passing through the membrane, as shown in Fig. 2-4.

ACKNOWLEDGMENTS

This work was supported by the Jet Propulsion Laboratory, California Institute of Technology, under contract with the National Aeronautics and Space Administration.

REFERENCES

1. Rudenko, O. V.; Soluyan, S. I. "Theoretical Foundations of Nonlinear Acoustics"; Plenum Press: New York, 1977.
2. Naugal'nykh, K. A. In "High Intensity Ultrasonic Fields"; Rozenberg, L. D., Ed.; Plenum Press: New York, 1971, pp. 355–376.

3. Beyer, R. T.; Letcher, S. V. "Physical Acoustics"; Academic Press: New York, 1969.
4. Beyer, R. T. "Nonlinear Acoustics;" Naval Ship Systems Command, U.S. Government Printing Office: Washington, DC, 1974.
5. Bjorno, L. *Ultrason. Int. Proc.*, **1975**, 110.
6. G. Stokes, *Trans. Camb. Phil. Soc.* **1945**, *8*, 287.
7. Fubini, E. *Alta Freq.* **1935**, *4*, 539.
8. Earnshaw, S. *Phil. Trans. R. Soc. London* **1860**, *150*, 133.
9. Riemann, B. "Collected Works"; Dover: New York, 1953.
10. Airy, G. B. "Encyclopedia Metropolitana"; London, 1845, Vol. 5.
11. Fay, R. *J. Acoust. Soc. Am.* **1931**, *3*, 222.
12. Mendousse, J. S. *J. Acoust. Soc. Am.* **1953**, *25*, 51.
13. Rudnick, I. *J. Acoust. Soc. Am.* **1958**, *30*, 565.
14. Fox, F. E.; Wallace, W. A. *J. Acoust. Soc. Am.* **1954**, *26*, 994.
15. Cook, B. D. *J. Acoust. Soc. Am.* **1962**, *34*, 941.
16. Blackstock, D. T. *J. Acoust. Soc. Am.* **1964**, *36*, 534.
17. Soluyan, S. I.; Khokhlov, R. V. *Sov. Phys. Acoust.* **1962**, *8*, 170.
18. Blackstock, D. T. *J. Acoust. Soc. Am.* **1964**, *36*, 217.
19. Krasilnikov, V. A.; Shklovskaya-Kordi, V. V.; Zarembo, L. K. *J. Acoust. Soc. Am.* **1957**, *29*, 642.
20. Barnes, R. P. Jr.; Beyer, R. T. *J. Acoust. Soc. Am.* **1964**, *36*, 1371.
21. Law, W. K.; Frizzell, L. A.; Dunn, F. *J. Acoust. Soc. Am.* **1981**, *69*, 1210.
22. Cobb, W. N. *J. Acoust. Soc. Am.* **1983**, *73*, 1525.
23. Apfel, R. E. *J. Acoust. Soc. Am.* **1983**, *74*, 1866.
24. Zhu, Z.; Roos, M. S.; Cobb, W. N.; Jensen, K. *J. Acoust. Soc. Am.* **1983**, *74*, 1518.
25. Muir, T. G.; Carstensen, E. L. *Ultrasound Med. Biol.* **1980**, *6*, 345.
26. Carstensen, E. L.; Law, W. K.; McKay, N. D.; Muir, T. G. *Ultrasound Med. Biol.* **1980**, *6*, 359.
27. Haran, M. E.; Cook, B. D. *J. Acoust. Soc. Am.* **1983**, *73*, 774.
28. Goss, S. A.; Fry, F. J. *IEEE Trans. Sonics Ultrason.* **1981** *SU-28*, 21.
29. Carstensen, E. L.; Becroft, S. A.; Law, W. K.; D. B. Barbee, *J. Acoust. Soc. Am.* **1981**, *70*, 302.
30. Ichida, N.; Sato, T.; Miwa, H.; Murakami, K. *IEEE Trans Sonics Ultrason.* **1984**, *SU-31*, 635.
31. Nakagawa, Y.; Nakagawa, M.; Yoneyama, M.; Kikuchi, M. *Proc. 1984 Ultrasonics Symposium* B. R. McAvoy, ed., IEEE: Piscataway, N.J., **1984**, 673.
32. Nyborg, W. L. In *Physical Acoustics*; Mason, W. P., Ed.; Academic Press: New York, **1965**, Vol. 2B, pp. 265–283.
33. Kukarkin, A. B.; Rudenko, O. V. *Sov. Phys. Acoust.* **1976**, *22*, 137.
34. Ostrovskii, L. A.; Popilova, I. A. *Sov. Phys. Acoust.* **1974**, *20*, 45.
35. Liebermann, L. N. *Phys. Rev.* **1949**, *75*, 1415.
36. Lord Rayleigh, *Phil. Trans. R. Soc. London* **1884**, *175*, 1.
37. Lord Rayleigh, "Theory of Sound": Dover: New York, 1945.
38. Fairbanks, H. V.; Cline, R. E. *IEEE Trans. Sonics Ultrason.* **1967**, *SU-14*, 175.
39. Fairbanks, H. V.; Hu, T. K.; Leonard, J. W. *Ultrasonics* **1970**, *8*, 165.
40. Schlichting, H. *Phys. Z.* **1932**, *33*, 327.
41. Schlichting, H. "Boundary Layer Theory"; McGraw-Hill: New York, 1955.
42. Holtsmark, J.; Johnsen, I.; Sikkeland, T.; Skavlem, S. J.; *J. Acoust. Soc. Am.* **1954**, *26*, 26.
43. Skavlem, S.; Tjotta, S. *J. Acoust. Soc. Am.* **1955**, *27*, 26.
44. Westervelt, P. J. *J. Acoust. Soc. Am.* **1953**, *25*, 1123.
45. Westervelt, P. J. *J. Acoust. Soc. Am.* **1955**, *27* 379.
46. Raney, W. P.; Corelli, J. C.; Westervelt, P. J. *J. Acoust. Soc. Am.* **1954**, *26*, 1006.
47. Lane, C. A. *J. Acoust. Soc. Am.* **1955**, *27*, 1082.
48. Rowe, W. E.; Nyborg, W. L. *J. Acoust. Soc. Am.* **1966**, *39*, 965.
49. Sorokodum, E. D.; Timoshenko, V. I. *Sov. Phys. Acoust.* **1973**, *19*, 597.
50. Kolb, J.; Nyborg, W. L. *J. Acoust. Soc. Am.* **1956**, *28*, 1237.
51. Elder, S. A. *J. Acoust. Soc. Am.* **1959**, *31*, 54.
52. Mason, W. P.; Wick, R. F. *J. Acoust. Soc. Am.* **1951**, *23*, 209.
53. Jackson, F. J. *J. Acoust. Soc. Am.* **1960**, *32*, 1387.
54. Jackson, F. J.; Nyborg, W. L. *J. Acoust. Soc. Am.* **1958**, *30*, 614.
55. Hughes, D. E.; Nyborg, W. L. *Science* **1962**, *138*, 108.
56. Fill, E. E. *J. Appl. Phys.* **1968**, *39*, 5816.
57. Williams, A. R.; Nyborg, W. L. *Ultrasonics* **1970**, *8*, 36.
58. Rooney, J. A. *Proc. IEEE Ultrason. Symp.*, J. deKlerk, ed., IEEE: New York, N.Y. Milwaukee, Wisconsin, 1974, p. 51.

59. Ingard, U.; Labate, S. *J. Acoust. Soc. Am.* **1950**, *22*, 211.
60. Thurston, G. B.; Martin, C. E. *J. Acoust. Soc. Am.* **1953**, *25*, 26.
61. Rooney, J. A. *Science* **1970**, *169*, 869.
62. Semenova, N. G. *Sov. Phys. Acoust.* **1974**, *20*, 65.
63. Hargrove, L. E. Jr.; Thurston, G. B. *J. Acoust. Soc. Am.* **1957**, *29*, 966.
64. Durelli, A. J.; Clark, J. A. *J. Strain Anal.* **1972**, *7*, 217.
65. Nagai, S.; Peters, A.; Candau, S. *Rev. Phys. Appl.* **1977**, *12*, 21.
66. Nyborg, W. L.; Gould, R. K.; Jackson, F. J.; Adams, C. E. *J. Acoust. Soc. Am.* **1959**, *31*, 706.
67. Bhadra, T. C.; Roy, B. *Proc. Ultrason. Int.* IPC Science and Technology, Ross Ltd., Guilford, Surrey, U.K., **1975**, 253.
68. Schmitt, F. O. *Protoplasma* **1929**, *7*, 332.
69. Dyer, H. J.; Nyborg, W. L. *IRE Trans. Med. Electron.* **1960**, *ME-7*, 163.
70. Gershoy, A.; Nyborg, W. L. *J. Acoust. Soc. Am.* **1973**, *54*, 1356.
71. El'piner, I. E.; Fairkin, I. M.; Basurmonova, O. K. *Biofizika* **1965**, *10*, 805.
72. Wilson, W. L.; Wiereinski, F. J.; Nyborg, W. L.; Schnitzler, R. M.; Sichel, F. J. *J. Acoust. Soc. Am.* **1966**, *40*, 1363.
73. Schnitzler, R. M. Ph.D. thesis, University of Vermont, Burlington, VT, 1969.
74. Williams, A. R. *J. Physiol.* **1976**, *257*, 23.
75. Williams, A. R. *Ultrasound Med. Biol.* **1977**, *3*, 191.
76. Williams, A. R. *J. Acoust. Soc. Am.* **1977**, *56*, 1640.
77. Williams, A. R.; Hughes, D. E.; Nyborg, W. L. *Science* **1970**, *169*, 871.
78. Rooney, J. A. *J. Acoust. Soc. Am.* **1972**, *52*, 1718.
79. Rooney, J. A. *J. Biol. Phys.* **1973**, *2*, 26.
80. Bundy, M. L. Ph.D. Thesis, University of Maine, Orono, Me, 1980.
81. Rooney, J. A.; Smith, C. W.; Carey, R. F. *J. Acoust. Soc. Am.* **1982**, *72*, 245.
82. Smith, C. W.; Rooney, J. A.; Carey, R. F. *Physica* **1981**, *107B*, 695.
83. Carey, R. F.; Rooney, J. A.; Smith, C. W. *Phys. Lett.* **1978**, *65A*, 311.
84. Khalatnikov, I. M. *Zh. Eksp. Teor, Fiz.* **1952**, *23*, 169.
85. Smith, C. W.; Rooney, J. A.; Carey, R. F. *Ultrasonics* **1983**, *21*, 43.
86. Weinstein, J. N.; Yoshikami, S.; Henkart, P.; Blumenthal, R.; Hagins, W. A.; *Science* **1977**, *195*, 489.
87. Genta, V. M.; Kaufman, D. G.; Kaufman, W. K. *Anal. Biochem.* **1975**, *67*, 279.
88. Hislop, T. W. *Ultrasonics* **1970**, *8*, 88.
89. Last, A. J. *Ultrasonics* **1969**, *7*, 131.
90. Crawford, A. E. *Research* **1953**, *6*, 106.
91. Greguss, P. *Ultrasonics* **1972**, *10*, 276.
92. Hinze, J. O. American Institute of Chemical Engineers J. **1955**, *1*, 289.
93. Goldsmith, H. L.; Mason, S. G. In "Rheology Theory and Applications"; Eirich, F. R., Ed.; Academic Press: New York, 1967, Vol. 4 pp. 85–104
94. Rumscheidt, F. D.; Mason, S. G. *J. Colloid Sci.* **1961**, *16*, 238.
95. Taylor, G. I. *Proc. R. Soc. London Ser. A* **1932**, *138*, 41.
96. Rajagopol, E. S. *Proc. Indian Acad. Sci. Sect A* **1959**, *49*, 333.
97. Neduzhii, S. A. *Sov. Phys. Acoust.* **1963**, *9*, 99.
98. Neduzhii, S. A. *Sov. Phys. Acoust.* **1961**, *7*, 79.
99. Neduzhii, S. A. *Sov. Phys. Acoust.* **1961**, *7*, 209.
100. Mullaney, P. F.; Cram, L. S.; Trujillo, T. T. *Acta Cytol.* **1971**, *15*, 217.
101. Williams, A. R.; Stafford, D. A.; Callely, A. G.; Hughes, D. E. *J. Appl. Bacteriol.* **1970**, *33*, 656.
102. Parry, J. S.; Cleary, B. K.; Williams, A. R.; Evans, D. M. D. *Acta Cytol.* **1971**, *15*, 163.
103. Williams, A. R.; Slade, J. S. *Ultrasonics* **1971**, *9*, 85.
104. Brakey, M. W.; Rooney, J. A. *Digest Int. Conf. Med. Biol. Eng.*; 11th, Ottawa, Ontario, Canada, 1976, p. 80.
105. Wood, W. R.; Loomis, A. L. *Phil. Mag.* **1927**, *4*, 417.
106. Topp, M. N.; Eisenklam, P. *Ultrasonics* **1972**, *10*, 127.
107. Eknadiosyants, O. K. In "Physical Principles of Ultrasonic Technology"; Rozenberg, L. D., Ed.; Plenum Press: New York, 1973, Vol. 1, pp. 121–132.
108. Sollner, K. *Trans. Faraday Soc.* **1936**, *32*, 1532.
109. Bisa, K.; Dirnogl, K.; Esche, R. *Siemens Z.* **1954**, *28*, 341.
110. Sorokin, V. I. *Sov. Phys. Acoust.* **1957**, *3*, 281.
111. Eisenmenger, W. *Acustica* **1959**, *9*, 327.
112. Peskin, R. L.; Raco, R. J. *J. Acoust. Soc. Am.* **1963**, *35*, 1378.

113. Eknadiosyants, O. K. *Sov. Phys. Acoust.* **1968**, *14*, 80.
114. Boguslavskii, Y. Y.; Eknadiosyants, O. K. *Sov. Phys. Acoust.* **1969**, *15*, 14.
115. Lang, R. J. *J. Acoust. Soc. Am.* **1962**, *34*, 6.
116. Fogler, H. S.; Timmerhaus, K. D. *J. Acoust. Soc. Am.* **1966**, *39*, 515.
117. Topp, M. N. *J. Aerosol Sci.* **1973**, *4*, 17.
118. Bassett, J. D.; Bright, W. W. *J. Aerosol Sci.* **1976**, *7*, 47.
119. Chiba, C. *Bull. Journal of Society of Mechanical Engineers* **1975**, *18*, 376.
120. Lierke, E. G.; Greisshamer, G. *Ultrasonics* **1967**, *5*, 224.
121. Wilcox, R. L.; Tate, R. W.; *AIChE J.* **1965**, *11*, 69.
122. Clamen, A.; Gauvin, W. H. *Can. J. Chem. Eng.* **1968**, *46*, 223.
123. Rayleigh, J. W. S. *Proc. London Math. Soc.* **1979**, *10*, 4.
124. Lindblad, N. R.; Schneider, J. M. *J. Sci. Instrum.* **1965**, *42*, 635.
125. Fulwyler, M. J. *Science* **1965**, *150*, 910.
126. Machara, N.; Nakane, S.; Yamamoto, K.; Iga, K. *Ultrasonics* **1984**, *22*, 259.
127. Herzenberg, L. A.; Sweet, R. G. *Sci. Am.* **1976**, *234*, 108.
128. Beasley, J. D. *Photogr. Sci. Eng.* **1977**, *21* (2), 89–92.
129. Stovel, R. T. *J. Histochem. Cytochem.* **1977**, *25*, 813.
130. Klopovsky, B. A.; Fridman, V. M. *Ultrasonics* **1976**, *14*, 107.
131. Post, E. J. *J. Acoust. Soc. Am.* **1953**, *25*, 55.
132. Dvorak, V. *Poggendorff's Ann.* **1876**, *157*, 42.
133. Eckart, C. *Phys. Rev.* **1948**, *73*, 68.
134. Lord Rayleigh, *Phil. Mag.* **1902**, *3*, 338.
135. Rooney, J. A.; Nyborg, W. L. *Am. J. Phys.* **1972**, *40*, 1825.
136. Zieniuk, J.; Chivers, R. C. *Ultrasonics* **1976**, *14*, 161.
137. Beyer, R. T. *J. Acoust. Soc. Am.* **1978**, *63*, 1025.
138. Lord Rayleigh, *Phil. Mag.* **1905**, *10*, 364.
139. Brillouin, L. *Ann Phys. (Paris)* **1925**, *4*, 528.
140. Beyer, R. T. *Am. J. Phys.* **1950**, *18*, 25.
141. Borgnis, F. E. *Rev. Mod. Phys.* **1953**, *25*, 633.
142. Gol'dberg Z. A. In "High Intensity Ultrasonic Fields"; Rozenberg, L. D., Ed.; Plenum Press: New York, 1971, pp. 73–133.
143. Chu, B.; Apfel, R. E. *J. Acoust. Soc. Am.* **1982**, *72*, 1673.
144. Chu, B.; Apfel, R. E. *J. Acoust. Soc. Am.* **1984**, *75*, 1003.
145. Nyborg, W. L.; Rooney, J. A. *J. Acoust. Soc. Am.* **1984**, *75*, 263.
146. Wemlen, A. *Med. Biol. Eng.* **1968**, *6*, 159.
147. Rooney, J. A. *Ultrasound Med. Biol.* **1973**, *1* 13.
148. Klein, E. *J. Acoust. Soc. Am.* **1938**, *10*, 105.
149. Henry, G. E. *Institute of Radio Engineers Trans. Ultrason. Eng.* **1957**, *6*, 17.
150. Kossoff, G. *Acoustica* **1962**, *12*, 84.
151. Hertz, G.; Mende, H. *Z. Phys.* **1939**, *114*, 354.
152. Herrey, E. M. J. *J. Acoust. Soc. Am.* **1955**, *27*, 891.
153. Cornhill, C. V. *Ultrasonics* **1982**, *20*, 282.
154. Hill, C. R. *Phys. Med. Biol.* **1970**, *15*, 241.
155. Cady, W. G.; Gitlings, C. E. *J. Acoust. Soc. Am.* **1953**, *25*, 892.
156. Kossoff, G. *J. Acoust. Soc. Am.* **1963**, *38*, 880.
157. Rooney, J. A. *J. Acoust. Soc. Am.* **1973**; *54*, 429.
158. Tarnocy, T. *Magy. Fiz. Foly.* **1954**, *2*, 159.
159. Wells, P. N. T.; Bullen, M. A.; Freundlich, H. F. *Ultrasonics* **1969**, *2*, 214.
160. Haran, M. E.; Cook, B. D.; Stewart, H. F. *J. Acoust. Soc. Am.* **1975**, *57*, 1436.
161. Greenspan, M.; Breckenridge, F. R.; Tschiegg, C. E. *J. Acoust. Soc. Am.* **1978**, *63*, 1031.
162. Bellin, J. L. S.; Beyer, R. T. *J. Acoust. Soc. Am.* **1960**, *32*, 339.
163. Hildebrand, B. P.; Brenden, B. B. "An Introduction to Acoustical Holography"; Plenum Press: New York, 1972, pp. 1–224.
164. Clark, A. V. *IEEE Trans. Sonics Ultrason.* **1981**, *Su-28*, 14.
165. Bjerknes, V. F. K. "Fields of Force"; Columbia Univ. Press: New York, 1906.
166. King, L. V. *Proc. R. Soc. London Ser. A* **1934**, *147*, 212.
167. Yosioka, K.; Kawasima, Y. *Acustica* **1955**, *5*, 17.
168. Chivers, R. C.; Anson, L. W. *Ultrasonics* **1955**, *20*, 25.
169. Gor'kov, L. P. *Soc. Phys. Dokl.* **1962**, *6*, 773.
170. Westervelt, P. J. *J. Acoust. Soc. Am.* **1951**, *23*, 312.

171. Embleton, T. F. W. *J. Acoust. Soc. Am.* **1954**; *26*, 40.
172. Nyborg, W. L. *J. Acoust. Soc. Am.* **1967**, *42*, 947.
173. Dystke, K. B. *J. Sound. Vib.* **1969**, *10*, 331.
174. Crum. L. A. *J. Acoust. Soc. Am.* **1971**, *50*, 157.
175. Hasegawa, T. *J. Acoust. Soc. Am.* **1979**, *65*, 32.
176. Hasegawa, T. *J. Acoust. Soc. Am.* **1979**, *65*, 41.
177. Hasegawa, T.; Yosioka, K. *J. Acoust. Soc. Am.* **1975**, *58*, 581.
178. Eller, A. I. *J. Acoust. Soc. Am.* **1968**, *43*, 170.
179. Embleton, T. F. W. *J. Acoust. Soc. Am.* **1954**, *26*, 46.
180. Hasegawa, T.; Yosioka, K. *J. Acoust. Soc. Am.* **1969**, *46*, 1139.
181. Breazeale, M. A.; Dunn, F. *J. Acoust. Soc. Am.* **1974**, *55*, 671.
182. Crum, L. A.; Eller, A. I. *J. Acoust. Soc. Am.* **1970**, *48*, 181.
183. Baxter, K.; Apfel, R. E.; Marston, P. L. *Rev. Sci. Instrum* **1978**, *49*, 224.
184. Apfel, R. E. *J. Acoust. Soc. Am.* **1971**, *49*, 145.
185. Apfel, R. E. *Nature (London) Phys. Sci.* **1971**, *233*, 119.
186. Apfel, R. E.; Harbison, J. P. *J. Acoust. Soc. Am.* **1975**, *57*, 1371.
187. Apfel, R. E. *J. Acoust. Soc. Am.* **1976**, *59*, 339.
188. Weiser, M. A. H.; Apfel, R. E.; Neppiras, E. A. *Acustica* **1984**, *56*, 114.
189. Minnaert, M. *Phil. Mag.* **1933**, *16*, 235.
190. Eatoch, B. C.; Nishi, R. Y.; Johnston, G. W. *J. Acoust. Soc. Am.* **1985**, *77*, 1692.
191. Strasberg, M. *J. Acoust. Soc. Am.* **1953**, *25*, 536.
192. Esche, R. *Acustica* **1952**, *2AB*, 208.
193. Mellen, R. H. *J. Acoust. Soc. Am.* **1954**, *26*, 356.
194. Negishi, K. *J. Phys. Soc. Jpn.* **1961**, *16*, 1450.
195. Neppiras, E. A. *J. Acoust. Soc. Am.* **1969**, *45*, 587.
196. Tucker, D. G. *J. Sound Vib.* **1965**, *2*, 429.
197. Adler, L.; Breazeale, M. A. *J. Acoust. Soc. Am.* **1970**, *48*, 1077.
198. Houghton, G. *J. Acoust. Soc. Am.* **1963**, *35*, 1387.
199. Eller, A.; Flynn, H. G. *J. Acoust. Soc. Am.* **1969**, *46*, 722.
200. Safar, M. H. *J. Phys. D: Appl. Phys.* **1970**, *3*, 635.
201. Storm, D. L. Ph.D. Thesis, University of Vermont, Burlington, Vt. 1970.
202. Neppiras, E. A. *J. Acoust. Soc. Am.* **1969**, *45*, 587.
203. Carey, R. F.; Rooney, J. A.; Smith, C. W. *J. Acoust. Soc. Am.* **1979**, *66*, 1801.
204. Smith, C. W.; Rooney, J. A.; Carey, R. F. *Phys. Lett.* **1979**, *70A*, 110.
205. Smith, C. W.; Tejwani, M. J. *Physics* **1983**, *7D*, 85.
206. Smith, C. W.; Tejwani, M. J.; Farris, D. A. *Phys. Rev. Lett.* **1982**, *48*, 492.
207. Carey, R. F.; Rooney, J. A.; Smith, C. W. *Ultrasonics* **1980**, *18*, 213.
208. Coakley, W. T.; Nyborg, W. L. In "Ultrasound: Its Applications in Medicine and Biology"; Fry, F. J., Ed.; Academic Press: New York, 1978, pp. 77–159.
209. Lauterborn, W. *J. Acoust. Soc. Am.* **1976**, *59*, 283.
210. Prosperetti, A. *J. Acoust. Soc. Am.* **1974**, *56*, 878.
211. Medwin, H. *J. Acoust. Soc. Am.* **1977**, *62*, 1041.
212. Miller, D. L.; Nyborg, W. L. *J. Acoust. Soc. Am.* **1983**, *73*, 1573.
213. Neppiras, E. A.; Nyborg, W. L.; Miller, D. L. *Ultrasonics* **1983**, *21*, 109.
214. Miller, D. L.; Neppiras, E. A. *J. Acoust. Soc. Am.* **1985**, *77*, 946.
215. Gramiak, R.; Shaw, P. M. *Radiology* **1971**, *100*, 415.
216. Mackay, R. S.; Rubissow, G. *IEEE Trans.* **1978**, *BME-25*, 537.
217. Gills, M. F.; Karagiones, M. T.; Peterson, P. L. *Science* **1968**, *16*, 579.
218. Nishi, R. Y. *Ultrasonics* **1972**, *10*, 173.
219. Belcher, E. O. *IEEE Trans.* **1980**, *BME-27*, 330.
220. Hills, B. A.; Grulke, D. C. *Ultrasonics* **1975**, *13*, 181.
221. Evans, A.; Walder, D. N. *Ultrasonics* **1970**, *8*, 216.
222. Klepper, J. R. *Proc. IEEE Ultrason. Symp.* **1982**, 668.
223. Miller, D. L. *Ultrasonics* **1981**, *19*, 217.
224. Tucker, D. G.; Welsby, V. G. *Lancet* **1968**, *1*, 1253.
225. Heyser, R. C. U.S. Patents Nos. 3,466,652, "Time Delay Spectrometer," and 4,279,019, "Method and Apparatus for Delay Analysis of Energy Transmitted Through a Medium."
226. Heyser, R. C.; LeCroissette, D. H. *Ultrasound Med. Biol.* **1974**, *1*, 119.

3

Industrial Applications of Ultrasound

ANDREW SHOH

Ultrasonic Research

GENERAL BACKGROUND

Macrosonic Applications

A U.S. Census Bureau report[1] estimates 1980 sales of ultrasonic power equipment at over $80 million. One should at least double that amount to obtain the world figure, which, allowing for the economic upturn in recent years, would place the current worldwide sales somewhere around $200 million. By now power ultrasound in industry has become well entrenched, and with its many uses in automotive, electronic, optical, semiconductor, biomedical, and other industries, it has become indispensable to modern manufacturing.

In some ways, however, the growth of industrial uses of power ultrasound has been unpredictable and asymmetrical. Ultrasonic plastic welding and cleaning, for instance, represent by far the bulk of applications and ultrasonic machining and soldering, which were small-scale established processes 20 years ago, still remain small. Some applications that were once in use, such as ultrasonic drying and air scrubbing, are now nearly extinct, and many others have never made it beyond the exploratory stage.

On the other hand, new opportunities have emerged for power ultrasound as a result of modern technologies, e.g., synthetic textiles and reinforced composite materials. Table 3-1 gives an overview of present industrial uses of power ultrasound.

Table 3-1. Macrosonic Applications

Applications in liquids	Applications in solids
Cleaning[a]	Plastic welding[a]
Soldering[b]	Metal welding[b]
Deburring[c]	Metal forming[c]
Erosion testing[c]	Impact grinding[b]
Cell disruption[b]	Rotary abrasive machining
Extraction from plants[c]	Metal cutting[c]
Emulsification[c]	Fatigue testing[c]
Dispersion of solids[c]	Curing[c]
Sterilization[d]	Trimming of composites[c]
Filtration[d]	Dental descaling[b]
Inhalation therapy[b]	
Fuel atomization[c]	Airborne applications
Drying of textiles[d]	Particle agglomeration
Crystal growth[c]	Defoaming
Metal grain refinement[d]	Drying
Degassing[b]	
Medical surgery[b]	

[a] Large established area.
[b] Small established area.
[c] Some industrial use.
[d] Experimental.

Process Parameters

Ultrasound is a form of mechanical energy, and its application to matter, under the right circumstances, can result in permanent physical changes. Because energy is a product of power and time, for a given power the length of ultrasonic exposure determines the total energy input into the material treated, which normally bears some relationship to the desired result. Cumulative energy input alone, however, does not insure such result. Most macrosonic effects exhibit a threshold phenomenon and do not occur unless ultrasonic intensity has reached a certain level. Therefore, it is meaningful to talk about macrosonic applications in terms of power density or watts of ultrasonic power per unit area or volume of the material treated. Although in some cases, such as airborne acoustic drying, this power density may be relatively low, in most macrosonic applications it is typically over 1 W cm^{-2}, and in high-intensity applications, such as ultrasonic metal welding, may be in the order of tens of thousands of watts per square centimeter.

By definition, ultrasound pertains to frequencies above human hearing (approximately over 18 kHz). Although macrosonic effects extend well into the megahertz range, most practical applications to date have been in the lower ultrasonic spectrum, between 20 and 60 kHz.

In general, power ultrasound is characterized by an ability to transmit substantial amounts of mechanical power at small mechanical movements. Ultrasonic motional amplitude is limited by the allowable stress in the ultrasonic transducer material and is frequency dependent. Table 3-2 illustrates the point.

Table 3-2. Ultrasonic Transducer Parameters as a Function of Frequency[a]

Parameter	$f(\omega)$
Length	$1/\omega$
Diameter	$1/\omega$
Output area	$1/\omega^2$
Displacement amplitude	$1/\omega$
Velocity amplitude	Constant
Acceleration	ω
Output power	$1/\omega$
Output power density	Constant
Transducer stress	Constant

[a] Parameters of an ultrasonic transducer of the same design and operating at the same stress in the transducer material are tabulated as a function of frequency. Transducer linear dimensions, displacement amplitude, and power capability vary inversely with the frequency, whereas ultrasonic acceleration increases directly with the frequency.

Transducer displacement, velocity, acceleration, and output power have been tabulated as a function of frequency for a constant stress in transducer material. It should be noted that the source velocity and power density are constant, whereas displacement decreases, and acceleration increases, directly with frequency. To illustrate typical values, a 20-kHz transducer operating at a peak displacement amplitude of 50 μm has a peak velocity of 6.28 m s^{-1} and a peak acceleration of $8 \times 10^4 g$. More completely then, power ultrasound is characterized by high vibrational frequencies (for reciprocating mechanical movement), small displacements, moderate point velocities, and extremely high accelerations.

Most macrosonic applications depend on compound acoustic phenomena occurring in matter, which, in turn, are caused by primary vibratory inputs. Thus, acoustic pressure causes cavitation and microstreaming in liquids, vibratory stress causes heating and fatiguing in solids, and ultrasonic acceleration is responsible for surface instability occurring at liquid–liquid and liquid–gas interfaces.

The ability of ultrasound to propagate through elastic media makes it possible to treat materials at some distance away from the source (far-field processing). Such treatment, however, may be nonuniform because of "standing wave" patterns that may form in the medium, causing alternate areas of high and low intensity at quarter-wavelength intervals.[2–7]

Ultrasonic Sources

The early industrial macrosonic applications made use of electronic as well as mechanical ultrasonic sources. The mechanical devices were pneumatic and liquid whistles. Pneumatic whistles were energized by compressed air and were used in airborne applications, such as acoustic defoaming and drying. In a liquid whistle, a resonant reed is brought into oscillation by a stream of pressurized liquid. Such equipment was used in high-intensity liquid processing, such as homogenizing.[8]

SONAR-related research during and following World War II has yielded manmade piezoelectric transducer materials as an alternative to natural quartz crystals, which were used in early ultrasonic work. Because of superior electromechanical coupling, piezoelectric ceramics have made it practical to generate power at lower ultrasonic frequencies, where cavitational effects, such as ultrasonic cleaning, can be achieved more effectively.

Over the years, piezoelectric ceramics have been greatly improved, in particular with regard to efficiency, power handling capacity, and ability to withstand higher temperatures. The representative material today is lead–zirconate–titanite (PZT) which has a polarization (Curie) temperature of 350°C and is capable of electromechanical conversion efficiency above 98%. The utilization of piezoelectric ceramics was further enhanced by sandwich-type transducer construction, where ceramic piezoelectric elements are clamped between metal members to produce a resonant (usually half-wave) structure. Such construction assures that the ceramic elements are always used in compression and therefore can withstand higher stresses, the end result being higher transducer amplitudes and more convenient means of coupling to the load or other resonant members.[9]

An alternate method of converting high frequency electric current to ultrasonic vibration is by means of magnetostrictive transducers which make use of materials that change dimension in response to magnetic field. Because of limitations that have not been resolved (magnetorestrictive metals are inefficient and magnetostrictive ferrites are brittle) they have been gradually replaced by piezoelectric transducers and are rarely used.

Concurrent with the development of piezoelectric materials has been the progress in electronic power supplies driving the ultrasonic transducers. Modern solid-state ultrasonic power supplies utilize transistors or thyristors (silicon-controlled rectifiers) and are efficient, compact, and reliable. Typical system efficiency, defined as a ratio of net acoustical power into the load to electrical input from power line, is over 80%.

APPLICATIONS IN LIQUIDS

Cleaning

Industrial ultrasonic cleaning was introduced in the early 1950s and for 25 years was the largest application of power ultrasound. Although the primary uses are in the electronic, optical, and medical industries, the process spans a large variety of applications, from large castings to watches. Typical examples include hospital glassware and surgical instruments, photographic lenses, filters, electronic printed circuit boards, semiconductors, ball bearings, automotive carburetors, engine parts, and heat exchangers.

Ultrasonic cleaning works best on sound-reflecting materials such as metal, glass, and plastic. It is less efficient on sound-absorbing materials such as rubber or cloth. In addition to saving time, it does jobs that are difficult or impossible by other methods, such as cleaning parts of complex geometries or having blind holes or crevices.

Ultrasonic cleaning has been described as "brushless scrubbing". The work of scrubbing is done by cavitational bubbles at or near the surface of the cleaned object. The explanation of the cleaning process mechanism varies somewhat with the nature of the contaminants, which can be either soluble or insoluble. Stable cavitation and microstreaming dissolve contaminants by bringing fresh solution to the surface cleaned. Insoluble contaminants may be dispersed or, with tougher deposits, eroded gradually by cavitational action. The contaminant coat may also be stripped by combined action of transient cavitation, which cracks the coat, and stable cavitational bubbles, which form between the coat and the surface and peel the coat away. Stable cavitation and microstreaming also contribute to emulsification of grease in aqueous solutions. Cleaning agents used in ultrasonic cleaning are selected on the basis of their ability to combine cavitational activity with chemical action. Aqueous solutions of surfactants (i.e., detergents) are most commonly used and can be of any pH.[10–12]

Industrial ultrasonic cleaners can be stand-alone units; likewise; the ultrasonic cleaning function may be built into larger systems, usually combining other pre- and postcleaning stages such as presoaking or vapor rinsing (see Figs. 3-1 and 3-2). Industrial stand-alone cleaners consist of rectangular stainless steel tanks of 5–150 L capacity, driven by electronic solid-state power supplies. The power ratings on standard models range from about 250 to 2000 W. At present, most operate at 40 kHz, although equipment at lower frequencies (in particular 25 kHz) is also in use. The more recent preference for 40 kHz originates from applications in the semiconductor industry, where lower frequencies may damage the product, and also from noise considerations. The tanks are usually equipped with temperature-controlled heaters. In larger systems, tanks of several hundred liters capacity, energized by a number of generators, are not uncommon. In recent years computer-controlled robotic devices have been introduced for automatic material handling. Small, inexpensive cleaners are available for light-duty uses, which combine the cleaning bath and electronics in one package. Used in laboratories, garages, and jewelry shops, such cleaners have power output of 25–150 W and operate at around 50–60 kHz (see Fig. 3-3).

Soldering

The principle of ultrasonic soldering is related to ultrasonic cleaning. Cavitation in molten solder erodes surface oxides and exposes clean metal to solder. This allows soldering without flux. Such fluxless soldering is mainly of importance

Figure 3-1. A typical configuration for washing, rinsing, and drying precision optics. Stainless steel tanks on the upper shelf with electronic power supplies below are representative of industrial ultrasonic cleaners. (Courtesy of Crestek, Inc.)

Figure 3-2. An automated system for cleaning hypodermic needles. (Courtesy of Crestek, Inc.)

for aluminum, which cannot be soldered with ordinary alkaline or acid fluxes. Fluxes used on aluminum are highly corrosive and attack the metal itself, which creates a problem after soldering when flux cannot be completely removed.

Other applications for ultrasonic soldering are in electrical and electronic components, where nickel, Kovar and other hard to tin metals are used. The advantages include elimination of flux for economic or corrosion reasons and faster or more complete tinning. Typical applications include termination of aluminum cables and other conductors, for which small, high-intensity solder pots are used, where two opposing horns are immersed in solder, and parts are placed near their oscillating surfaces (see Fig. 3-4). Larger structures, such as aluminum window frames and heat exchangers, can also be soldered ultrasonically. For heat exchangers large solder tanks are used, where transducers are attached to the underside of the tank, and the whole volume of solder is ultrasonically activated,

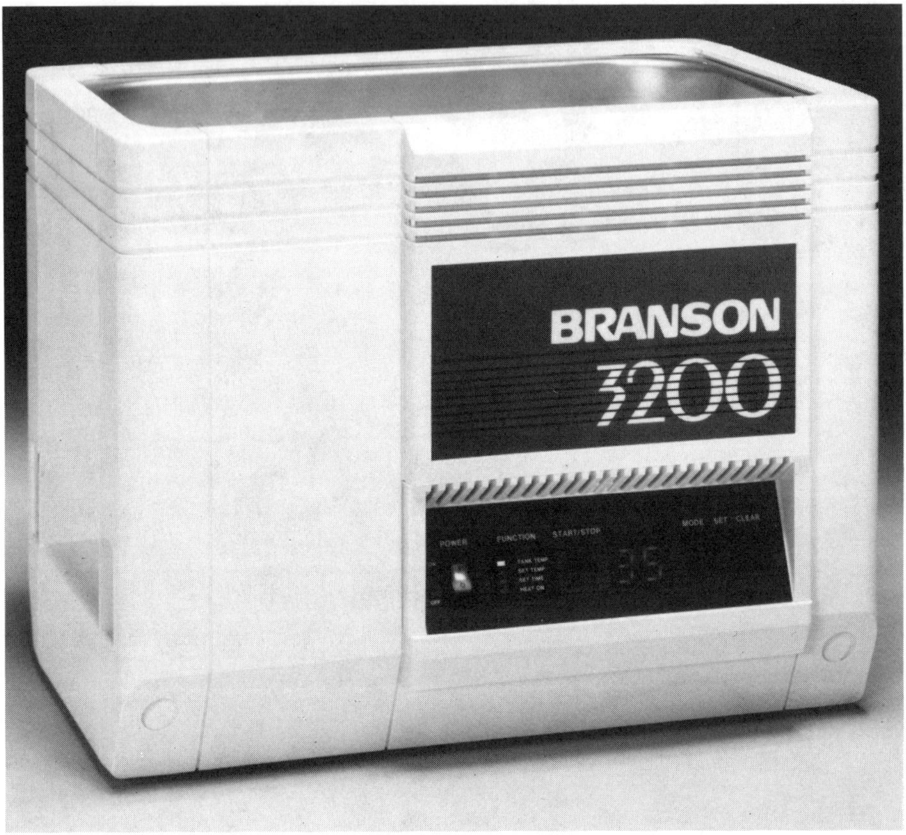

Figure 3-3. 150-Watt benchtop cleaner with built-in power supply and microprocessor-controlled temperature control. (Courtesy of Branson Ultrasonics Corporation.)

similar to the cleaning bath in ultrasonic cleaning.[13] Such ultrasonic solder tanks operate at temperatures up to 450°C.

Deburring

Ultrasonic deburring is an extension of high-intensity ultrasonic cleaning. It is used infrequently, usually on delicate parts that can be damaged by conventional deburring methods, such as tumbling or sandblasting.[14] The work of deburring is done by cavitational erosion. Although conventional cleaning tanks can be used, the process works better in pressurized liquid because cavitational shock intensity under pressure is higher. Deburring action can be further enhanced by the use of an abrasive suspension in water–glycerine solution. Equal parts of water, glycerine, and abrasive are suggested, and the size of abrasive particles should be optimized for the ultrasonic frequency used.

Figure 3-4. Ultrasonic soldering tank for tinning small parts. High-intensity cavitation in molten solder is produced between two opposing horns. (Courtesy of Branson Ultrasonics Corporation.)

Erosion Testing

Testing the resistance of materials to cavitational erosion by ultrasound has an advantage of control and convenience compared to other methods, such as venturi throat or rotating disk. The subject is of interest to studies of material erosion by water impact, such as encountered with turbine blades, boat propellers, and hydraulic machinery.[15,16]

In ultrasonic erosion testing the specimen may be the ultrasonic horn, or material to be tested may be bonded to the horn. Material samples may be also placed near the ultrasonic horn. The easily measured parameters are ultrasonic horn amplitude, frequency, and power. Normally 20-kHz equipment is used.

Cell Disruption

Of the ultrasonic extraction processes, biological cell disintegration is the most important at present and is used in research as well as low-volume production. High-intensity cavitation punctures the cell walls and releases the cell content. The process is used to extract active antigens for making vaccines and in the study of lipids, enzymes, and viruses.[17] The advantages of the ultrasonic treatment

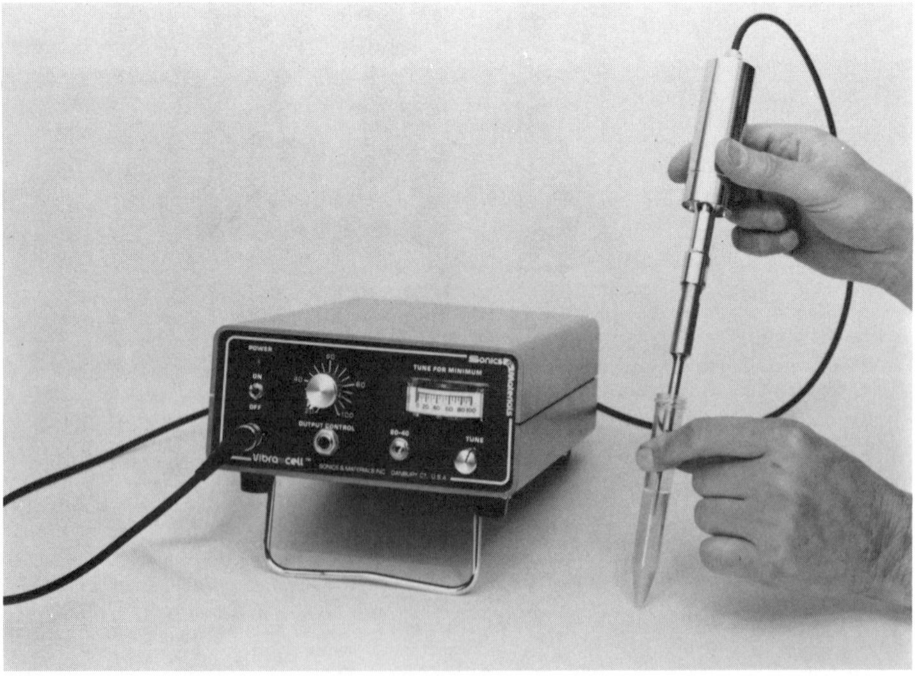

Figure 3-5. 40-Watt, 20-KHz biological cell disrupter for laboratory use. (Courtesy of Sonics & Materials, Inc.)

include relative safety for materials being extracted. Typical extraction times range from a few seconds to tens of minutes. Pulsed operation, available on some equipment, allows treatment at high ultrasonic intensities while maintaining lower average temperatures. Twenty-kilohertz, high-amplitude horns are used at power levels from 10 to 300 W (see Fig. 3-5). Solutions are usually treated in test tubes, but more elaborate continuous-flow processing equipment is available for higher throughout. Optional cooling may be incorporated to reduce the damage to the extracts.

Extraction from Plants

There have been numerous reports of ultrasonic extraction from vegetable matter, including alkaloids, glucosides, and scents; fruit juices; and essential oils from hops.[18–23] The advantages sought are higher yield and reduced processing time. Ultrasonic extraction can be done at lower temperatures, and there is less damage to extracts and less loss of volatile components in boiling. Although much of the investigatory work on ultrasonic extraction was reported in the 1960s, the process has not caught on and is not being used on any significant scale.

Emulsification

It has been demonstrated that stable emulsions can be produced ultrasonically without the use of surfactants or with a lower surfactant content.[24–26] This would appear to be of value to processes where surfactants can be eliminated to save cost or where the product cannot tolerate additives. In spite of an early start—ultrasonic emulsification of mineral and essential oils, antibiotic dispersions, and lotions was done in the early 1960s using liquid whistles—there is not much demand, at least in the United States and Western Europe, for ultrasonic equipment for these uses today. One explanation is that chemists prefer chemical solutions to their problems, and the progress in surfactants has made it possible to produce emulsions with simple and inexpensive mixing equipment.

An interesting opportunity for ultrasonic emulsification has emerged in the energy field for preparation of coal–oil mixtures. It has been established that oil with up to 40% coal powder content can be burned in existing oil-buring installations after some modification of the equipment. However, continuous agitation of the coal–oil mixture is necessary to prevent coal particles from settling, which makes the fuel unsuitable for storage or transportation. When a small amount of water is added to the mixture followed by ultrasonic agitation, a stable suspension results. A pilot plant has been built to produce sample quantities of such fuel for evaluation, using high-power, continuous-flow ultrasonic equipment operating at 10 kHz and capable of flow rates up to 160 L min^{-1}. The outcome of this experiment is uncertain in view of the current oil glut and general de-emphasis of synfuel effort.[27,28]

Dispersion of Solids

Concurrent with the introduction of ultrasonic emulsification was the effort to apply ultrasound to other mixing processes. It was found that ultrasound could be effective in improving liquid–solid mixtures by breaking up and dispersing clumps of solid particles to obtain uniform dispersions.[29] Many industrial uses were reported, including manufacture of dye pigment preparations; zinc oxide, barium meal, and other ointments; insecticides; and dispersions of china clay, mica, and titanium oxides and other materials used in the rubber and paper industries. Liquid whistles were used for these applications, which may explain why the need for ultrasonic equipment has not grown in this area, as one would expect considerable equipment wear as a result of abrasive particle impingement. Present-day ultrasonic horns can be hard faced, thus being better equipped to deal with the wear problem, and ultrasonic dispersion may have some potential in today's industry.

Ultrasonic dispersion of magnetic oxides used in manufacturing audio, video, and computer tapes and disks has been found to improve the quality by producing coatings with fewer imperfections, which cause information dropouts. This could be a significant factor, considering the continuing drive for higher information density of magnetic storage devices.

A technique for ultrasonically mixing solid epoxy curing agents into liquid epoxy resins has been reported. This procedure allows curing agents to be mixed without prior melting, and also the mixing of agents with very high melting temperatures without premature curing.[30]

A large area of unresolved potential is ultrasonic treatment of mineral slurries. There is evidence that ultrasonic agitation reduces soaking time and improves yields in floatation and separation processes. Possible uses include particle size reduction, removal of coatings blocking access to reagents, particle wetting, and slime removal. Precipitation of ultrafine coal and metal dust, and separation of sulfur and ash from coal have been considered.[31,32]

Sterilization

Destruction of bacteria by ultrasound has been studied with a view toward pasteurization of milk and as an alternative or supplement to traditional sterilization methods.[33–37] It may be concluded that complete destruction of bacteria by ultrasound alone is difficult, but that ultrasonically enhanced conventional sterilization can be accomplished in a shorter time and at lower temperatures. This is primarily because dispersion of clumps of bacteria by ultrasound allows the conventional reaction to proceed faster.

Filtration

Most filters require maintenance, such as periodic replacement or cleaning. Additionally, progressive clogging reduces filtration rates. Substantial increases in filtration rates were achieved with ultrasonically enhanced filtering, where the filtered element was vibrated through direct contact with the horn, or the horn was placed in the liquid near the filter.[35–41] Slower buildups of deposit were demonstrated, resulting in less need for cleaning, and a drier cake. The filters used in the experiments included sintered brass, stainless steel, sandstone, and other materials with pore sizes ranging from a few to over 100 μm. The materials filtered included contaminated motor oil and coal slurries. Most of the work was done at 20 kHz using relatively low power levels. Filtering of high-viscosity slurries, unsuitable for conventional treatment, may also be possible.

Atomization

Ultrasonic atomization of liquids has been used in medical inhalation therapy, air humidification, and on a very small scale in fuel atomization. The advantages include small, predictable droplet size, narrow droplet size distribution, and an ability to atomize high-viscosity liquids, which cannot be handled by nozzles.[42,43] For a given liquid, the droplet size is determined by the ultrasonic frequency and gets smaller as the frequency is increased. The efficiency of ultrasonic atomization in terms of ultrasonic energy per unit volume of dispersed liquid improves as the frequency is decreased.

Medical nebulizers used in inhalation therapy marked the first significant commercial use. Nebulizers operate at frequencies of 1–3 MHz and produce droplet sizes of 1–5 μm. An added advantage of the ultrasonic method is an absence of gas, which makes it suitable for anesthetic systems. In recent years, inexpensive ultrasonic humidifiers for home use have appeared on the market.

Ultrasonic atomization of fuel for both commercial and home heating purposes has been researched in considerable depth, with expectations of an improved fuel utilization efficiency because of more complete combustion made possible by reduced fuel droplet size. So far there has been little commercial progress, possibly because improvements in conventional fuel burning equipment have made the more expensive ultrasonic method a less attractive option. Similarly, commercialization of ultrasonic equipment for automotive carburetion is not in evidence at this time. Ultrasonic fuel atomization has found some uses in small special-purpose electric power generators where the rate of fuel dispersion is too low for nozzles.

It has been demonstrated that fine solid particles can be produced by ultrasonic atomization. Atomized molten metal upon cooling results in metal powder.[44,45] Glass and other materials have been pulverized. Fine, uniform crystals can be produced by ultrasonically atomizing and then cooling a supersaturated solution.

In ultrasonic atomization applications mentioned so far, the focus has been on producing a product of unusual quality. Experiments show that the process may be also useful for bulk moisture removal. Continuous drying of fabric is a common operation in textile processing plants and at present is done by squeezing out part of the moisture by mechanical means and evaporating the rest by heat.

It has been shown that a high-intensity ultrasonic horn in direct contact with the fabric can remove a higher percentage of moisture than mechanical rollers, thus reducing the final drying requirements. Trials indicate that ultrasonic moisture removal works best on thin, nonabsorbing materials. While these results are impressive, other alternatives for fabric moisture removal enhancement are possible, such as vacuum slots, and the acceptance of the ultrasonic method for this application is not certain.

Crystal Growth

Exposure of supersaturated solutions to an ultrasonic field speeds up crystallization and results in smaller crystals.[46–50] Large installations exist in the pharmaceutical industry for continuous harvesting of crystals of small, controlled size. Crystals are grown in a fluidized bed in a long vertical column having the bottom lined with high-amplitude ultrasonic horns. An upward liquid flow is maintained in the column. As crystals increase in size, they descend against the current, and a gradation develops, with the smallest crystals at the top and the largest at the bottom of the column. At the bottom of the column, oversized crystals are ultrasonically shattered, and at some higher location those of desired size are collected, while small fragments rise to the top to provide seed for new crystals.[51]

Crystallization of Melts

It is known that crystallization rates during the solidification phase proceed faster under the influence of ultrasound, and the resultant grain structure in the solidified material is finer.[52] Lower ultrasonic frequencies, in the 20–30 kHz range, are most effective, and breakup of larger crystals by cavitation has been suggested as the primary mechanism of the process. Of the many materials investigated, metals have been of most interest with regard to potential improvement of their mechanical characteristics. It has been shown that ductility, impact toughness, specific elongation and other characteristics can be significantly improved as a result of ultrasonic treatment. To date, however, the process has made little commercial progress. The main reasons are the unfavorable attitude toward ultrasonic equipment acceptance because of the high temperatures involved, and the requirement for a uniform treatment of a large volume of material.

Degassing

Gasses dissolved in a liquid can be expelled effectively with ultrasound, and the process is used in the industry for degassing of carbonated drinks, beer, photographic solutions, and other liquids.[53,54] Ultrasonic cleaning tanks can be used, or liquid containers may be brought into contact with the horn, as is done in beer fobbing. Special equipment for continuous-flow degassing is another possibility. Equipment in the 20–40 kHz range is normally used. Ultrasonic degassing of high-temperature melting materials, such as metals and glass, has been demonstrated with an aim of reducing porosity upon solidification, but commercial implementations have not followed.[55]

APPLICATIONS IN SOLIDS

Plastic Welding

Few industrial processes have had the rapid acceptance of ultrasonic plastic welding. Introduced commercially in 1963, within a few years it captured a large number of operations previously done with adhesives, solvents and mechanical fasteners. Today the process constitutes the preferred method of plastic assembly in all industries where thermoplastic materials are used and has surpassed ultrasonic cleaning in industrially developed countries. Representative applications include automotive lenses, air ducts, dashboards, and trim; numerous household appliances; radio and TV cases; tape recording cassettes; photographic cameras, lenses and film cartridges; computer tape reels and disk enclosures; medical devices; toys; and textiles. The reason for this is that the process possesses some inherent desirable characteristics, which include speed, energy efficiency, absence of comsumables, and adaptability to automation. Typical welds are performed in less than a second.[56,57]

Ultrasonic welding is suitable for most thermoplastics (see Table 3-3). It works best on materials having low mechanical damping, such as styrenes. In such materials, because of their ability to transmit vibration with low attenuation, "far-field" welding is possible, in which welds can be made at some distance away from the contact with the ultrasonic tool. Energy efficiency results from the fact that ultrasonically induced heat can be limited to the welding zone, with minimal heating of the rest of the welded object. This results in fast welding and low part distortion. Because heat is generated in the material internally, rather than by conduction from a tool, welding can generally be accomplished at lower temperatures, causing less material degradation and therefore yielding stronger welds.

In ultrasonic welding, plastic parts to be welded are clamped under some pressure between an ultrasonic tool (horn) and a stationary fixture (see Fig. 3-6). The direction of vibration is perpendicular to the weld plane. To concentrate ultrasonic stress, and therefore heating, in the weld zone, the contact area between the two parts is reduced by shaping into the interface a projection called an "energy director." Upon vibration, the energy director melts and spreads across the interface, fusing the parts together. After ultrasonic exposure the parts remain under pressure for a fraction of a second to allow the plastic to solidify before unclamping.

To reduce ultrasonic plastic welding to practice, significant progress has occurred in the development of acoustic tooling, transducers, electronic power supplies, and controls. Plastic welding requires rather high power densities, in

Table 3-3. Weldability[a] of Thermoplastics by Ultrasound[b]

Material	Spot welding	Staking swaging	Inserting	Field of welding Near	Field of welding Far
Amorphous					
ABS	E	E	E	E	G
ABS/polycarbonate	G	G	G	G	F
ABS/PVC	G	G	F	G	F
Acrylic	G	F	G	G	F
Acrylic multipolymer-XT polymer	G	G	G	G	F
Acrylic/PVC	G	G	F	G	F
Acrylic, impact modified	F	F	P	F	P
BDS	G	G	G	G	F
Cellulosics (CA, CAB, CAP)	P	G	E	P	—
Modified phenylene oxide	E	E	E	E	G
Polyarylate	F	F	G	G	F
Polycarbonate	G	F	G	G	F
Polyetherimide	G	G	E	E	G
Polystryene, G.P.	F	F	G	E	E
Polystyrene, impact modified	F	F	G	G	P
PVC, rigid	F	G	E	P	P
PVC, flexible	P	—	—	P	—
SAN, NAS, ASA	F	F	G	E	E
Styrene-maleic anhydride	E	E	E	E	G
Sulfone polymers	F	F	G	G	F
Crystalline					
Acetal copolymer	F	F	G	G	F
Acetal homopolymer	F	F	G	G	F
Fluoropolymers	—	—	—	P	—
Nylon	F	F	G	G	F
PC-PET	G	G	E	E	G
Polyester (PBT)	F	F	G	G	F
Polyester (PET)	F	F	G	G	F
Polyetheretherketone	G	G	E	E	G
Polyethylene (LDPE, HDPE)	G	F	G	P	P
Polyethylene (UHMW)	—	—	—	—	—
Polymethylpentene	G	F	E	F	P
Polyphenylene sulfide	F	P	G	G	F
Polypropylene	E	E	G	F	P

[a] E, excellent; G, good; F, fair; P, poor; —, not suitable for ultrasonic assembly.
[b] Courtesy of Sonics & Materials, Inc.

the order of 1000 W cm^{-2}. To deliver such power densities, ultrasonic horns must operate at high amplitudes. Additionally, the geometry of the horn output face must conform to the shape of the welded part and the ultrasonic motion across the horn face must be fairly uniform. Because horns are resonant sections, tuned precisely to the operating frequency of the welder, this imposes a set of tough conditions on horn design, the problem becoming more difficult with increasing horn size (see Fig. 3-7). Depending on the application, at 20 kHz, to-

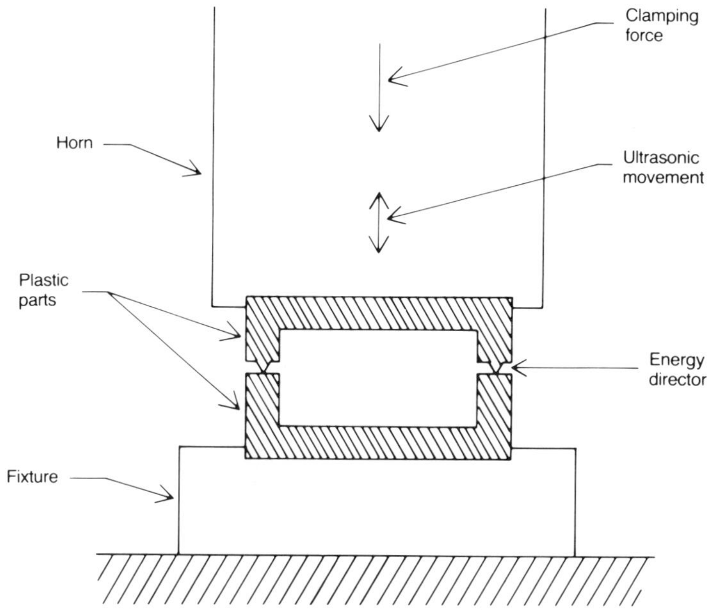

Figure 3-6. Cross-section of an ultrasonic plastic weld.

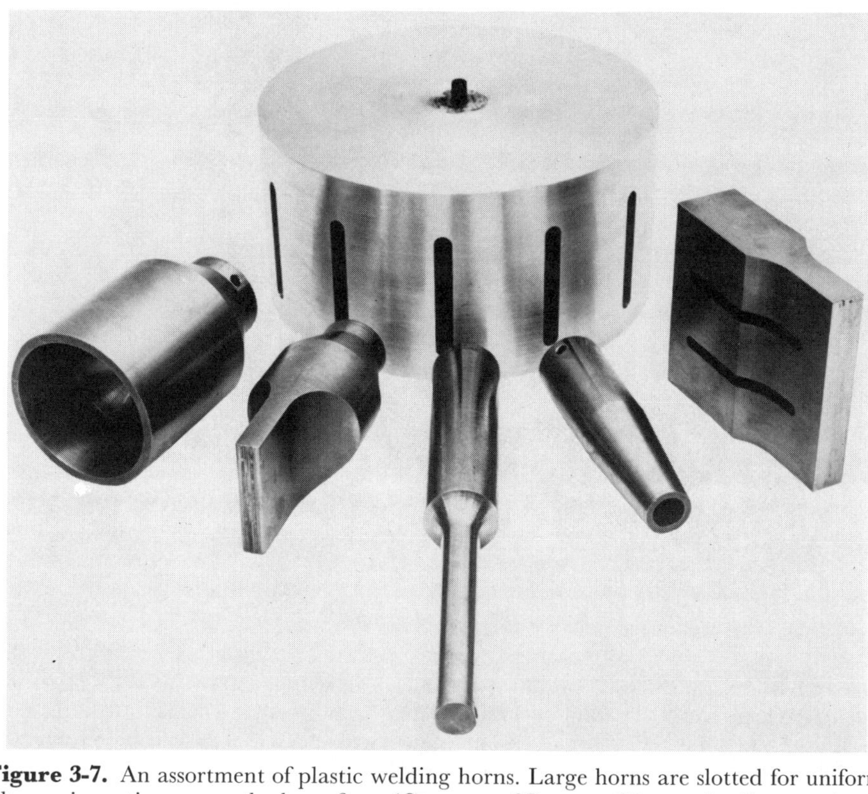

Figure 3-7. An assortment of plastic welding horns. Large horns are slotted for uniform ultrasonic motion across the horn face. (Courtesy of Branson Ultrasonics Corporation.)

day's linear horns are limited in width to about 20–25 cm, round and rectangular shapes to somewhat smaller sizes. When part dimensions are too large for a single horn, two or more horns can be used. Ultrasonic welding horns are usually made of aluminum or titanium, because of the high strength to weight ratios and low acoustic losses of these metals.

Because the loading on a transducer during welding is not constant and for part of the time may be absent altogether, it is necessary that the equipment

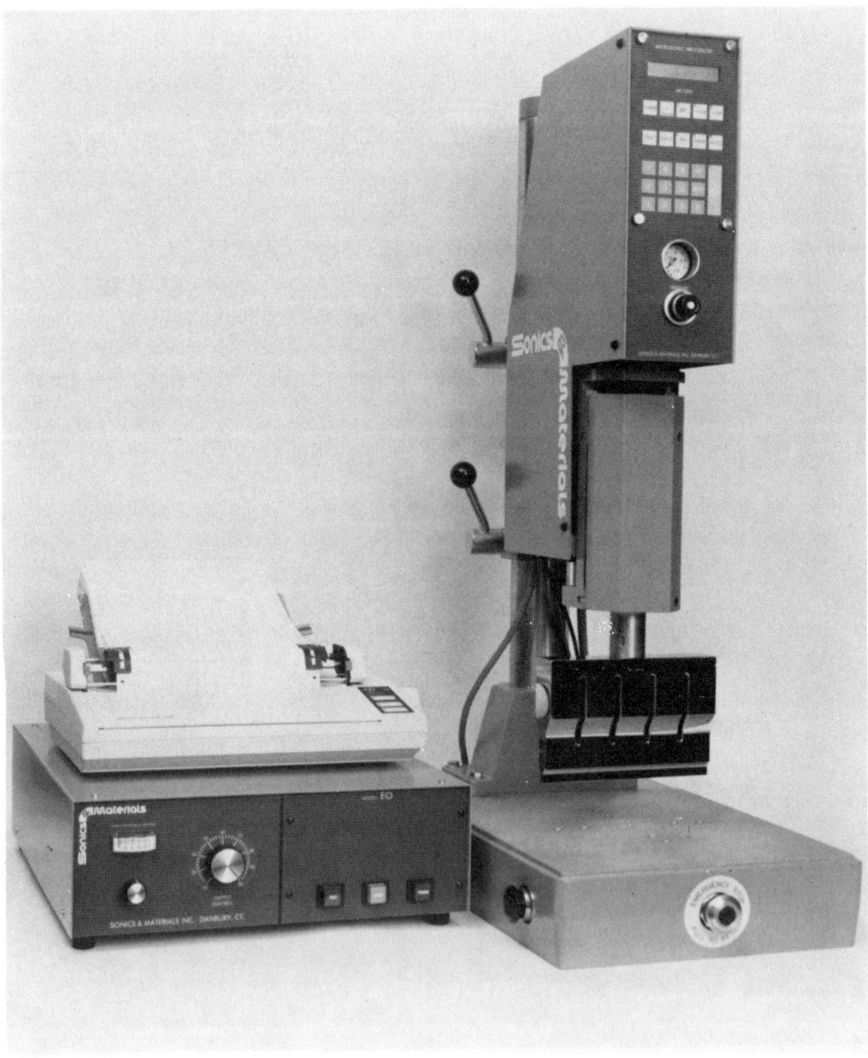

Figure 3-8. Microprocessor-controlled plastic welder with a 1500-Watt, 20-KHz power supply packaged in a separate enclosure. Time or cumulative energy can be used to control or to monitor ultrasonic exposure, and data for each weld can be output to a printer. (Courtesy of Sonics & Materials, Inc.)

be able to accommodate such changes without damage. This has led to the "constant-amplitude" concept now commonly used in welding equipment. In this method the vibratory stroke of the ultrasonic tool stays constant regardless of changes in loading, and power into the workpiece varies upon demand. Automatic frequency control is another feature necessary for attendance-free operation of ultrasonic welders.

Ultrasonic exposure in plastic welding is usually controlled by timing. In recent years, the industry trend toward higher speed, shorter weld times, automation, and better quality control has facilitated introduction of more sophisticated controls based on cumulative energy. Whereas the concept is not new, present day progress in microelectronics has made its implementation practical. Ultrasonic welders with elaborate controls are now available at a reasonable price, performing on-line weld quality inspection and automatic ultrasonic exposure correction to insure equal energy input into each weld (see Fig. 3-8). A readout of process parameters is available for printout, magnetic storage, or output to other computers.

Welding of rigid thermoplastic parts represents the majority of applications. Other areas include spot welding, staking, and insertion of metal into plastic. Continuous welding of thermoplastic textiles was introduced around 1970 and deserves a special consideration.[58] In this method the fabric is fed between a stationary horn and a rotating anvil (see Fig. 3-9). The anvil is power driven and is usually engraved with a "stitch pattern" in the form of solid or broken

Figure 3-9. Typical ultrasonic equipment setup using a horn and a power-driven drum for continuous bonding of thermoplastic textiles. Material is bonded in the pattern engraved in the drum. Arrays of adjacent horns are used to span wider webs. (Courtesy of Branson Ultrasonics Corporation.)

lines, dots, etc.. Welding of synthetic fabrics can be done by this arrangement at speeds exceeding conventional sewing. One such major application is fabric laminating and quilting. Large machines have been built for this purpose, where the rotary anvil is a roller extending the full width of the quilt (335 cm) and an array of adjacent ultrasonic horns is used to cover that span. Typically, a filler is sandwiched between the "face" and the "back" to form a quilt, or three or four nonwoven layers may be laminated together, as in production of furniture backing materials. Depending on the material, such machines run at 8–10 m s^{-1}, or about 10 times faster than the conventional quilting process.

A more advanced development based on rotary anvil principle is a textile manufacturing process where thermoplastic fibres are ultrasonically welded to each other to produce a nonwoven fabric.[59] While the process is in its infancy, its potential may be large in view of the projected growth for nonwoven fabrics. Although relatively slow at this time, ultrasonic bonding, because of its low heat input and absence of additives, has an advantage over calendering and binders, the traditional methods of manufacturing nonwoven goods.

Metal Welding

Ultrasonic welding of metals has been practiced in the industry for almost three decades on applications that are not suitable for conventional metal welding processes.[60,61] Because ultrasonic metal welding is relatively "cold," such problems as embrittlement resulting from recrystallization and the formation of intermetallic compounds at dissimilar metal joints can be reduced or eliminated. Ultrasonic welding is also independent of electrical resistivity and has an advantage for joining high-conductivity metals, such as electrical grade aluminum and copper, which can be troublesome with conventional resistance welding. Welding of thin foil and of parts of grossly unequal heat capacity can also be done ultrasonically. Although forming of butt joints has been demonstrated on specially designed equipment, ultrasonic metal welding is mainly suitable for lap welds where the thickness of at least one of the parts is limited to about 1 cm. Welders in the power range of 300–5000 W are used, operating between 10 and 20 kHz. Equipment for continuous welding of foil and sheet is also available.

An important application of the ultrasonic metal welding technique is microbonding, or welding of miniature semiconductor leads and chips, and is used extensively in the semiconductor manufacturing industry. Welds are performed by ultrasonic vibration alone, or ultrasound is used in conjunction with thermal bonding. Stand-alone manual bonders are used or ultrasonic components are built into automated systems operating at rates over 10 welds per second. Low-power equipment operating around 60 kHz is used.

Ultrasonic metal welding is done in shear mode, where parts are clamped between a stationary anvil and a laterally oscillating horn and are caused to slip with respect to each other at the joint. Ultrasonic scrubbing breaks up and disperses surface oxides and other contamination and exposed metal surfaces fuse together under pressure. In this respect there is some similarity to friction welding and compression welding used in the industry.

Metal Forming

Much work has been done on ultrasonic metal forming, from fundamental process investigations to adaptations of ultrasonic equipment to metal-forming machinery.[62–66] Wire drawing, deep drawing, bending, rivet forming, and other forming operations have been investigated. The advantages sought are reductions in forming forces, higher formability, and improvements in the quality of the formed product, such as better surface finish and absence of tearing. To date, the process has found only occasional use, usually involving hard to form metals, for exotic applications, such as drawing of thin walled tubular shapes or thin wire.

Machining of Hard, Brittle Materials

Ultrasonic impact grinding is used on hard to machine materials, such as ceramics, ferrites, glass, and gemstones.[67–71] An abrasive slurry, usually boron carbide, silicon carbide, or aluminum oxide, is fed between the vibrating horn and the work piece, and a negative impression of the horn face results in the finished part. Machining of irregular, three-dimensional shapes is possible, also simultaneous drilling of clusters of holes. The method is slow and is limited to small-scale production in manufacturing of optical devices, such as lasers, and also for production of jewelry.

An alternate method for similar applications is ultrasonic rotary abrasive machining, where axial vibratory motion is superimposed on a rotating tool. Drilling is most common, although other machining operations are possible. Diamond-impregnated tools, usually liquid cooled through the center, are used. Faster rates compared to conventional rotary abrasive machinery are possible because the tool is continuously cleaned by cavitation, which prevents the tool from loading. Also, lower resultant forces on the tool make more accurate machining possible, such as drilling of deep, small-diameter holes; 20-kHz equipment is used (see Fig. 3-10).

Metal Cutting

Reduction in cutting forces, faster cutting rates, better surface finish, and improved tool life have been demonstrated for various metal cutting operations in the presence of ultrasound, by vibrating either the cutting tool or the work piece.[71] Single and multiple cutting point operations were tried, including drilling, tapping, and turning. On balance, the benefits so far have proved marginal, considering the added expense and encumberment of conventional machinery with ultrasonics, and the process has not had much following in the industry.

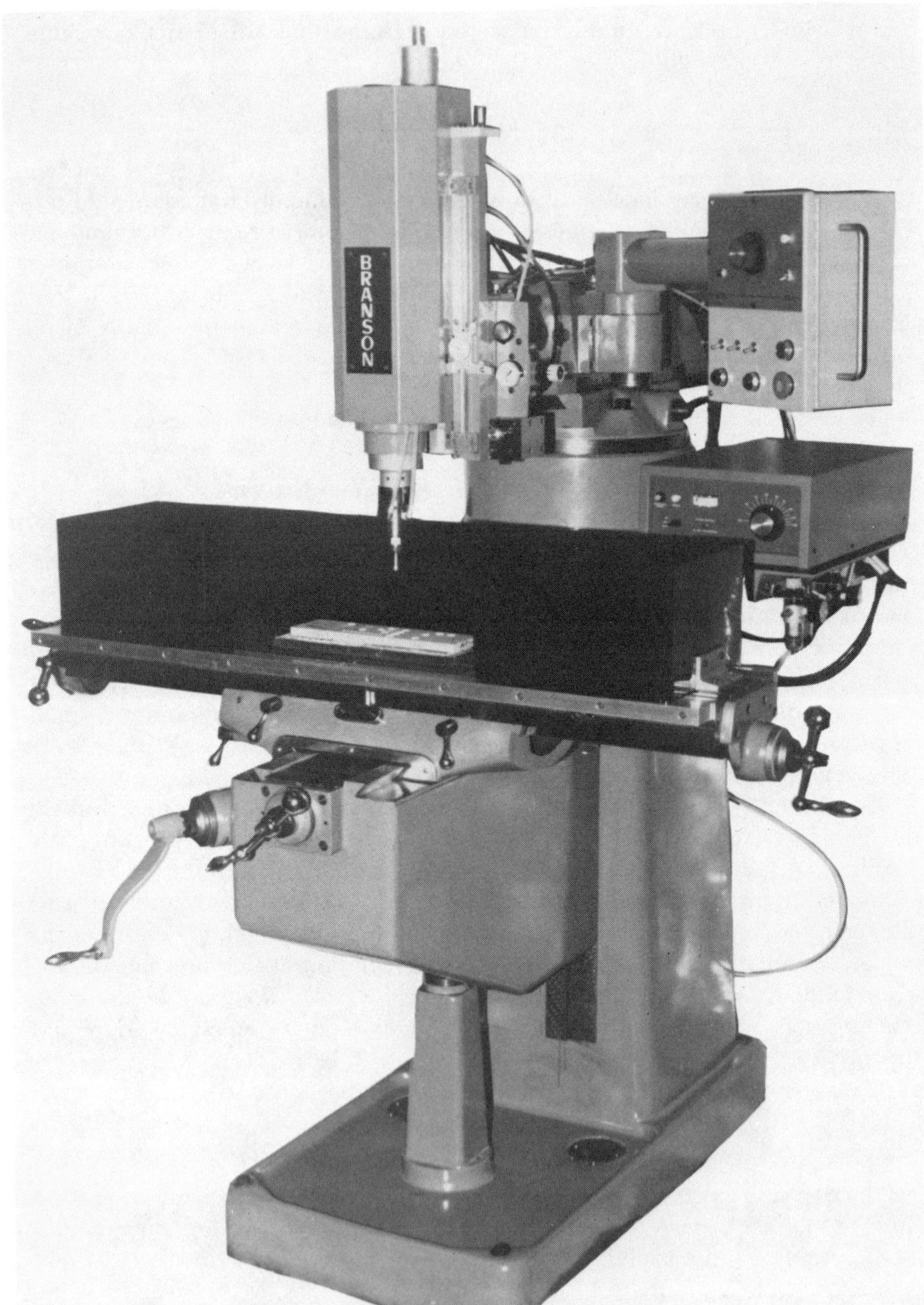

Figure 3-10. Ultrasonic machine tool for abrasive rotary machining of hard and brittle materials. Diamond-impregnated core drills, water-cooled through the center, are used at speeds of several thousand rpm. (Courtesy of Branson Ultrasonics Corporation.)

Fatigue Testing

In addition to providing direct data on fatigue endurance of materials at ultrasonic frequencies, ultrasonic fatigue testing is much faster than low-frequency testing. Thus, comparing 20-kHz to 50-Hz testing, a 400-fold saving in testing time is possible, and 10^8 cycles can be accumulated in under 2 h compared to almost a month at the lower frequency. A considerable amount of data exists for different materials. Since for the same stress levels fatigue life at high and low frequencies is generally different, much of the effort has been in establishing a correlation between high- and low-frequency testing.[71–73]

Fatigue strength of adhesive bonds can be tested ultrasonically by bonding a small mass to an ultrasonic horn. The method works well, provided the adhesive does not get overheated as a result of ultrasonic fatigue.

Most ultrasonic fatigue testing has been at 20 kHz. Constant-amplitude ultrasonic welding equipment is particularly suited for this purpose.

Curing

The heat generated in a material by ultrasonic stain can be used to speed up an endothermic reaction, such as the curing of epoxies.[74] Because of attenuation and the nonuniformity caused by the standing wave pattern, this process is effective only on relatively small volumes of shallow depth (about 2 cm at 20 kHz) and is rarely used.

Friction Reduction

Friction reduction caused by ultrasonic acceleration has been demonstrated in many experiments, and attempts have been made to put this effect to use in ultrasonically activated sieves, chutes, surgical instruments, razors, etc.[75] The most promising of industrial applications are ultrasonically vibrated blades for cutting of modern fibrous composite materials, such as graphite–epoxy laminates, which cut poorly by conventional means. Ultrasonically vibrated cutters cut with less effort, produce cleaner cuts, and require less frequent sharpening. A number of such devices are in use in the aerospace industry.

AIRBORNE APPLICATIONS

Coagulation

Ultrasonic agglomeration of particles suspended in gasses has been a subject of some early studies, and air-driven sirens were promoted in the 1950s for air scrubbing of industrial pollutants. Electrostatic precipitators proved more efficient and no ultrasonic devices are used for this purpose today.[76,77]

Defoaming

Some ultrasonic defoaming is done in beverage packaging, where filled containers are passed under a high-intensity horn before closure. The advantage of the ultrasonic method is that it does not require contact with the liquid, which is desirable from the sanitary standpoint.[78]

Drying

In 1950s ultrasonic airborne drying was promoted using air-driven whistles. Because in the presence of sound drying can be accomplished at lower temperatures, the method appeared well suited for drying temperature-sensitive powders, such as foodstuffs and pharmaceuticals. Possibly because of the drawbacks of mechanical whistles, few of these devices are in use today. High ultrasonic intensitites in gasses can now be produced more efficiently,[79] which may rekindle some of the interest in ultrasonic airborne drying.[80–84]

REFERENCES

1. *1980 Current Industrial Report*, MA-36N, Department of Commerce, U.S. Bureau of Census, Washington, DC, p. 7.
2. Rosenberg, L. D., Ed., "High Intensity Ultrasonic Fields"; Plenum Press: New York, 1971, p. 249.
3. Neppiras, E. A. *Phys. Rep.* **1980**, *61*, 159.
4. Akulichev, V. A. In "High Intensity Ultrasonic Fields"; Plenum Press: New York, 1971, pp. 203–419.
5. Nyborg, W. L. In "Physical Acoustics"; Mason, W. P., Ed.; Academic Press: New York, 1965, Vol. 2, Part B.
6. Nyborg, W. L. In "Proc. First Int. Symp. High Power Ultrasonics"; I.P.C. Science and Technology Press: Guildford, Surrey, UK, 1970, pp. 124–135.
7. Niemczevski, B. *Ultrasonics* **1980**, *18*, 107–110.
8. Brown, B; Goodman, J. E. "High-Intensity Ultrasonics"; Iliffe Books: London, 1965, pp. 56–74.
9. Neppiras, E. A. In "Ultrasonics International 1973 Conference Proceedings"; I.P.C. Science and Technology Press: Guildford, Surrey, U.K., pp. 295–302.
10. Agranat, B. A.; Bashkirov, V. I; Kitaigorodskii, Y. I. In "Physical Principles of Ultrasonic Technology"; Rosenberg, L. D., Ed.; Plenum Press: New York, 1973, Vol. 1, pp. 247–376, 515.
11. Brown, B; Goodman, J. E. "High-Intensity Ultrasonics"; Iliffe Books: London, 1965, pp. 116–135.
12. Pohlman, R.; Werden, B; Marziniak, R. *Ultrasonics* **1972**, *10*, 156
13. Hunicke, R. L. In "Ultrasonics Int. 1975 Conf. Proc."; I.P.C. Science and Technology Press: Guildford, Surrey, UK, pp. 32–38.
14. Agranat, B. A.; Bashkirov, V. I; Kitaigorodskii, Y. I. In "Physical Principles of Ultrasonic Technology"; Rosenberg, L. D., Ed., Plenum Press: New York, 1973, Vol. 1, pp. 319–330, 515.
15. "*Erosion by Cavitation*", a symposium; ASTM Spec. Tech. Publ. No. 408; American Society for Testing and Materials: Philadelphia 1967, pp. 159–219, 239–283.
16. Hansson, I; Morch K. A; Preece, C. M. In "Ultrasonics Int. 1977 Conf. Proc. I.P.C. Science and Technology Press: Guildford, Surrey, UK, pp. 159–219.
17. McIntosh, A. F.; Munro, R. F. *Proc. Biochem.* **1971**, *6*, 22, 23, 27.
18. Ovadia, M. E.; Skauen, D. M. *J. Pharm. Sci.*, **1965**, *54*, 1013.
19. Patel, I. C.; Skauen, D. M. *J. Pharm. Sci.*, **1969**, *58*, 1135.

20. Velichkov, V.; Georgiev, E. *Nauch. Tr., Vissh. Inst. Khranit. Vkusova Prom. Plovdiv*, **1969**, *16*(2), 177; *Chem. Abstr.* **1973**, *78*, 20,103.
21. Srinivasulu, C.; Srivastava, S. C.; Mahapatra, S. N. In *Ultrasonics Ind. 1973 Conf. Proc.*, I.P.C. Science and Technology Press: Guildford, Surrey, UK, pp. 25–27.
22. Solonko, V. M.; et al. *Farm. Kh.* **1975**, 30(1) 84–86; *Biol. Abstr. 1976*, *61*(9), 50,696.
23. Brown, B; Goodman, J. E. "High-Intensity Ultrasonics"; Iliffe Books: London, 1965, pp. 222–223.
24. Nosov, V. A. "Ultrasonics in the Chemical Industry"; Consultants Bureau: New York, 1965, Vol. 2, pp. 40–43, 164.
25. Higgins, D. M.; Skauen, D. M. *J. Pharm. Sci.* **1972**, *61*, 1567.
26. Brown, B; Goodman, J. E. "High-Intensity Ultrasonics"; Iliffe Books: London, **1965**, pp. 141–159.
27. O'Neill, J. K.; Kenneberg, D. J.; Viola, M. A. In "Proc. Third Int. Symp. Coal-Oil Mixture Combustion;" Pittsburgh Energy Technology Center: Pittsburgh, Vol. II.
28. Hawkins, G. T. In "Proc. Third Int. Symp. Coal-Oil Mixture Combustion"; Pittsburgh Energy Technology Center: Pittsburgh, Vol. II.
29. Nosov, V. A. "Ultrasonics in the Chemical Industry"; Consultants Bureau: New York, 1965, Vol. 2, pp. 43–47, 164.
30. Hodges, W. T.; St. Clair, T. L. *SAMPE Quart.* **1983**, *14*, April.
31. Agranat, B. A.; Kirillov, O. D.; Preobrazhensky, N. A.; Khavsky, N. N.; Yakubovich, I.A. "Ultrazvuk v Gidrometallurgil"; Metallurgia: Moscow, 1969, pp. 56–159.
32. Kowalski, W.; Kowalska, E. *Ultrasonics*, **1978**, *16*, 84–86.
33. Jacobs, S. E.; Thornley, M. J. *J. Appl. Bacteriol.* **1954**, *17*, 383.
34. Rubleson, G. R.; Murray, T. M.; Pollard, M. *Appl. Microbiol.* **1975**, p. 340.
35. Sierra, G. U.S. Patent 3,697,222, October 10, 1972.
36. Uko, T. S.; et al. *Yakazaigaku*, **1973**, *33*, 101; *Chem. Abstr.* **1974**, *80*, 30,691.
37. Bybel, D.; et al. U.S. Patent 4,076,617, February 28, 1978.
38. Nosov, V. A. "Ultrasonics in the Chemical Industry"; Consultants Bureau: New York, 1965, Vol. 2, pp. 47–48, 164
39. Fairbanks, H. V. In "Ultrasonics Int. 1973 Conf. Proc."; I.P.C. Science and Technology Press: Guildford, Surrey, UK, 1973, pp. 11–15.
40. A. Semmelink, In "Ultrasonics Int. 1973 Conf. Proc."; I.P.C. Science and Technology Press: Guildford, Surrey, UK, 1973, pp. 7–10.
41. Bjorno, L.; Gram, S.; Steenstrup, P. R. *Ultrasonics* **1978**, *16*, 103.
42. Eknadiosyants, O. K.; In "Physical Principles of Ultrasonic Technology"; Rosenberg, L. D., Ed.; Plenum Press; New York, 1973, Vol. 2, pp. 3–88.
43. Topp, M. N.; Eisendiam, P. *Ultrasonics* **1972**, *10*, 127.
44. Pohlman, R. G. "Ultrasonics Int. 1973 Conf. Proc."; I.P.C. Science and Technology Press: Guildford, Surrey, UK, pp. 52–55.
45. Pohlman, R.; Heisler, K.; Cichos, M. *Ultrasonics*, **1974**, *12*, 11.
46. Agranat, B. A.; Bashkirov, V. I.; Kitaigorodskii, Y. I. In "Physical Principles of Ultrasonic Technology"; Rosenberg, L. D., Ed.; Plenum Press: New York, 1973, Vol. 2, 348–378, 515.
47. Nosov, V. A. In "Ultrasonics in the Chemical Industry"; Consultants Bureau: New York, 1965, Vol. 2, pp. 32–34, 164.
48. Skauen, D. M. *J. Pharm. Sci.* **1967**, *56*(11), 1373.
49. Hem, S. L. *Ultrasonics* **1967**, *5*, 202.
50. Kortnev, A. V.; Martynovskaya, N. V. *Sb. Mosk. Inst. Stali Splavov* **1974**, *77*, 98–100; *Chem. Abstr. 82*, 49,945.
51. Middler, M. U.S. Patent 3,510,266, May 6, 1970.
52. Abrams, O. V.; Teumin, I. I. In "Physical Principles of Ultrasonic Technology"; Rosenberg, L. D., Ed.; Plenum Press: New York, 1973, Vol. 2, pp. 145–273.
53. Kapustina, O. A. In "Physical Principles of Ultrasonic Technology"; Rosenberg, L. D., Ed.; Plenum Press: New York, Vol. 1, pp. 379–509.
54. Brown, B; Goodman, J. E. "High-Intensity Ultrasonics"; Iliffe Books: London, 1965, pp. 210–213.
55. Brown, B; Goodman, J. E. "High-Intensity Ultrasonics"; Iliffe Books: London, 1965, pp. 496–509.
56. Shoh, A. *Ultrasonics* **1976**, *14*, 209.
57. "Ultrasonic assembly of thermoplastic mouldings and semifinished products"; Fachferband Electroschweissgeraete, Stresemannallee 19, 6 Frankfurt/M. 70.

58. Shoh, A. In "Advanced Forming/Bonding Conference"; San Antonio, Texas, November 13–15, 1983; Marketing/Technology Service, Inc., 3125 West Main St., Kalamazoo, MI 49007.
59. Hill, B. R. U.S. Patent 4,311,540, January 19, 1983.
60. Mitskevich, A. M. In "Physical Principles of Ultrasonic Technology"; Rosenberg, L. D., Ed.; Plenum Press: New York, 1973, Vol. 1, pp. 99–243.
61. Hulst, A. P. *Ultrasonics* **1972**, *10*, 256.
62. Langenecker, B.; Vodep, O. In "Ultrasonics Int. 1975 Conf. Proc."; I.P.C. Science and Technology Press; Guildford, Surrey, UK, 1975, pp. 202–205.
63. Dawson, G. R. In "Ultrasonics Int. 1975 Conf. Proc."; I.P.C. Science and Technology Press,: Guildford, Surrey, UK, 1975, pp. 206–209.
64. Herbertz, J. In "Ultrasonics Int. 1977 Conf. Proc."; I.P.C. Science and Technology Press: Guildford, Surrey, UK, 1977, pp. 323–328.
65. Anatasiu, N. E. *Ultrasonics* **1980**, *18*(6), 255.
66. Hansson, I; Thoelen, A. *Ultrasonics* **1978**, *16*, 57.
67. Kazantsev, V. F. In "Physical Principles of Ultrasonic Technology"; Rosenberg, L. D., Ed.; Plenum Press: New York, Vol. 1, pp. 3–98.
68. Markov, A. I. "Ultrasonic Machining of Intractable Materials"; Illife Books: London, 1966.
69. Kremer, et al., *EDM Digest*, **1982**, *4(2)*.
70. Fairbanks, H. V. In "Ultrasonics Int. 1975 Conf. Proc."; I.P.C. Science and Technology Press: Guildford, Surrey, UK, 1975, pp. 107–109.
71. Neppiras, E. A. *Proc. Am. Soc. Test. Mater.* **1959**, *59*, 691
72. Bajons, P.; Kromp, K.; Kromp, W.; Langer, H.; Weiss, B.; Stickler, R. In "Ultrasonics Int. 1975 Conf. Proc."; I.P.C. Science and Technology Press: Guildford, Surrey, UK, 1975, pp. 95–101.
73. Kuzmenko, V. A. *Ultrasonics* **1975**, *13*, 21.
74. Saksena, T. K.; Babbar, N. K. *Ultrasonics* **1979**, *17*, 122.
75. Balamuth, L. U.S. Patent 3,794,040, Februrary 26, 1974.
76. Shirokova, N. L. In "Physical Principles of Ultrasonic Technology"; Rosenberg, L. D., Ed.; Plenum Press, New York, 1973, Vol. 2, pp. 477–539.
77. Nosov, V. A. "Ultrasonics in the Chemical Industry"; Consultants Bureau: New York, 1965, Vol. 2, pp. 51–56, 164.
78. Matzner, E. UK Patent Application GB 2,051,754 A, January 23, 1980.
79. Garreton, G.; et al. In "9th Int. Congr. Acoustics"; Madrid, 4/9/-VII-1977.
80. Borisov, Y. Y.; Gyukina, N. M. In "Physical Principles of Ultrasonic Technology"; Rosenberg, L. D., Ed.; Plenum Press, New York, 1973, Vol. 2, pp. 381–474.
81. Nosov, V. A. "Ultrasonics in the Chemical Industry"; Consultants Bureau: New York, 1965, Vol. 2, pp. 57–61, 164.
82. Seya, K. In "Proc. First Int. Symp. High Power Ultrasonics"; I.P.C. Science and Technology Press: Guildford, Surrey, UK, 1970, pp. 136–39.
83. Fairbanks, H. V. In "Ultrasonics Int. 1975 Conf. Proc."; I.P.C. Science and Technology Press; Guildford, Surrey, UK, pp. 43–45.
84. Otsuku, T.; Purdum, H.; Fairbanks, H. In "Ultrasonics Int. 1977 Conf. Proc."; I.P.C. Science and Technology Press: Guildford, Surrey, UK, pp. 91–93.

4

Homogeneous Sonochemistry

KENNETH S. SUSLICK

School of Chemical Sciences, University of Illinois at Urbana-Champaign

INTRODUCTION, INTENT, AND NOMENCLATURE

The chemical[1] and biological[2] effects of ultrasound were first reported by Loomis more than 50 years ago. Within 15 years of that work, widespread industrial applications of ultrasound were found, as described in Chapter 3. In spite of early work in the area of sonochemistry, interest within the chemical community remained exceedingly modest until the past few years. With the advent of inexpensive and reliable sources of ultrasound, however, increasing use of sonochemistry in a variety of reactions is being reported.

The purpose of this chapter is to act as a critical and comprehensive review for those interested in the chemical effects of ultrasound, with a focus on homogeneous media. The next chapter deals in detail with the effects of ultrasound on heterogeneous reactions. Both more introductory[3] and more specialized reviews[4,5] have recently been published.

An overview of the physics of cavitation begins this review, but the reader is referred to the first two chapters for detailed discussions of acoustic cavitation and other nonlinear phenomena. The use of chemical dosimetry of ultrasonic irradiation and its applications to determining the local conditions generated during cavitational collapse follows. The influence of experimental variables on sonochemical reactions is discussed in detail. Finally, the review of the literature is divided between aqueous and nonaqueous media; this division has strong historical origins but also represents major differences in the types of chemistry observed.

A number of terms in this area are likely to be unfamiliar to most chemists. *Cavitation* is the formation of gas bubbles (or cavities) in a liquid and occurs when the pressure within the liquid drops sufficiently lower than the vapor pressure of the liquid. Cavitation can occur from a variety of causes: turbulent flow, laser heating, electrical discharge, boiling, radiolysis, or acoustic irradiation. We will be concerned exclusively with *acoustic cavitation*. When sound passes through a liquid, it consists of expansion (negative pressure) waves and compression (positive pressure) waves. These cause preexisting bubbles to grow and recompress. Acoustic cavitation can lead, as discussed later, to an implosive collapse of such cavities with associated high-energy chemistry. The importance of acoustic cavitation extends well beyond its chemical effects and is relevant to studies of heat transport, liquid tensile strengths, and superheating and boiling phenomena. Furthermore, since ultrasound is heavily used both for medical treatment (hyperthermia for soft tissue traumas) and diagnosis (sonography of fetal development), the biological and chemical effects of ultrasound are of immediate importance to the health services community, as discussed in the last two chapters.

There has been no consensus on the chemical symbol to be used to indicate ultrasonic irradiation. Several that are in common usage include:

$$-)-)-)\rightarrow, \quad \xrightarrow{\text{u.s.}}, \quad \text{and} \quad \xrightarrow{)))}$$

The first of these has some historical precedent in the early Soviet work in this area. This first symbol, however, has two disadvantages: (1) the breaking of the arrow could be mistaken for the symbol often used for "no reaction observed" (-/-/→); (2) it does not lend itself well to use in multistep reactions (e.g., ultrasonic irradiation followed by acidic workup). The second arrow has difficulties of interlingual ambiguity and chauvinism. The third symbol has none of these problems and will be used exclusively in this book to indicate ultrasonic irradiation of a solution leading to a sonochemical reaction. We urge all workers in the area of sonochemistry to adopt its use.

There are also a few matters of terminology that need clarification. The terms that have been used for ultrasonic irradiation include insonation, sonication, sonification, sonolysis, and ultrasonication. In the interest of simplification, we suggest that "sonication" and "ultrasonic irradiation" be the phrases of first choice. Sonolysis should refer specifically to the breaking of chemical bonds by ultrasonic irradiation, in analogy to photolysis. "Sonocatalysis" is restricted in its use here only to the creation of catalytically competent intermediates by ultrasonic irradiation; we do not refer to a simple sonochemical rate enhancement of an already ongoing reaction by this term.

MECHANISTIC AND EXPERIMENTAL CONSIDERATIONS

Ultrasound spans the frequencies of roughly 20 kHz to 10 MHz (human hearing has an upper limit of <18 kHz). Since the velocity of sound in liquids is ≈ 1500 m s^{-1}, ultrasound has acoustic wavelengths of roughly 7.5 to 0.015 cm. Clearly no direct coupling of the acoustic field with chemical species on a molecular level can account for sonochemistry. Instead, the chemical effects of ultrasound derive from several different physical mechanisms, depending on the nature of the system.

Acoustic Cavitation in Homogeneous Liquids

The most important of these is acoustic cavitation, which involves at least three discrete stages: nucleation, bubble growth, and, under proper conditions, implosive collapse. The dynamics of cavity growth and collapse are strikingly dependent on local environment, and one must consider separately cavitation in a homogeneous liquid and cavitation near a liquid–solid interface. The latter is discussed in the following chapter.

The question of nucleation of cavitation arises from the very high tensile strength of pure liquids expected from the attractive intermolecular forces that maintain the liquid state. The calculated tensile strength of water, for example, is in excess of -1000 atm.[6] In practice however, the measured threshold for initiation of cavitation is never more than a small fraction of that: tap water cavitates at a negative acoustic pressure of a few atmospheres. Indeed, if the observed tensile strengths of liquids did approach their theoretical limits, the acoustic intensities required to initiate cavitation would be well beyond that generally available, and no sonochemistry would be observed in homogeneous media!

Thus, cavitation is initiated at nucleation sites where the tensile strength is dramatically lowered. An obvious site would be small gas bubbles present in the liquid. Free gas bubbles, however, present a double bind: small ones of the size needed for acoustic cavitation (a few microns in radius) redissolve in a few seconds, whereas larger ones rapidly rise to the surface.[7] The nucleation mechanism generally accepted at this time involves gas entrapped in small-angle crevices of particulate contaminants,[8–10] as discussed in Chapter 1. As the crevice-stabilized nucleus is subjected to large, negative acoustic pressures, the bubble volume grows, releasing small free bubbles into solution or undergoing violent collapse itself. Those actions which remove such nucleation sites (e.g., ultrafiltration to remove particulates, evacuation or pressurization to flood the crevices) thus increase the cavitation threshold. In liquids undergoing cavitation, it should be noted that after the inital cycle of cavitation, the implosive collapse of bubbles generates microcavities that can then serve as nucleation sites for the next cycle.[11]

Flynn proposed the generally accepted division of cavitation phenomenon in homogeneous liquids into (1) transient cavitation, in which a short-lived bubble undergoes large excursions of size in a few acoustic cycles and may terminate in a violent collapse, and (2) stable cavitation, in which a bubble oscillates many times with limited change about its equilibrium radius.[12] In practice, however, this distinction is of limited utility, because both stable and transient cavitation may occur simultaneously in a solution and a bubble undergoing stable cavitation may change to transient cavitation. The oscillatory behavior of cavities in an acoustic field has been well described by a variety of mathematical models and is discussed in Chapter 1. One can easily calculate from such equations that at 20 kHz, a typical frequency of laboratory ultrasonic irradiations, the resonance size of cavities before implosive collapse will be $\approx 170\ \mu m$, and 1 at MHz, $\approx 3.3\ \mu m$. The dynamic process of bubble collapse has been observed by Lauterborn and others by ultrahigh-speed photography (10^5 frames per second) of laser generated cavitation,[13] and the comparison between theory and experiment is remarkably good.

Generation of Ultrasound and Reactor Design

A variety of devices has been used for ultrasonic irradiation of solutions. There are three general designs in common laboratory use: the ultrasonic cleaning bath, the "cup-horn" sonicator, and the direct immersion ultrasonic horn. In all cases the original source of the ultrasound is a piezoelectric material, usually a lead zirconate titanate ceramic (PZT) or occasionally quartz, which is subjected to a high-voltage alternating current with an ultrasonic frequency (roughly 15 kHz to 10 MHz). The piezoelectric source expands and contracts in this electric field and is attached to the wall of a cleaning bath or to an amplifying horn.

The ultrasonic cleaning bath is clearly the most accessible source of laboratory ultrasound and has been used successfully for a variety of liquid–solid heterogeneous sonochemical studies. There are, however, several potential drawbacks to its use. There is no means of control of the acoustic intensity, which will vary from bath to bath and over the lifetime of a single cleaning bath. In addition, their acoustic frequencies are not well controlled and differ from one manufacturer to another, and reproducibility from one bath to another may therefore suffer. Reproducible positioning of the reaction flask in the bath is critical, since standing waves in the bath create nodal spots where cavitation does not occur.[14,15] Similarly, the height of the bath liquid and of the solution within the reaction vessel are extremely important.[16,17] Temperature control is often neglected with this apparatus. Since the bath temperature can rise $>25°C$ during the course of a long irradiation, this can significantly influence both the intensity of the cavitational collapse and the rate of background thermal reactivity. Thermostating is best done using coolant passed through copper coils suspending in the bath (not in contact with the walls). The temperature inside the reaction

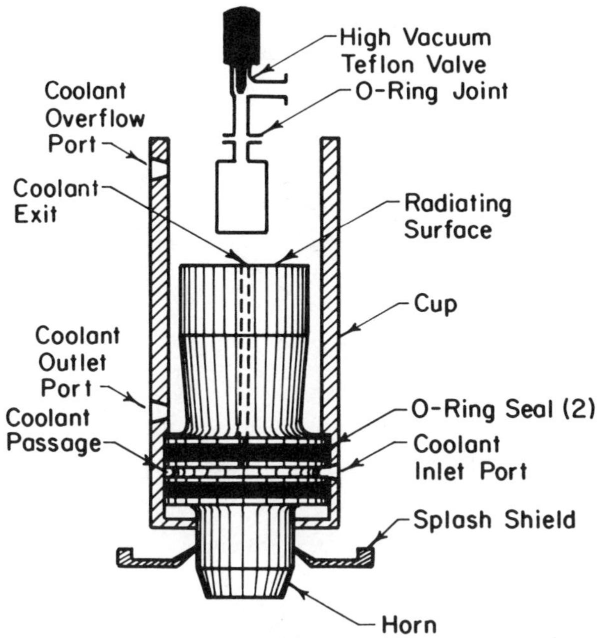

Figure 4-1. Cup–horn sonicator (modification of a design from Heat Systems-Ultrasonics, Inc.[18,19]).

vessel must be measured directly since it is often warmer than that in the bath itself. Finally, and most critically, the acoustic intensities present in most cleaning baths are only marginal for the generation of cavitation in homogeneous liquids. When solids are present, the weakened tensile strength of the liquid at the interface allows cavitation at thresholds well below those of simple solutions. Even in the case of heterogeneous sonochemistry, however, the ultrasonic cleaning bath must be viewed as an apparatus of limit capability.

The cup-horn configuration, shown in Fig. 4-1, was originally designed for cell disruption but has been adopted for sonochemical studies as well.[18] It has greater acoustic intensities, better frequency control, and potentially better thermostating than the cleaning bath. Again, however, it is very sensitive to the liquid levels and to shape of the reaction vessel. In addition, the reaction vessel faces a size restriction of $\approx$5-cm diameter. Since the ultrasonic radiating surface is not in direct contact with the reaction soluttion, the acoustic intensities are much lower than those of the direct immersion horn, and so homogeneous sonochemistry is often quite sluggish. On the other hand, there is no possibility of contamination from erosion of the titanium horn.

The most intense source of ultrasound generally used in the chemical laboratory is the direct immersion ultrasonic horn, which has been adapted for inert atmosphere work, as shown in Fig. 4-2, or for moderate pressures (<10 atm). These devices are available from several manufacturers[19] at modest cost and are used primarily for cell disruption. A variety of sizes of power supplies and titanium

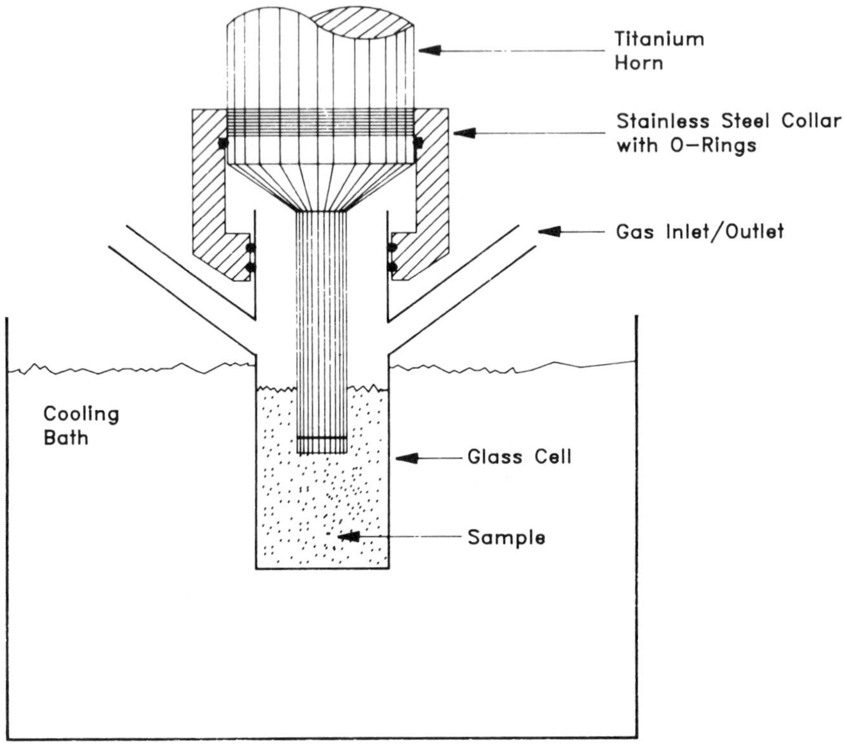

Figure 4-2. Direct immersion ultrasonic horn equipped for inert atmosphere work (design of K. S. Suslick[211]).

horns are available, thus allowing flexibility in sample size. The acoustic intensities are easily and reproducibly variable; the acoustic frequency is well controlled, albeit fixed (typically at 20 kHz). Since power levels are quite high, countercooling of the reaction solution is essential to provide temperature control; cooling of the piezoelectric ceramic may also be necessary, depending on the configuration. One potential disadvantage in corrosive media is the erosion of the titanium tip; this is generally a very slow process without chemical consequences, given the high tensile strength and low reactivity of Ti metal. This configuration may be used for both homogeneous and heterogeneous sonochemistry.

A rough, but useful, comparison between typical sonochemical and photochemical efficiencies is shown in Table 4-1. As shown, homogeneous sonochemistry is typically more efficient than photochemistry, and heterogeneous sonochemistry is several orders of magnitude better. Unlike photochemistry, whose energy inefficiency is inherent in the production of photons, ultrasound can be produced with nearly perfect efficiency from electric power. Still, a primary limitation of sonochemistry remains its energy inefficiency because of the small fraction of acoustic power involved in the cavitation events. This may be significantly improved, however, if a more efficient means of utilizing the sound

Table 4-1. Comparisons among Sonochemical and Photochemical Apparatuses

	Photochemistry	Homogeneous sonochemistry	Heterogeneous sonochemistry
Source	250-W quartz-halogen	200-W cell disrupter (at 60% power)	200-W cleaning bath
Approximate cost (1987)	\$1800	\$1900	\$700
Typical rates (μmol min^{-1})	7	10	500
Electrical efficiency (mmol $(kWh)^{-1}$)	2	5	200

field to generate cavitation can be found. Along these lines, Margulis and coworkers have recently compared cavitation generated by a high-speed turbine with a magnetostrictive ultrasonic transducer for the generation of both homogeneous and heterogeneous sonochemistry.[20,21] Although the scale and total rate were slower with the turbine, its electrical efficiency (moles of product per kilowatt hour of electrical input) was as much as tenfold higher.

Large-scale ultrasonic sources are readily accessible technology, as discussed in Chapter 3. Liquid processing rates of 200 L min^{-1} are routinely accessible from a variety of modular, in-line designs with acoustic power of several kilowatts per unit.[19] The industrial uses of these units include (1) degassing of liquids, (2) dispersion of solids into liquids, (3) emulsification of immiscible liquids, and (4) large-scale cell disruption. Although these units are of limited use for most laboratory research, they are of importance in eventual industrial application of sonochemical reactions.

The Nature of the Sonochemical Reaction Site

Given the detailed understanding of the dynamics of cavitation, the relevant question for the chemist lies in the actual mechanisms responsible for sonochemical reactions in homogeneous media. Historically, there have been two separate proposals: "hot-spot" pyrolysis[22,23] and electrical discharge[24,25]. The implosive collapse of a bubble will obviously produce adiabatic heating of its contents: estimates of the conditions so induced are in the thousands of degrees and thousands of atmospheres, as discussed shortly.

The several proposals of electrical discharge during cavitation (including recent suggestions[26,27]) have not been well developed on a molecular level and recently have been thoroughly rebutted as inconsistent with observed sonochemical reactivities and sonoluminescent behavior.[28,29] A very recent proposal for a new electrical discharge mechanism has been made by Margulis[30] with special reference to low-frequency (<100 Hz) induced cavitation; the separation of smaller fragmentation bubbles from local surface deformations of large cavitating bubbles under these conditions is the suggested source of very high local electric fields leading to discharge. Experimental conformation of this mechanism for sonochemical systems will be an important area for further exploration.

The high temperatures and pressures created during transient cavitation are difficult both to calculate and to determine experimentally. The simplest models of collapse, which neglect heat transport and the effects of condensable vapor, predict maximum temperatures and pressures as high as 10,000K and 10,000 atm.[31,32] More realistic estimates from increasingly sophisticated hydrodynamic models yield estimates of ≈ 5000K and ≈ 1000 atm, with effective residence times of <100 ns, but these are very sensitive to initial assumptions of the boundary conditions.[33–35] A detailed hypothesis for the diffusion and spatial distribution of radicals produced in and near collapsing cavities has also been proposed.[36]

However, few experimental determinations of the nature of the sonochemical reaction site or of cavitational conditions have been made. Sonoluminescence spectra have been used by Verrall and Sehgal to estimate the local conditions at the site of sonoluminescence, as discussed in Chapter 6. It is not yet established, however, whether sonoluminescence arises in the same location as most sonochemistry; excited-state Na emission created up sonication of aqueous NaCl solutions, for example, is not likely to have been generated in the gas phase of a collapsing bubble because Na^+ has virtually no vapor pressure.

The important question as to the nature of the sonochemical reaction site has recently been answered. The sonochemistry of volatile metal carbonyls (which are discussed later in detail) has recently been used as a dosimeter of acoustic cavitation by Suslick et al.[18,37] In order to determine whether the sonochemical reactions were occurring in the gas phase or the liquid phase of the cavitation event, the first-order rate coefficients of sonochemical ligand substitutions were measured as a function of metal carbonyl vapor pressure. If the sonochemical reaction site is gas phase only, then a plot of reaction rate coefficient versus dosimeter vapor pressure will be linear with a zero intercept; if the site is liquid phase only, then the plot will have a nonzero intercept with a zero slope (i.e., the rate coefficient will be independent of dosimeter vapor pressure).

As discussed later, the efficacy of cavitational collapse and the temperatures so generated are strongly dependent on the vapor pressure of the solvent system.[38,39] Therefore, sonochemical substitutions at various ambient temperatures were done in solutions of two *n*-alkanes that had been mixed in the proper proportion to keep the total system vapor pressure constant (at 5.0 torr (6.7 mBar)). In all cases the observed sonochemical rate coefficient increases linearly with increasing dosimeter vapor pressure and has a nonzero intercept, as shown in Fig. 4-3. The linear dependence of the observed rate coefficients on metal carbonyl vapor pressure is expected for reactions occurring in the gas phase: as the dosimeter vapor pressure increases, its concentration within the gas-phase cavity

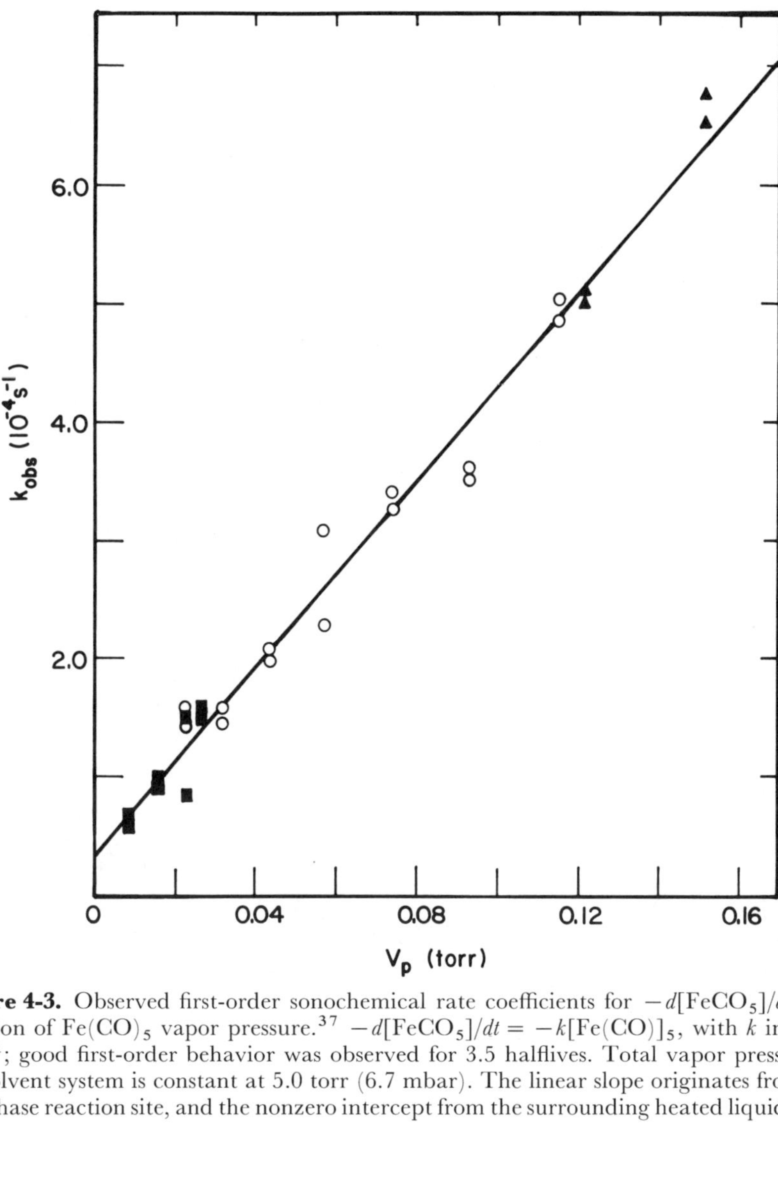

Figure 4-3. Observed first-order sonochemical rate coefficients for $-d[FeCO_5]/dt$ as a function of $Fe(CO)_5$ vapor pressure.[37] $-d[FeCO_5]/dt = -k[Fe(CO)]_5$, with k in units of s^{-1}; good first-order behavior was observed for 3.5 halflives. Total vapor pressure of the solvent system is constant at 5.0 torr (6.7 mbar). The linear slope originates from the gas-phase reaction site, and the nonzero intercept from the surrounding heated liquid shell.

increases linearly, thus increasing the observed sonochemical rate coefficients. In addition, the nonzero intercept indicates that there is a vapor-pressure-independent component of the overall rate. There must be, therefore, an additional reaction site occurring within the liquid phase, presumably in the thin liquid shell surrounding the collapsing cavity.

With these data, one may also estimate the effective temperatures reached in each site by the use of comparative rate thermometry, a technique developed for similar use in shock tube chemistry.[40] If the chemical reaction rate coefficients follow Arrenhius behavior ($k = A \exp(E_a/RT)$, where A is the frequency factor, E_a is the activation energy, and R is the gas constant), then a plot of ln (k_i/A_i) against E_{ai} for a series of reactions i gives a line with slope $1/RT_{\text{eff}}$, as shown in Fig. 4-4. Using the sonochemical kinetic data for ligand substitution of metal carbonyls, in combination with the activation parameters (A and E_a) recently determined by high-temperature gas-phase laser pyrolysis,[41] the effective temperatures of both sonochemical reaction sites have been experimentally estimated recently by Suslick and co-workers:[42,43] the gas-phase reaction zone effective temperature is 5200 ± 650K, and the liquid-phase effective temperature is ≈ 1900K. Using a simple thermal conduction model, the liquid reaction zone is estimated to be ≈ 200 nm thick and to have a lifetime of less than 2 μs, as shown in Fig. 4-5.

The sonochemical estimates of cavitational temperature in the liquid shell are comparable to those determined by sonoluminescence (Chapter 6), which may mean that the site of sonoluminescence (at least for alkali metal ions) is in the heated liquid shell surrounding the collapsing bubble. Given the differences in irradiation conditions used in these various studies, however, it is not yet possible to propose this as more than a working hypothesis. Regardless of the details,

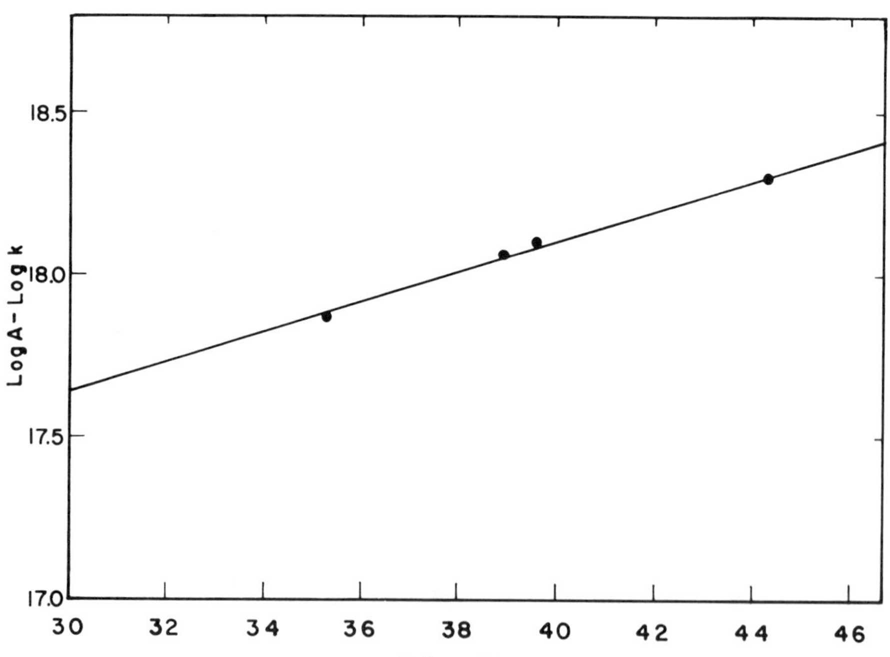

Figure 4-4. Comparative rate thermometry of the gas-phase reaction site. In order of increasing E_a, these data are for $Cr(CO)_6$, $Mo(CO)_6$, $Fe(CO)_5$, and $W(CO)_6$. The slope of the lease-squares fit line (solid) is $1/RT_{\text{eff}}$.

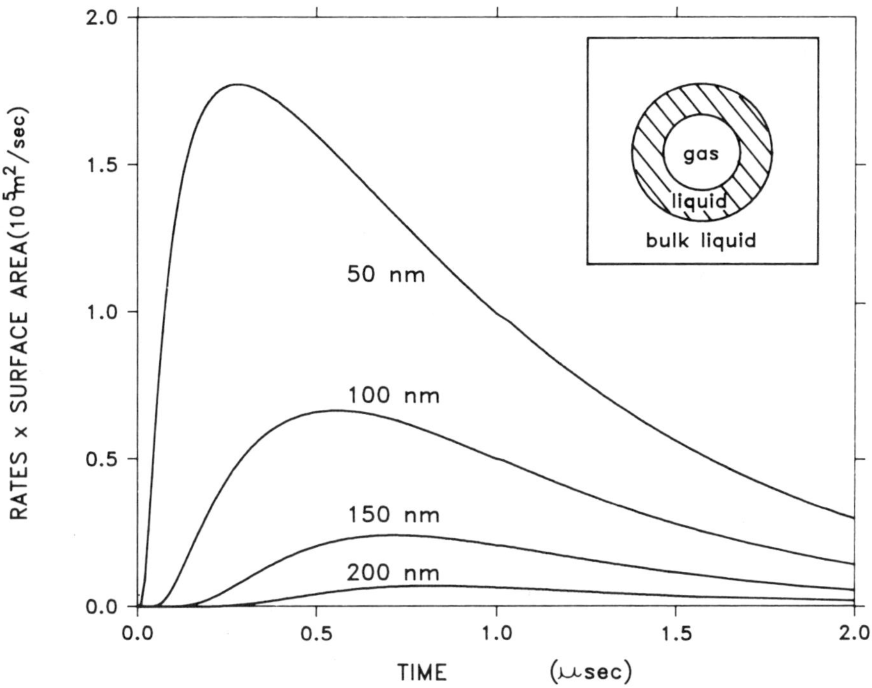

Figure 4-5. Temporal and spatial evolution of reaction rates in the liquid-phase reaction zone.[42] Rates were calculated as a function of time and distance from the bubble surface assuming only conductive heat transport from a sphere with radius 150 μm at 5200K, embedded in an infinite matrix at 300K.

however, it is clear that cavitational collapse generates hot spots with effective temperatures of several thousand degrees.

Experimental Influences on Sonochemistry

Sonochemistry is strongly affected by a variety of external parameters, including acoustic frequency, acoustic intensity, bulk temperature, static pressure, choice of ambient gas, and choice of solvent. These are important considerations in the effective use of ultrasound to influence chemical reactivity and are also easily understandable in terms of the cavitational hot-spot mechanism. A summary of these effects is given in Table 4-2.

The frequency of the sound field has relatively little effect on the nature of most sonochemistry. Unlike photochemistry, there is no direct coupling of the irradiating field with the molecular species in sonochemistry. The effect of changing sonic frequency is simply one of altering the resonant size of the cavitation event. The overall chemistry is therefore little influenced over the range where cavitation can occur (from tens of hertz to a few megahertz[12]); the observed

Table 4-2. The Effects of Extrinsic Variables on Sonochemistry

Extrinsic variable	Physical property	Effect
Acoustic frequency	Period of collapse	Resonant bubble size
Acoustic intensity	Reaction zone size	Cavitation events per volume
Bulk temperature	Liquid vapor pressure	Bubble content, intensity of collapse
	Thermal activation	Enhanced secondary reaction rates
Static pressure	Total applied pressure	Intensity of collapse
	Gas solubility	Bubble content
Ambient gas	Polytropic ratio	Intensity of collapse
	Thermal conductivity	Intensity of collapse
	Chemical reactivity	Primary or secondary sonochemistry
	Gas solubility	Bubble content
Choice of liquid	Vapor pressure	Intensity of collapse
	Surface tension	Transient cavitation threshold
	Viscosity	Transient cavitation threshold
	Chemical reactivity	Primary or secondary sonochemistry

sonochemical rates may change, but controlled comparisons of efficiency are lacking at this time and are likely to prove difficult. The observed sonochemistry of aqueous solutions, for example, is unchanged over this entire range.[44] Cavitation induced by very low-frequency sound (<100 Hz) has an observational advantage because of its large spatial (≈ 1 cm) and temporal (≈ 0.1 s) scale, which permits more easily detailed insights into bubble dynamics.[45] At very high frequencies (above a few megahertz) cavitation ceases, and sonochemistry is generally not observed.[46,47] The observed thresholds for cavitation in homogeneous liquids are strongly frequency dependent (cf. Chapter 1); since homogeneous sonochemistry is generally studied at acoustic intensities well above the threshold, however, this is not a major concern.

Acoustic intensity has a dramatic influence on the observed rates of sonochemical reactions. Below a threshold value, the amplitude of the sound field is too small to induce nucleation or bubble growth. Above the cavitation threshold, increased intensity of irradiation (from an immersion horn, for example) increases the effective size of the zone of liquid undergoing cavitation and thereby increases the observed sonochemical rate. Furthermore, as the acoustic pressures increase, the range of bubble sizes that undergo transient cavitation increases;[10] this too increases the observed sonochemical rate. It is often observed experimentally, however, that as one continues to increase acoustic amplitude, eventually rates begin to diminish again.[48] Possible explanations for this behavior include bubble shrouding of the sonic horn and overgrowth of bubbles. At high intensities, the cavitation of the liquid near the radiating surface becomes so intense as to produce a shroud of bubbles that diminishes the penetration of the sound into the liquid. Also at high intensities, bubble growth may become so rapid, that the bubble grows beyond the size range of transient cavitation before implosive collapse may occur.[49]

The effect of the bulk solution temperature lies primarily in its influence on the bubble content before collapse. As the ambient temperature is increased, in general, sonochemical reaction rates are slowed. From a different perspective, sonochemical reactions usually have apparent negative activation energies! This reflects the dramatic influence that solvent vapor pressure has on the cavitation event: the greater the solvent vapor pressure found within a bubble prior to collapse, the less effective the collapse. In fact, one can quantitate this relationship rather well. Three different derivations from three different approaches and different assumptions (one of which is given in Chapter 6 by Verrall and Sehgal)[39,50] all predict that the effective peak temperature generated during cavitational collapse will be inversely proportional to P_v, the vapor pressure of the system. Assuming Arrhenius behavior $[k = A \exp(-E_a/RT)]$, one expects that the sonochemical rate coefficient (k_{obs}) should follow:

$$\ln k_{obs} = C_0 \ln A - (E_a/C_1)P_v/T_0 \tag{4-1}$$

where T_0 is the ambient temperature of the bulk liquid; C_0 and C_1 contain various constants, including the number of cavitation events per volume per time and parameters that determine the efficiency of the cavitational collapse (e.g., the polytropic ratio). These derivations are only rough approximations of very complex events. Nonetheless, the linear correlation of $\ln k_{obs}$ and P_v is the experimentally observed behavior in a wide range of sonochemical systems in a variety of solvents;[38,39,50] an example is shown in Fig. 4-6. In these results, temperature, solvent vapor pressure, and gas solubility have been varied independently, which

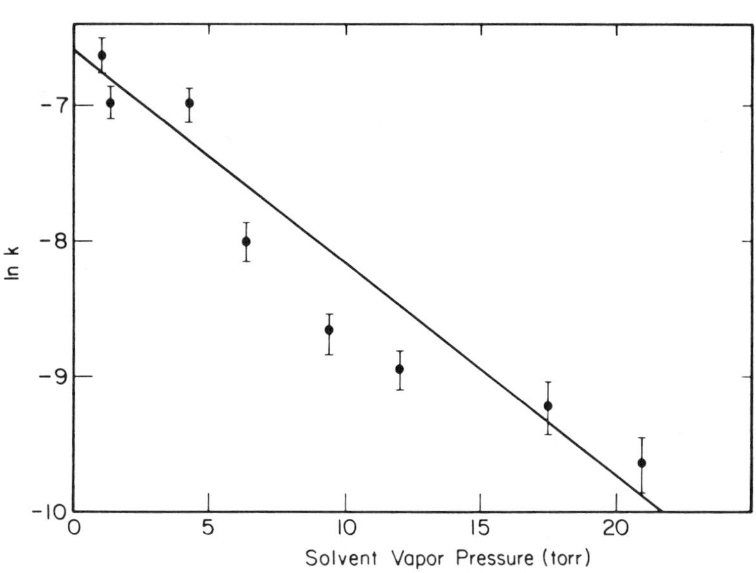

Figure 4-6. The ln k of the observed first-order rate coefficient for $-d[Fe(CO)_5]/dt$ vs. total system vapor pressure at 25°C.

rules out gas solubility as an important variable in sonochemistry when ultrasonic intensities are well above the threshold level. When secondary reactions are being observed (e.g., as in secondary chemical corrosion that occurs after initial acoustic erosion of a passivated surface), then temperature can play its usual role in thermally activated chemical reactions. This explains the occasional observation of increasing rates of corrosion associated with cavitation with increasing temperature.[51]

Sonochemical yields as a function of increasing static pressure have been reported by different researchers[52,53] to increase, to decrease, and to increase to some point and then decrease. One would expect cavitational collapse to increase in intensity with increasing external pressure, because the total imposed pressure at the initiation of collapse would be increased. Given a fixed acoustic intensity, however, nucleation of cavities can no longer occur above some limiting applied pressure, since the acoustic field must overcome the combined tensile strength of the liquid and the ambient pressure. In contrast, as one reduces the ambient pressure, eventually one deactivates the gas-filled crevices that serve as nucleation sites (discussed earlier) and therefore also diminishes observed sonochemistry. Further experimental difficulties occur when one attempts to maintain a pressure vessel at constant temperature while under ultrasonic irradiation. It is perhaps not surprising, then, that the experimental results are conflicting on this question. In reactions that involve the ambient gas directly, enhanced solubility also plays a role in the overall observed rates.

The choice of ambient gas also has a major impact on sonochemical reactivity. The maximum temperature reached during cavitation is strongly dependent on the polytropic ratio (C_p/C_v) of the ambient gas,[22,23] which defines the amount of heat released during the adiabatic compression of that gas. This can have a dramatic impact: all other factors being equal the difference between cavitation in the presence of xenon ($C_p/C_v = 1.67$) and a freon ($C_p/C_v \approx 1.1$), for example, would yield a ratio of maximum temperatures of sevenfold! Sonochemical rates are also significantly influenced by the ambient gas's thermal conductivity, as shown in Fig. 4-7, so even the noble gases affect cavitation differently.[54,55] The role of thermal transport during cavitational collapse has long been recognized as evidence in favor of the hot-spot mechanism of sonochemistry,[54] and recent calculations underscore its effect on conditions generated during cavitational collapse.[34,35] In addition, sonochemical reactions often involve the gases present in the cavitation event.[56] For example, H_2, N_2, O_2, and CO_2 are not inert during cavitation and undergo a variety of redox and radical reactions, as discussed in the later section, Sonochemistry of Gases in Water. Another relevant parameter, gas solubility, has been observed to affect the concentration of cavitation nuclei,[57] and in this way, may play a role in determining the observed cavitation threshold.

The choice of the solvent has a profound influence as well over the observed sonochemistry. The effect of vapor pressure has already been mentioned. It is the principal determinant of the conditions formed during cavitation. Other liquid properties, such as surface tension and viscosity, alter the threshold of cavitation, but this is generally a minor concern. The chemical reactivity of the

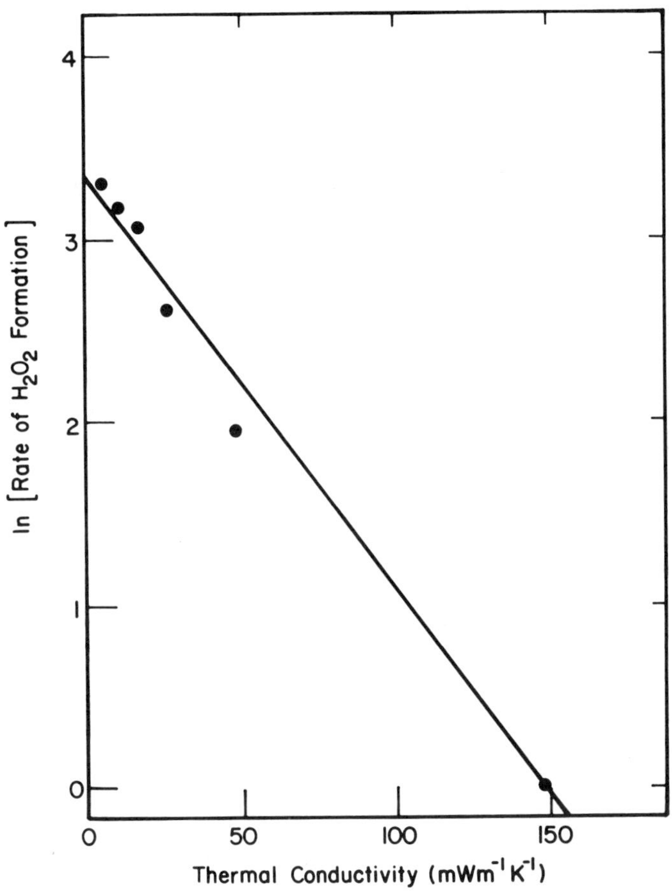

Figure 4-7. Sonochemical rates as a function of ambient gas thermal conductivity (replotted data from R. O. Prudhomme[142]).

solvent is often much more important. As discussed in later sections, aqueous sonochemistry is dominated by secondary reactions of $OH^{\bullet}$ and $H^{\bullet}$ formed from the sonolysis of water vapor in the cavitation zone. No solvent is inert under the high-temperature conditions of cavitation: even linear alkanes undergo pyrolysis like cracking during high-intensity sonication.[39] This problem may be minimized, however, by using robust solvents (avoiding halocarbons, in particular) which have low vapor pressures in order to reduce their concentration in the vapor phase of the cavitation event. Similarly, one must anticipate secondary solvent reactivity in the trapping of high-energy species produced during cavitation.

Thus, the parameters of acoustic intensity, temperature, ambient gas, and solvent choice have strong influences on sonochemical reactions. It is clear that one can fine tune the energetics of cavitation by the use of these variables, and hence exercise control on the rates and reaction pathways followed by the associated chemistry. Most notably, the thermal conductivity of the ambient gas (e.g., a variable He/Ar atmosphere) and the overall solvent vapor pressure provide easy mechanisms for experimental control of the peak temperatures generated during the cavitational collapse.

AQUEOUS SONOCHEMISTRY

Sonolysis of Water

History

The early studies of the chemical effects of ultrasound have been thoroughly reviewed,[58,59] but a brief introduction of the history of sonochemistry (which was essentially all in aqueous solutions until 1980) is worthwhile. The sonolysis of water to produce H_2O_2 was first reported by Schmitt et al.[60] in 1929. This early study led to exhaustive further work by a large number of chemists,[53,56,58,61–67]; the work of Weissler[68,69] and of Griffing[22,23,54] especially began to delineate the experimental parameters influencing sonochemistry. Sonochemical oxidations of many inorganic anions, including I^-, Br^-, and S^{2-}, were observed by many early workers. Reductions were also observed, as for example by Beuthe,[70] in the sonochemical reduction of mercuric chloride by aqueous oxalate solutions. The reaction of N_2 in aqueous solutions to give nitric and nitrous acids was recognized in 1936 by Schultes and Gohr.[71] Virtanen and Ellsfolk further expanded these observations.[72] These findings were paralleled by the first observation of sonoluminescence in 1935 by Frenzel and Schultes.[73] The first studies of organic sonochemistry in water were made by Schmitt et al.[60] on C—Cl bond cleavage and by Porter and Young[74] for enhancement of the Curtius rearrangement of benzazide ($C_6H_5CON_3$ to N_2 and C_6H_5NCO). Miyagawa reported in 1942 the use of ultrasound to increase the rates of acetate hydrolysis and fat saponification, as well as the rate of mellowing whiskey, sake, and shoyu (soy sauce).[75]

The principal products from the ultrasonic irradiation of water are H_2O_2 and H_2, and various data support the hypothesis of the intermediacy of hydroxyl radicals and hydrogen atoms:

$$H_2O \xrightarrow{)))} OH^\bullet + H^\bullet \longrightarrow H_2O_2 + H_2 \qquad (4\text{-}2)$$

The wide range of oxidations and reductions that occurs with aqueous sonochemistry is often a consequence of secondary reactions of these high-energy intermediates.

Contemporary Work

Spin trapping experiments by Riesz and co-workers[76] have recently provided elegant evidence for the intermediacy of $OH^\bullet$ and $H^\bullet$ during the sonolysis of water, although other data have also suggested the formation of a variety of other high-energy species, depending on conditions, including e_{aq}^{-}[77,78] and $(HO_2)^{\bullet}$.[79] In Riesz's definitive work, standard diamagnetic nitrone and nitroso compounds were used as reactive traps for short-lived hyrdoxyl radicals and hydrogen atoms in order to generate relatively long lived nitroxide radicals, which were observed (and assigned) by conventional electron spin resonance. An example of such a spin trapping reaction is shown below:

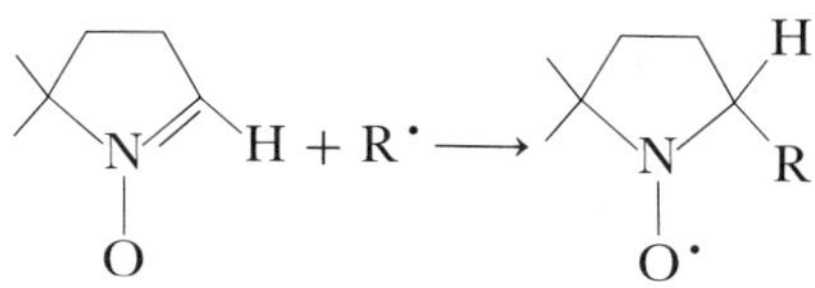

(4-3)

In the first studies by Makino, Mossoba, and Riesz, the aqueous solutions containing low concentrations of the spin traps were sonicated in a cleaning bath at room temperature under Ar.[80,81] Comparisons of the observed hyperfine coupling constants of the spin adducts formed after ultrasonic irradiation permitted unambiguous identification of both $OH^{\bullet}$ and $H^{\bullet}$ adducts. Sonication of D_2O gave confirmatory proof of $OD^{\bullet}$ and $D^{\bullet}$ adducts. That these species were formed directly during sonication was confirmed by competitive scavenging studies with ethanol and sodium formate, which decreased the signals observed for the $OH^{\bullet}$ adduct and created new signals for the $[C_2H_4OH]^{\bullet}$ and $[CO_2]^{-}$ adducts, respectively.

A variety of sources of ultrasound were investigated by Riesz and co-workers, with special effort made to create well-characterized acoustic exposures comparable to those used in human clinical applications. For example, Ar-saturated water solutions exposed to 25-kHz ultrasound from dental cleaning equipment generated hydroxyl radicals and hydrogen atoms.[82] Of most interest, perhaps, is the medical diagnostic use of high-frequency, pulsed ultrasound for imaging (sonography), which is presently used routinely, for example, in a majority of pregnancies. Using 10-μs pulses of 1.0-MHz frequency with spatial peak temporal average intensities of 90 W cm^{-2}, these same radicals were observed.[83] These exposure conditions approximate, but do not duplicate, diagnostic exposures used in medical ultrasonic imaging, which usually use even higher frequencies and shorter pulse durations. Pulse repetition rates are also an important parameter to quantify. These results are consistent with Flynn's prediction that cavitational collapse can occur even under high-frequency, short-pulsed ultrasonic irradiation.[84] Christman's et al. most recent paper[85] concludes:

> the possibility of cavitation *in vivo* due to diagnostic exposure [of ultrasound] is either unknown or considered unlikely by most physicians and other health professionals. Our recent experimental results suggest that perhaps this confidence that cavitation cannot occur for diagnostic pulsed ultrasound in not warranted. [p. 34]

Further detailed discussion of the clinical and biological implications of ultrasonic exposures is made in the final two chapters of this book.

Sonochemistry of Gases in Water

In the presence of O_2, early isotope labeling studies by DelDuca et al. showed that significant amounts of the peroxide derive from the O_2, without O—O bond cleavage,[86] which probably involves the reduction of O_2 with $H^\bullet$. Other work by Anbar and Pecht[87] demonstrated that isotopic scrambling of ^{18}O from labeled water or O_2. In an large series of recent papers,[88] Henglein has reexamined in detail the isotopic scrambling that occurs upon sonolysis of water in the presence of O_2 or D_2. The former interconversions[89] appear to involve the following reactions, which have precedent in combustion chemistry.

$$^{18}O_2 \xrightarrow{)))} 2\ ^{18}O \tag{4-4}$$

$$^{18}O + {}^{16}OH_2 \longrightarrow {}^{18}OH + {}^{16}OH \tag{4-5}$$

$$^{16}OH + {}^{18}OH \longrightarrow {}^{16}O + {}^{18}OH_2 \tag{4-6}$$

Henglein also demonstrated that $OH^\bullet$ and $H^\bullet$, the initial products of cavitational collapse, could be trapped by D_2 to form HDO and HD, respectively.[90] As increasing amounts of D_2 were present in the Ar atmosphere, the H_2O_2 production rapidly dropped off, and HD production increased to a maximum at about 60% D_2/Ar. As more D_2 is present, the temperature in the cavitation bubbles decreases substantially because of enhanced thermal transport; under these conditions, all sonochemistry decreases. These workers were able to estimate that in the absence of scavengers, under Ar, roughly 80% of the H atoms and OH radicals initially produced during the ultrasonic irradiation of water recombine. Similar conclusions had been reached earlier by Margulis.[91,92] For a sense of scale, under typical laboratory conditions of ultrasonic irradiation with an immersion horn at $\approx 50\ W\ cm^{-2}$, the rate of peroxide formation will be 10–50 $\mu M\ min^{-1}$.

The sonication of aqueous solutions of ozone leads to its very rapid destruction.[93,94] The rate of O_2 formation from O_3 is roughly 100-fold that of H_2O_2 production from the sonolysis of water. It was proposed that this was not a chain reaction but involved the initial cleavage of O_3 to form O_2 and O atom, which then attacked a second molecule of O_3. Given the weakness of the bonds in ozone, this mechanism is plausible, but not yet demonstrated. The early report[95] of ozone production upon sonication of water has not been reproduced and appears to be in error.

The reaction of dissolved N_2 under ultrasonic irradiation of water has been repeatedly studied[56,96–98] since the first reports by Schultes and Gohr. The principal products after sonication of water under N_2 are HNO_2 and HNO_3, with smaller amounts of NH_3 and N_2O. Both Margulis and Henglein have recently reported $^{14}N_2/^{15}N_2$ scrambling in this reaction. Margulis[99] assumed without evidence that the mechanism involved direct bond cleavage of N_2. In contrast, however, Henglein proposed that the scrambling occurred not via direct sonolysis of N_2, but through the attack of O atoms on N_2 to give NO and N

Table 4-3. Aqueous Sonochemistry of Gases

Substrate present	Principal products	Ref.
HD	H_2, D_2	90
$H_2 + N_2$	NH_3	96
$H_2 + CO$	HCHO	97
$^{14}N_2 + {}^{15}N_2$	$^{14,15}N_2$	99, 100
N_2	HNO_2, HNO_3, NH_2OH, N_2O, NH_3	56, 96–98
N_2 + (CO, CH_4 or HCHO)	Amino acids	97, 101
$^{18}O_2 + {}^{16}OH_2$	$^{16,18}O_2$, $^{16}O_2$	86–88
O_3	O_2	93, 94
N_2O	N_2, O_2	100

atoms, as shown below, which are known reactions from combustion and shock tube studies:

$$H_2O \xrightarrow{)))} H^{\bullet} + OH^{\bullet} \tag{4-7}$$

$$2\ OH^{\bullet} \longrightarrow O + H_2O \tag{4-8}$$

$$N_2 + O \longrightarrow N + NO \tag{4-9}$$

$$^{14}N + {}^{15}NO \qquad {}^{14,15}N_2 + O \tag{4-10}$$

$$^{14}NO + {}^{15}NO \longrightarrow {}^{14,15}N_2O + O \tag{4-11}$$

The low yields of N_2O are attributed to its own facile sonolysis[100] to N_2 and O_2.

Irradiation of water under mixtures of H_2 and CO yields formaldehyde; irradiation of N_2 with CO, CH_4, or HCHO gives a variety of amino acids in low yields.[97,101] It has, in fact, been suggested that cavitation induced by turbulent flow in ocean waves may have been a significant source of prebiotic organic compounds in the "primordial soup". A summary of the aqueous sonochemistry of gases is given in Table 4-3.

Sonochemistry of Inorganic Compounds in Water

Given the facile homolytic cleavage of water during ultrasonic irradiation, a wide range of secondary sonochemistry in aqueous solutions can be expected and indeed has been repeatedly observed. One of the first detailed studies was by Solov'eva, who demonstrated that iron(II) oxidation by O_2 was greatly accelerated upon sonication of aqueous solutions.[102] An extensive list of oxidations and reductions has been reported of various inorganic species. Cited in Table 4-4 are those aqueous sonochemical reactions in which products have been well characterized; in some cases, trace amounts of other products were reported in addition to the major products listed.

Table 4-4. Aqueous Sonochemistry of Inorganic Compounds

Substrate present	Principal products	Ref.
Br^-, Cl^-	Br_2, Cl_2	61–67, 110
Ce^{4+}	Ce^{3+}	111
$[Co(NH_3)N_3]^{2+}$	$Co^{2+} + N_3^{\bullet}$	109
Fe^{2+}	Fe^{3+}	112, 108
$[Fe(III)(C_2O_4)_3]^{3-}$	Fe^{2+}	113
H_2S	$H_2 + S_8$	114
I^-	I_3^-	18, 46, 63
$[MnO_4]^-$	MnO_2	70
$[NO_3]^-$	$[NO_2]^-$	115
OsO_4	OsO_2	116
$[PO_3]^{2-}$	$[PO_4]^{2-}$	95
Tl^+	Tl^{3+}	117

A great deal of effort in some quarters has been made to examine in detail the energetics of aqueous sonochemistry[103–105] in order to generate an equivalent to the primary radical and molecular yields (i.e., the number of radicals or molecules produced per 100 eV of ionizing radiation absorbed: so-called *g* values) used so successfully in radiolysis. For example, the ratio of rates of oxidation of Fe^{2+} and of reduction of Ce^{4+} is dependent on the linear energy transfer (energy transferred by radiolytic particle per length of penetration through water, in eV Å^{-1}). Comparison to the same sonochemical oxidations leads to an analogy between secondary aqueous sonochemistry and heavy fission product radiolysis (e.g., α-particles from ^{210}Po).[104,106,107] These attempts are of limited use, however, given the sensitivity of sonochemical yields to experimental conditions (e.g., nature of the dissolved gas, concentration of particulate matter for nucleation sites, past history of the liquid, total vapor pressure, hydrophobicity of added solutes, among many others as discussed earlier). The addition of small amounts of alcohols, for example, greatly increases the apparent sonochemical *g* values of Fe^{2+} oxidation through their chemical participation in a radical-chain oxidation initiated by aqueous sonolysis.[108]

Sonochemical redox reactions occur for both simple inorganic ions and for transition-metal complexes as well. For example, Rehorek and Janzen have recently examined[109] the sonochemical degradation of $[Co(NH_3)_5N_3]^{2+}$. Using ESR spin trapping they were able to demonstrate the transient formation of $N_3^{\bullet}$. Given the high liability of Co(II) complexes, the following mechanism was proposed:

$$H_2O \xrightarrow{)))} H^{\bullet} + OH^{\bullet} \quad (4\text{-}12)$$

$$H^{\bullet} + [Co(NH_3)_5N_3]^{2+} \longrightarrow Co^{2+} + [NH_4]^+ + 4\,NH_3 + N_3^- \quad (4\text{-}13)$$

$$OH^{\bullet} + N_3^- \longrightarrow OH^- + N_3^{\bullet} \quad (4\text{-}14)$$

Sonochemistry of Organic Compounds in Water

Various organic compounds have been sonicated either as aqueous solutions or as suspensions, generally with a wide range of highly degraded products being formed. Since extremely reactive intermediates are formed at respectable rates from the sonolysis of water itself, it is not surprising to see this general lack of control of sonochemistry in aqueous media. Given the nature of the cavitation event, the high vapor pressure of water (relative to inorganic species or to dilute organic compounds) condemns aqueous sonochemistry to domination by secondary chemical reactions unrelated to the direct processes that such dissolved substrates might have undergone, had they been the major species found in the collapsing bubble.

Weissler[68,69] discovered that aqueous solutions of carbon tetrachloride, even in the absence of O_2, liberated Cl_2 upon ultrasonic irradiation and greatly accelerated the formation of I_3^- from aqueous iodide solutions. He conclude that the fundamental reaction was between water and CCl_4:

$$CCl_4 + H_2O \xrightarrow{)))} Cl_2 + CO + 2\ HCl \qquad (4\text{-}15)$$

$$3\ I^- + Cl_2 \longrightarrow I_3^- + 2\ Cl^- \qquad (4\text{-}16)$$

This became a standard chemical dosimeter for sonochemical apparatus, since I_3^- has a high absorbance in the near ultraviolet and can be detected even more easily upon addition of starch. This dosimeter has serious difficulties with reproducibility, however, for the concentration of CCl_4 is hard to control because of its low aqueous solubility and high volatility; given the sensitivity of modern spectrophotometers, the rate enhancement provided by CCl_4 for I_3^- formation is no longer needed.[18]

A wide range of organic compounds were subjected to ultrasonic irradiation by Spurlock and others, as cited in Table 4-5 below. Various degradation products were identified; these products principally arise from secondary reactions of $OH^\bullet$ and $H^\bullet$. For example, irradiation of aqueous suspensions of thioethers gives mostly sulfoxides and sulfonic acids; of aldehydes gives mostly carboxylic acids; of carboxylic acids gives CO, CO_2 and hydrocarbons; and of amines gives aldehydes, other amines, *N*-oxides, alcohols, and hydrocarbons.

A rather interesting use of ultrasound to initiate a radical-chain isomerization was first reported by El'piner and co-workers[118] in 1965. When maleic acid was sonicated as a concentrated solution in water with a trace of Br_2 present, very rapid isomerization to fumaric acid occurred with its subsequent precipitation. Alkyl bromides also catalyze the reaction. The mechanism is thought to be analogous to the photochemical isomerization and involves the sonolytic production of $Br^\bullet$:

$$Br_2 \xrightarrow{)))} 2\,Br^{\cdot} \qquad (4\text{-}17)$$

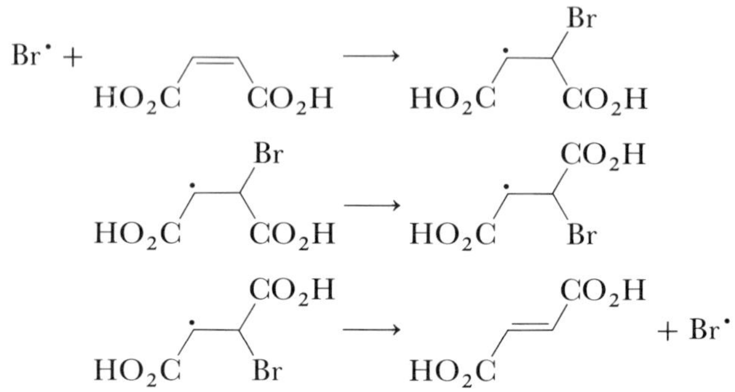

(4-18)

(4-19)

(4-20)

Even though the lifetime of $Br^{\cdot}$ in solution is quite brief, sonocatalytic isomerization is efficient. In more recent studies, Margulis reexamined this reaction under low-frequency sonic irradiation with similar results.[119] In other recent work, Magulis and co-workers found that *S*-alkylation of thiocarbamates with alkyl chlorides showed ≈75-fold rate enhancements in the presence of ultrasound compared to vigorous stirring; the origin of this acceleration remains obscure.[120]

Verrall et al. extensively investigated the aqueous sonochemistry of nucleic acid bases.[121] Many products are produced and their formation generally explained in terms of the expected secondary processes involving $OH^{\cdot}$ and $H^{\cdot}$. Mead et al.[122,123] found strong similarities between the sonochemical products of aerated thymine solutions and those after γ-irradiation. In other work the acoustic thresholds for decomposition of these bases were examined[124] but were found to be quite dependent on the external parameters, such as rate of gas bubbling during sonication.[125] The susceptibility to sonochemical reaction of these bases vary significantly in the order thymine > uracil > cytosine > guanine > adenine. Wang and Gupta recently examined the effects of pulsed ultrasound on these same compounds and found that the effective sonochemical rates of degradation were greater under pulsing than under a temporal averaged equivalent continuous irradiation.[126]

In a very recent and important study, Henglein et al. examined the trapping of $OH^{\cdot}$ as a function of added organic compounds.[127] They determined the concentration of scavenger necessary to reduce the rate of H_2O_2 production by one half. It was revealed that the vapor pressure of the scavenger was not the determining factor, but in fact the hydrophobicity (i.e., its solubility in water versus its solubility in a nonpolar organic liquid) was the important parameter, as shown in Fig. 4-8. Here a somewhat artificial hydrophobicity factor was defined as the ratio of the number of hydrophobic groups to that of hydrophilic groups. From these data, it appears that the trapping of $OH^{\cdot}$ is occurring selectively at the interface between the cavitation bubble and the liquid where hydrophobic solutes concentrate. This is a major difference between free-radical chemistry under sonolysis in water and that created in homogeneous solutions by photochemical or radiolytic means.

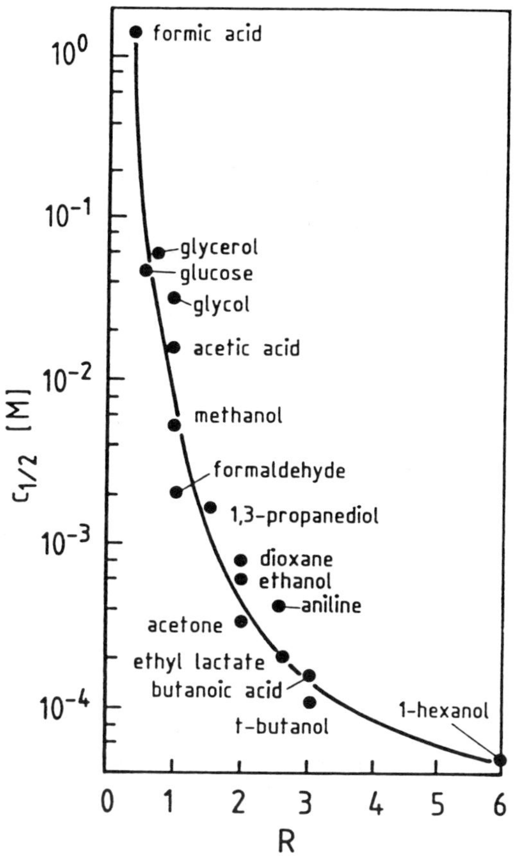

Figure 4-8. The Effect of solute hydrophobicity on $OH^\bullet$ trapping in aqueous sonolysis.[127] Here, $C_{1/2}$ is the solute concentration at which the yield of H_2O_2 is decreased by 50%; R, a measure of the solute hydrophobicity, is defined as the ratio of hydrophobic to hydrophilic groups in each solute.

Finally, there have been generally small rate enhancements[128–130] (<20%, although there is one report of an 10-fold increase[131]) of solvolysis reactions (shown below) observed during the sonication of esters and halocarbons in mixed aqueous media:

$$Cl{-}C(CH_3)_3 + H_2O \xrightarrow{)))} HCl + HO{-}C(CH_3)_3 \qquad (4\text{-}21)$$

These have been interpreted in terms of a disruption of the solvent structure[132] by the ultrasonic irradiation; the details of this proposed mechanism remain to be fully understood. The rate enhancement of solvolysis increases with decreasing temperature and increasing ethanol content, although the expected relationship between the log of the sonochemical rate constant and the vapor pressure seems not to apply.[133] There are, however, two significant experimental concerns with some of this work. First, the rate of solvolysis has not been measured directly but rather through the change in conductivity of the solutions through the formation of HCl. Because the sonolysis of aerated water produces nitric and nitrous

Table 4-5. Aqueous Sonochemistry of Organic Compounds

Substrate	Principal products	Ref.
CCl_4	Cl_2, CO_2, HCl, C_2Cl_6, HOCl	22, 63, 134–136
CH_3I	CH_4, I_2, CH_3OH, HI, C_2H_6	137
R_2CHCl	R_2CHOH, HCl	129–133
$Cl_3CCH(OH)_2$	HCl	138, 139
C_6H_5Br	Br^-, C_2H_2	140, 141
Maleic acid + Br_2	Fumaric acid	118, 119
CS_2	S, H_2S	142
$(C_4H_9)_2S$	$(C_4H_9)_2SO$, polymer	143
$[R_2NC(=O)S]^-$ + R′Cl	$R_2NC(=O)SR'$	120
RCHO	CO, CH_4, C_2H_4, $C_2H_4O_2$, RCO_2H	144
$[HCO_2]^-$	CO_2	145
C_5H_5N	HCN, H_2C_2, H_2C_4	141, 146
C_6H_5OH	$C_6H_4(OH)_2$	147
$C_6H_5CO_2H$	$C_6H_4(OH)(CO_2H)$	148
$C_6H_{11}OH$	C_2H_2	149
RCO_2H	CO, CH_4	144
ClH_2COO_2H	H_2, CO, CH_4^-, Cl^-	103
RCO_2R'	RCO_2H, R′OH	128, 75, 150–153
RCH_2NH_3	H_2, CH_4, NH_3, RCHO, RCH_2OH	154
$(CH_2NH_2)_2$	NH_3	87, 117
Thymine	Hydroxylated products	58, 121–126
Uracil	Hydroxylated products	58, 121–126
Various amino acids	H_2, CO, NH_3, RNH_2, HCHO	155
Cysteine	Cystine	155

acids, as discussed earlier, increases in solution conductivity may not be a reliable measure of rates of solvolysis. Second, in the comparison of observed rate increase versus solvent vapor pressure, the significant contribution of the halocarbon (2-chloro-2-methylpropane) to the total system vapor pressure appear to have been neglected.

A summary of aqueous sonochemists reactions in organic compounds appears in Table 4-5.

NONAQUEOUS SONOCHEMISTRY

Sonolysis of Organic Liquids

Until recently, very few examples of homogeneous nonaqueous sonications had been reported. These included the very slow degradation of acetonitrile and carbon tetrachloride,[156] the initiation of explosions of tetranitromethane and nitroglycerine,[157] the sevenfold acceleration[158] of the Curtius rearrangement of

$C_6H_5CON_3$ to C_6H_5NCO and N_2, and the depolymerization of high molecular weight polymers (discussed later). The pioneer study by Weissler et al. on the sonochemistry of CH_3CN (which gave small amounts of N_2, CH_4 and H_2 through the cleavage of both C—C and C—H bonds) and CCl_4 (which gave Cl_2) demonstrated that cavitational effects *could* occur outside of water. In general, however, sonochemistry had not been observed in most common, volatile organic solvents (or aqueous solutions with volatile organics). This led to the commonly stated assumption that sonochemical events could be produced only in high tensile strength or high dielectric liquids such as water.[159,160] As noted earlier, the problem is simply that many organic liquids have high vapor pressures, which greatly diminish the intensity of cavitational collapse. A listing of the nonaqueous sonochemistry of organic compounds is given in Table 4-6.

It is now clearly demonstrated through the use of free-radical traps that all organic liquids undergo cavitation and generate bond homolysis if the ambient temperature is sufficiently low (i.e., in order to reduce the solvent system's vapor pressure).[38,39,161,162] Within a single class of solvents, the rate of radical production is primarily determined by the solvent vapor pressure. As the vapor pressure decreases, the efficacy of cavitation collapse increases, the peak temperature reached during collapse increases, and the rate of radical production increases. There are secondary considerations (e.g., bond strengths, solvent viscosity) when

Table 4-6. Nonaqueous Sonochemistry of Organic Compounds

Substrate	Principal products	Ref.
Alkanes	H_2, CH_4, C_2H_2. 1-alkenes	39
CH_3CN	H_2, CH_4, N_2	156
$CH_3CN + O_2$	CO, CO_2, H_2O, N_2	156
$CH_3COCOCH_3$	$(H_3C)_2CO$	168
$RCHO + O_2$	RCO_2H, RCO_3H, CO_2, CO, CH_4	171, 172
$CHCl_3$	HCl, CCl_4, CH_2Cl_2, C_2Cl_5H, C_2Cl_4, C_2Cl_6, C_2HCl_3	166, 167
CCl_4	Cl_2, C_2Cl_6	156
C_6H_6	H_2, CH_4, C_2H_2, $C_6H_5CH_3$, $(C_6H_5)_2$	165
$C_6H_5CH_3$	H_2, CH_4, C_6H_6, $C_6H_4(CH_3)_2$, $[C_6H_5(CH_2)]_2$	165
1,3-$C_6H_4(CH_3)_2$	H_2, CH_4, $C_6H_5CH_3$, $C_6H_3(CH_3)_3$, $[(H_3C)C_6H_4(CH_2)]_2$	165
1,3,5-$C_6H_3(CH_3)_3$	H_2, CH_4, C_5H_6, $C_6H_n(CH_3)_{6-n}$, $[(H_3C)_2C_6H_3(CH_2)]_2$	165
$C_6H_5N_2^+ X^-$	N_2, $[C_6H_5]^\bullet$, $X^\bullet$	170
$C_6H_5CHCH_2$	Polymerization	192, 196, 197
$H_2CC(CH_3)(CO_2H)$	Polymerization	185
$H_2CC(CH_3)(CO_2CH_3)$	Polymerization	196, 197
$H_2CCH(CONH_2)$	Polymerization	193
Many polymers	Depolymerization	179, 180

comparisons are made between different solvent classes; nevertheless, vapor pressure remains the most important determinant.[39]

The sonolysis of alkanes is quite similar to very high temperature pyrolysis, yielding the products shown in Fig. 4-9. The principal products (H_2, CH_4, and 1-alkenes) are those expected from the well-understood Rice radical-chain mech-

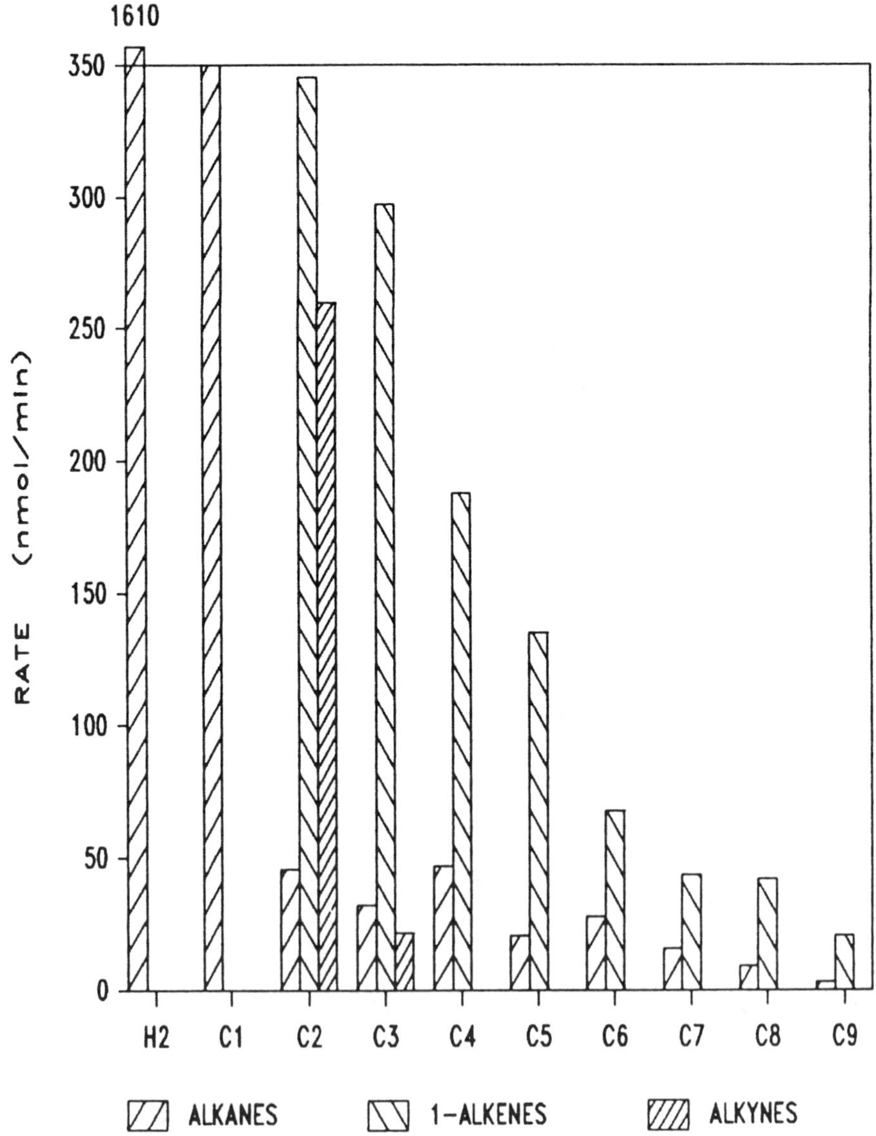

Figure 4-9. Product distribution from *n*-decane sonolysis under Ar at 25°C. Rates are of total gas- and liquid-phase products: 20-mL gas volume and 10-mL liquid volume (plotted from data in Ref. 39).

anism,[39] given as a partial reaction scheme in Fig. 4-10. Acetylene production is a secondary process also observed with extremely high temperature pyrolysis and shock tube chemistry.

Nonaqueous sonochemistry of arenes has been examined as well. It was originally reported that a variety of substituted benzenes gave deeply colored, uncharacterized tars upon sonication.[163,164] Suslick and Doktycz[165] recently have completed a comprehensive analysis of the soluble products (by capillary gas chromatography–mass spectrometry) and find that the tars represent only a small fraction (<20%) of the total sonochemistry observed. The principal products from the sonolysis of benzene, for example, are H_2, CH_4, $C_6H_5CH_3$, and $(C_6H_5)_2$; from alkyl-substituted benzene (toluene, o-xylene, and mesitylene) the product distribution is similar, with additional products from methyl group scramblings being formed in high yield.

The sonochemistry of halocarbons has been thoroughly examined; in general, rates of sonolysis of halocarbons are comparable to those of water. Spin trapping experiments by Rosenthal et al. on CCl_4 and CCl_3Br demonstrated the formation of $Cl^{\bullet}$ upon sonication, which leads to the Cl_2 finally observed. Suslick and Schubert reported[166] the use of $X^{\bullet}$ produced on the irradiation of various

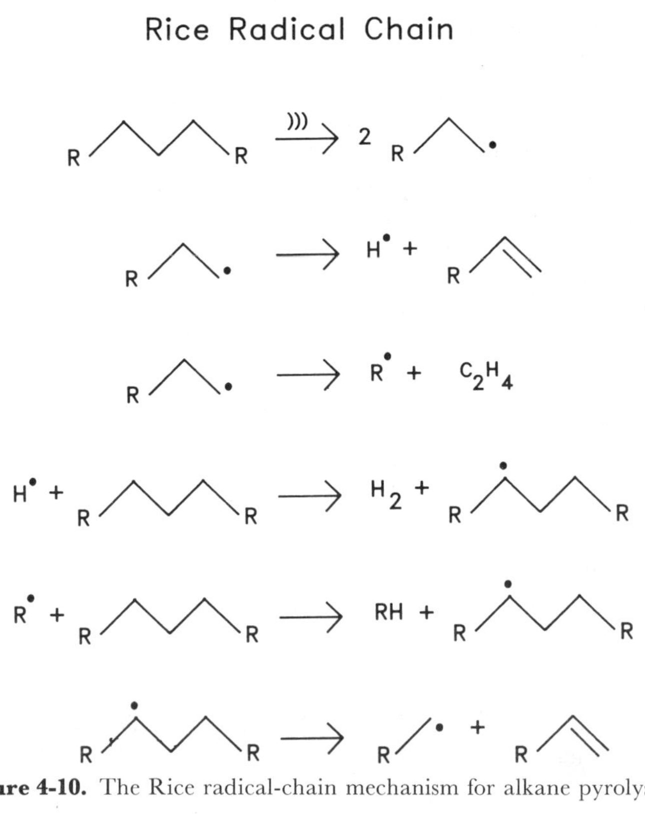

Figure 4-10. The Rice radical-chain mechanism for alkane pyrolysis.[39]

halocarbons for the halogenation of $M_2(CO)_{10}$ (M = Mn, Re), as discussed latter in detail. Henglein and Fischer have recently examined[167] the full product distribution of chloroform and find a large number of halocarbon products (notably HCl, and in decreasing amounts CCl_4, CH_2Cl_2, C_2Cl_5H, C_2Cl_4, C_2Cl_6, and C_2HCl_3), arising from both radical and carbene intermediates:

$$CHCl_3 \xrightarrow{)))} HCl + CCl_2 \tag{4-22}$$

$$CHCl_3 \xrightarrow{)))} Cl_2 + CHCl \tag{4-23}$$

$$CHCl_3 \xrightarrow{)))} H^{\bullet} + CCl_3 \tag{4-24}$$

$$CHCl_3 \xrightarrow{)))} Cl^{\bullet} + CHCl_2 \tag{4-25}$$

From the products observed, one may conclude that the carbene pathways predominate. In the presence of cyclohexene, the rate of sonolysis of chloroform increases because of trapping of intermediates that otherwise recombine.

Other recent reports compare the sonolysis and pyrolysis of biacetyl (which gives primarily acetone)[168] and the sonolysis and radiolysis of menthone.[169] Spin trapping of aromatic diazonium salts ($C_6H_5N_2^+ X^-$ where X = BF_4, PF_6, SCN, etc.) during sonication indicates the intermediacy of phenyl radicals and $X^{\bullet}$.[170] The use of ultrasound under an O_2 atmosphere to oxidize aldehydes to the carboxylic acids and (principally) peracids has also been reported, with a rate increase a few times over nonirradiated controls.[171,172]

By the proper choice of solvent and experimental conditions (e.g., low volatility, highly stable liquids at low temperature) the rates of degradation of nonaqueous liquids can be made quite slow. This is of considerable advantage, because one may then observe the primary sonochemistry of dissolved (and usually volatile) substrates rather than secondary reactions with solvent fragments. In general, the examination of sonochemical reactions in aqueous solutions has produced results difficult to interpret because of the complexity of the secondary reactions that so readily occur. In order to avoid such difficulties, one may hope to see the increased use of low-volatility organic liquids in future sonochemical studies.

Polymer Sonochemistry

The effects of ultrasound on polymers in solution and on the process of polymerization have been studied for quite some time. The first reports by Szalay,[173] Flosdorf and Chambers,[174] and Szent-Gyorgyi[175] in 1933 and by Schmid and Rommel[176,177] in 1939 showed that ultrasonic irradiation depolymerized starch, gum arabic, gelatin, and polystyrene. This work was followed by detailed examinations of sonochemical polymer degradation by many workers. The first use of ultrasound to initiate polymerization was made by Demann and Asbach.[178] In this chapter, only the most recent work is discussed since thorough reviews of the early studies have been published.[179,180]

The rate of ultrasonic depolymerization decreases with decreasing molecular weight of the polymer; there is a limiting molecular weight below which no further cleavage occurs. The rate of degradation also depends on the duration of irradiation, concentration in solution, nature of polymer and solvent, and the intensity of the ultrasound, in accord with the effect of experimental parameters on cavitation, as discussed earlier. One notable exception, however, is the fact that polymer degradation is independent of the ambient (inert) gas, which indicates that the cleavage is not occurring inside the collapsing bubble by thermal processes, since polyatomic gases greatly reduce the cavitational temperature. Ultrasonic degradation of polymers in solution creates a nonrandom scission, with a well-defined preference for rupture near the middle of the chain.

The transient existence of macromolecular radicals formed during the ultrasonic degradation of polymers is well established. Henglein early reported the use of radical traps (both stable free radicals and radiolabeled I_2) to detect free-radical fragments produced in the ultrasonic decomposition of poly(methyl methacrylate); in the absence of O_2, two atoms of iodine are incorporated into the polymer per C—C bond broken.[181,182] More recent spin trapping studies have confirmed the generality of these observations.[183,184]

It is generally accepted that ultrasonic depolymerization is caused by hydrodynamic forces, which are created either by the cavitation even itself or by shock waves so generated and can cause intense accelerations ($\approx 10^5$g at 500 kHz).[179,180] Secondary reactions with high-energy species produced from solvent sonolysis may also contribute to polymer degradation,[185] but not substantially to chain cleavage under most conditions. No fundamental differences in mechanism are observed in the cleavage of synthetic polymers in organic liquids compared to naturally occurring polymers or proteins in aqueous media, although the former have been examined in more detail.

The exact nature of the processes that result in bond cleavage and depolymerization have been a matter of some dispute, with various models being proposed.[186–190] It is evident that the hydrodynamic shear forces created by acoustic microstreaming are sufficient to cleave very large macromolecules (with molecular weights $> 10^7$). For moderate size macromolecules, however, the principal source of the mechanical cleavage is the frictional force generated by the relative movement of the solvent and polymer molecules.

Numerous studies have been reported on the use of ultrasound to initiate radical polymerization, often of aqueous emulsions of various monomers[191–195] where the production of radicals from the sonolysis of water is well established. Much less work had been done on polymerization in nonaqueous media. In recent years, Kruus and co-workers have reported the formation of darkly colored, but uncharacterized, polymeric tars upon the ultrasonic irradiation of various substituted benzenes.[50,163,164] As discussed earlier, Suslick and Doktycz[165] have examined the products formed upon irradiation of some of these aromatics and have found that polymer is actually only a small part of the total reaction product. The nature of the polymeric produc.s has not yet been clarified, and this may prove difficult because of its inhomogeneity, insolubility, and change in properties upon aging.

In more recent studies, Kruus has reexamined in some detail the use of ultrasound to initiate polymerization of styrene and of methyl methacrylate in organic solvents;[196,197] relatively low molecular weight polymers are produced. The kinetics of polymerization as a function of experimental conditions are consistent with initiation arising from the localized hot spot produced by cavitational collapse. Other workers[198] have recently used ultrasound to initiate copolymerization of styrene and methyl methacrylate in the presence of tris(acetylacetonato) aluminum(III). A quite general scheme has been developed to use the macromolecular radicals formed during ultrasonic depolymerization as a means of forming block copolymers, either by initiating chain polymerization[199–202] or by cross-radical recombination from mixtures of different homopolymers.[203,204]

Organometallic Sonochemistry

The effects of high-intensity ultrasound on organometallic systems is an area of only recent investigation; consequently, a limited range of complexes and reactions have been examined. Still, a variety of novel reactivity patterns are beginning to emerge that are distinct from either normal thermal or photochemical activation.[205] Most of the reactions that have been reported are stoichiometric in terms of the metal or metal complex, but a few examples of true sonocatalysis have also appeared.

In 1981, Suslick, Schubert, and Goodale made the first report on the sonochemistry of organometallic complexes and demonstrated the effects of ultrasound on the iron carbonyls in alkane solutions.[206] The transition-metal carbonyls were chosen for these initial studies because their thermal and photochemical reactivities have been well characterized. The comparison among the thermal, photochemical, and sonochemical reactions of $Fe(CO)_5$ provides an excellent example of the unique chemistry that homogeneous cavitation can induce. Because of the mechanistic insights this system has provided, for our present discussion we focus upon it as an archetype. Thermolysis of $Fe(CO)_5$, for example, gives pyrophoric, finely divided iron powder;[207] ultraviolet photolysis[208] yields $Fe_2(CO)_9$, via the intermediate $Fe(CO)_4$; multiphoton infrared photolysis in the gas phase[41,209] yields isolated Fe atoms. Multiple ligand dissociation, generating $Fe(CO)_3$, $Fe(CO)_2$, etc., is not available from ordinary thermal or photochemical processes but does occur in matrix-isolated and gas-phase laser photolyses.[210] These observations reflect the dual difficulties inherent in creating controlled multiple ligand dissociation: first, to deliver sufficient energy in a utilizable form and, second, to quench the highly energetic intermediates before complete ligand loss occurs.

During sonolysis in alkane solvents in the absence of alternate ligands, the unusual clusterification of $Fe(CO)_5$ to $Fe_3(CO)_{12}$ is observed, together with the formation of finely divided iron.[206,211] The rate of decomposition is clearly first order, and the log of the observed first-order rate coefficient is linear with the solvent vapor pressure. This is consistent with a simple dissociation process activated by the intense local heating generated by acoustic cavitation. As discussed earlier, the intensity of the cavitational collapse and the maximum temperature reached during such collapse decreases with increasing solvent vapor pressure.

Given this method for controlling the conditions generated during cavitation, one would also expect to see the ratio of products vary as a function of solvent vapor pressure. This proves to be the case: the ratio of products can be varied over a 100-fold range. The production of $Fe_3(CO)_{12}$ is strongly favored by increasing solvent volatility, as expected, since the sonochemical production of metallic iron requires greater activation energy than the production of $Fe_3(CO)_{12}$.

The proposed chemical mechanism by which $Fe_3(CO)_{12}$ is formed during the sonolysis of $Fe(CO)_5$ is:

$$Fe(CO)_5 \xrightarrow{)))} Fe(CO)_{5-n} + n\,CO \quad (n = 1\text{–}5) \tag{4-26}$$

$$Fe(CO)_3 + Fe(CO)_5 \longrightarrow Fe_2(CO)_8 \tag{4-27}$$

$$Fe_2(CO)_8 + Fe(CO)_5 \longrightarrow Fe_3(CO)_{12} + CO \tag{4-28}$$

$Fe_2(CO)_9$ is not generated during the synthesis of $Fe_3(CO)_{12}$, and sonolysis of $Fe_2(CO)_9$ yields only $Fe(CO)_5$ and finely divided iron. The production of $Fe_3(CO)_{12}$ arises from initial multiple dissociative loss of CO from $Fe(CO)_5$ during cavitation, followed by secondary reactions with excess $Fe(CO)_5$. Ligand trapping studies confirm the formation of $Fe(CO)_3$ but cannot rule out the dimerization of $Fe(CO)_4$ in the localized cavitation site. The reaction of the putative $Fe_2(CO)_8$ with $Fe(CO)_5$ may proceed through initial dissociation in analogy to the matrix isolation reactivity[210,212] of $Fe(C_4H_4)_2(CO)_4$ and of $Fe_2(CO)_9$.

In addition to clusterification, sonochemical ligand (L) substitution also occurs for $Fe(CO)_5$ and, in fact, for most metal carbonyls. This has proved useful as a mechanistic probe of the reactive species formed during cavitation. Sonication of $Fe(CO)_5$ in the presence of phosphines or phosphites produces $Fe(CO)_{5-n}L_n$, $n = 1$, 2, and 3. The ratio of these products is independent of length of sonication; the multiply substituted products increase with increasing initial [L]; $Fe(CO)_4L$ is not sonochemically converted to $Fe(CO)_3L_2$ on the time scale of its production from $Fe(CO)_5$. These observations are consistent with the same primary sonochemical event responsible for clusterification:

$$Fe(CO)_5 \xrightarrow{)))} Fe(CO)_{5-n} + n\,CO \quad (n = 1\text{–}5) \tag{4-29}$$

$$Fe(CO)_4 + L \longrightarrow Fe(CO)_4L \tag{4-30}$$

$$Fe(CO)_3 + L \longrightarrow Fe(CO)_3L \tag{4-31}$$

$$Fe(CO)_3 + CO \longrightarrow Fe(CO)_4 \tag{4-32}$$

$$Fe(CO)_3L + L \longrightarrow Fe(CO)_3L_2 \tag{4-33}$$

Table 4-7. Homogeneous Organometallic Sonochemistry

Reactants	Products	Ref.
Clusterification		
$Fe(CO)_5$	$Fe_3(CO)_{12}$, Fe	206, 211
$Fe_2(CO)_9$	Fe, $Fe(CO)_5$	206, 211
Ligand substitution[a]		
$Cr(CO)_6$ + L	$Cr(CO)_5L$, $Cr(CO)_4L_2$, $Cr(CO)_3L_3$	42, 43, 211
$Mo(CO)_6$ + L	$Mo(CO)_5L$, $Mo(CO)_4L_2$	42, 43, 211
$W(CO)_6$ + L	$W(CO)_5L$, $W(CO)_4L_2$	42, 43, 211
$Fe(CO)_5$ + L	$Fe(CO)_4L$, $Fe(CO)_3L_2$, $Fe(CO)_2L_3$	42, 43, 211
$FeCp(CO)_2I$ + L	FeCp(CO)(L)I	215
$Fe_3(CO)_{12}$ + L	$Fe(CO)_4L$, $Fe(CO)_3L_2$	206, 211
$Mn_2(CO)_{10}$ + L	$Mn_2(CO)_8L_2$	37, 166
$Co(Cp)_2$ + CO	$CoCp(CO)_2$	217
Sn_2R_6	$[SnR_3]^\bullet$, $R^\bullet$	219
Secondary reactions		
$M_2(CO)_{10} + R_3CX$	$M(CO)_5X$ (M = Mn, Re; X = Cl, Br)	166
$Co_2(CO)_8$ + alkane	$Co_2(CO)_6(C_2H_2)$, $Co_4(CO)_{10}(C_2H_2)$	215
Sonocatalytic reactions		
$Fe_x(CO)_y$ + 1-alkene	*cis-,trans*-2-alkene	206, 211
$Ru_x(CO)_y$ + 1-alkene	*cis-,trans*-2-alkene	206, 211
$Mo(CO)_6$ + 1-alkene	*cis-,trans*-2-alkene	206, 211
$Co_2(CO)_8$ + 1-alkene	*cis-,trans*-2-alkene	206, 211
$Mo(CO)_6$ + alkene + O_2	1:1 enol + epoxide	225

[a] L = various phosphines and phosphites.

Sonochemical ligand substitution readily occurs with a variety of other metal carbonyls, as shown in Table 4-7. In all cases, multiple ligand substitution originates directly from the parent carbonyl. The rates of sonochemical ligand substitution of the various metal carbonyls follow their relative volatilities, as predicted from the nature of the cavitational collapse.

The sonolysis of $Mn_2(CO)_{10}$ makes for an interesting comparison,[166] since either metal–metal (as in photolysis)[213] or metal–carbon (as in moderate-temperature thermolysis)[214] bond breakage can occur. Ligand substitution can occur from either route, producing the axially disubstituted $Mn_2(CO)_8L_2$. Using benzyl chloride as a trap for the intermediacy of $[Mn(CO)_5]^\bullet$, the sonochemical substitution of $Mn_2(CO)_{10}$ has been shown to follow the thermal, rather than the photochemical, pathway of dissociative CO loss.

Upon sonication in halocarbon solvents, metal carbonyls undergo facile halogenations.[166] The rates of halogenation are solvent dependent, but independent of choice of metal carbonyl or its concentration, and represent the products of secondary reactions occurring from the sonolytic decomposition of the halocarbon solvent:

$$R_3CX \xrightarrow{)))} [R_3C]^{\bullet} + X^{\bullet} \quad (4\text{-}34)$$

$$2[R_3C]^{\bullet} \longrightarrow R_3CCR_3 \quad (4\text{-}35)$$

$$2\,X^{\bullet} \longrightarrow X_2 \quad (4\text{-}36)$$

$$M_2(CO)_{10} + 2\,X^{\bullet} \longrightarrow 2\,M(CO)_5X \quad (4\text{-}37)$$

$$M_2(CO)_{10} + X_2 \longrightarrow 2\,M(CO)_5X \quad (4\text{-}38)$$

Alkanes and other halogen atom traps suppress the halogenation of the metal carbonyls.

Another example of a secondary sonochemical reaction is the very slow production of acetylene complexes of cobalt carbonyls upon lengthy sonolysis of $Co_2(CO)_8$ in *n*-alkanes (C_5H_{12} through $C_{10}H_{22}$).[215] The principal products are $Co_2(CO)_6(H_2C_2)$ and $Co_4(CO)_{10}(H_2C_2)$ and small amounts of $Co_4(CO)_{12}$. The $Co_4(CO)_{12}$ is an expected product, since it is easily formed upon pyrolysis of $Co_2(CO)_8$. The acetylene of the former complexes originates from the solvent, as confirmed by isotope labeling. Their formation is initially quite surprising, until one notes that their rates of formation are comparable to the slow rate of H_2C_2 formation from sonolysis of the alkane,[39] and that cobalt carbonyls undergo facile thermal reactions with alkynes.[216] Therefore, the origin of this sonochemical alkane activation is not in some high-energy organometallic fragment, but in the secondary trapping of acetylene sonochemically produced from the alkane.

The sonochemistry of non-carbonyl organometallics has not yet been well developed. Complexes that contain both CO and Cp undergo CO substitution upon sonolysis.[215] In preliminary studies of the metallocenes,[217] $Co(Cp)_2$ has been found to undergo facile ligand substitution during sonication under CO to yield $Co(Cp)(CO)_2$. This reaction under simple thermal conditions[218] requires 200 atm CO at 90–150°C and gives low yields; with ultrasonic irradiation, excellent yields are obtained at 3 atm and 20°. These results are in keeping with the high-temperature/high-pressure conditions generated during acoustic cavitation and suggest a useful analogy between sonochemistry and bomb reactions.

A recent report describes the sonochemical decomposition of organotin compounds.[219] Trapping of intermediate radicals by nitrosodurene and analysis by ESR demonstrated alkyl–tin bond cleavage during sonication in benzene solutions. The following reaction scheme was suggested to explain the observed ESR spectra:

$$R_4Sn \xrightarrow{)))} R^{\bullet} + [R_3Sn]^{\bullet} \quad (4\text{-}39)$$

$$R^{\bullet} + ONC_6H(CH_3)_4 \longrightarrow {}^{\bullet}\{ON(R)[C_6H(CH_3)_4]\} \quad (4\text{-}40)$$

$$[R_3Sn]^{\bullet} + C_6H_6 \longrightarrow [R_3SnC_6H_6]^{\bullet} \quad (4\text{-}41)$$

$$[R_3SnC_6H_6]^{\bullet} + ONC_6H(CH_3)_4 + \text{ox.} \longrightarrow {}^{\bullet}\{ON(C_6H_4SnR_3)[C_6H(CH_3)_4]\} + \text{red.} \quad (4\text{-}42)$$

The yields, rates, source of oxidant, or final products in the absence of spin traps were not determined.

Sonocatalysis

Having demonstrated that ultrasound can induce ligand dissociation, the initiation of homogeneous catalysis by ultrasound becomes practical. The potential advantages of such sonocatalysis include (1) the use of low ambient temperatures to preserve thermally sensitive substrates and to enhance selectivity, (2) the ability to generate high-energy species unobtainable from photolysis or simple pyrolysis, (3) the mimicry, on a microscopic scale, of bomb reaction conditions, and (4) possible ease of scaleup. The transient, coordinatively unsaturated species produced from the sonolysis of metal carbonyls are likely candidates, since similar species produced photochemically are among the most active catalysts known.[208,220] The thermal[221,222] and photochemical[208] isomerization of terminal olefins by metal carbonyls have been extensively studied and provide a useful test case for applications of sonocatalysis.

As shown in Fig. 4-11, a variety of metal carbonyls, upon sonication, catalyze the isomerization of 1-pentene to *cis*- and trans-2-pentene.[211] Initial turnover rates are $\approx 1-100$ mol 1-pentene isomerized per mole of precatalyst per hour and represent rate enhancements of $\approx 10^5$ over thermal controls. The relative sonocatalytic and photocatalytic activities of these carbonyls are in general accord. An exception is $Ru_3(CO)_{12}$, which is relatively more active as a sonocatalyst and shows very different trans/cis ratios under sonolysis; it appears that the catalytic agent produced sonochemically in this case is not that produced photochemically. A variety of terminal alkenes serve as substrates for sonocatalytic isomerization, although increasing steric hindrance, as in 2-ethylpent-1-ene and allylbenzene, significantly diminishes the observed rates. Alkenes without β-hydrogens do not serve as substrates, in contrast to the radical-chain isomerization of maleic acid, which occurs upon ultrasonic irradiation of aqueous Br_2 solutions.[118,119]

The exact nature of the catalytic species generated during sonolysis remains unknown. Results are consistent with the generally accepted mechanism for alkene isomerization in analogous thermal[223] and photochemical[224] systems. This involves the formation of a hydrido-π-allyl intermediate and alkene rearrangement via hydride migration to form the thermodynamically more stable 2-alkene complex, as shown in a general sense in Eq. (4-43) to (4-47):

$$M(CO)_n \xrightarrow{)))} M(CO)_m + (n-m)CO \tag{4-43}$$

$$M(CO)_m + \text{1-alkene} \longrightarrow M(CO)_x(\text{1-alkene}) + (m-x)CO \tag{4-44}$$

$$M(CO)_x(\text{1-alkene}) \longrightarrow M(CO)_x(H)(\pi\text{-allyl}) \tag{4-45}$$

$$M(CO)_x(H)(\pi\text{-allyl}) \longrightarrow M(CO)_x(\text{2-alkene}) \tag{4-46}$$

$$M(CO)_x(\text{2-alkene}) + \text{1-alkene} \longrightarrow M(CO)_x(\text{1-alkene}) + \text{2-alkene} \tag{4-47}$$

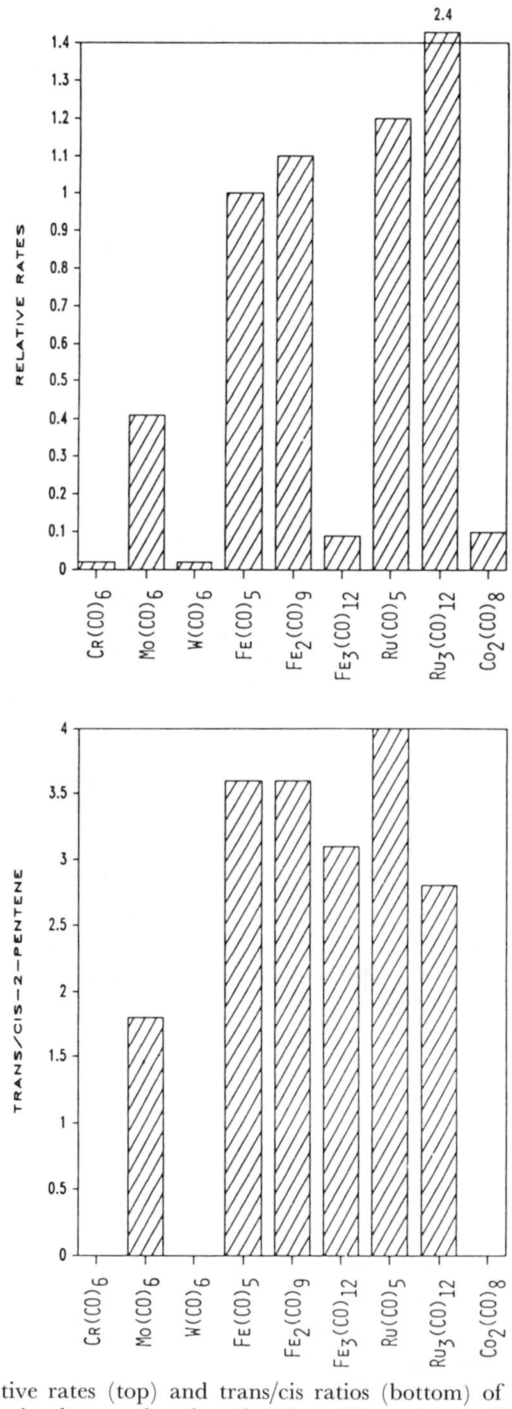

Figure 4-11. Relative rates (top) and trans/cis ratios (bottom) of the sonocatalysis of 1-pentene isomerization by metal carbonyls (plotted from data in Ref. 211).

In keeping with this scheme, sonication of $Fe(CO)_5$ in the presence of 1-pentene and CO does produce $Fe(CO)_4$(pentene), as determined by Fourier Transform infrared spectral stripping.[211]

Recently, Suslick and Hammerton[205] have demonstrated another sort of homogeneous sonocatalysis in the sonochemical oxidation of alkenes by O_2. Upon sonication of alkenes under O_2 in the presence of $Mo(CO)_6$, 1-enols and epoxides are formed in 1 : 1 ratios. Radical trapping and kinetic studies suggest a mechanism involving initial allylic C—H bond cleavage (caused by the cavitational collapse), and subsequent well-known autoxidation and epoxidation steps. The following scheme is consistent with their observations:

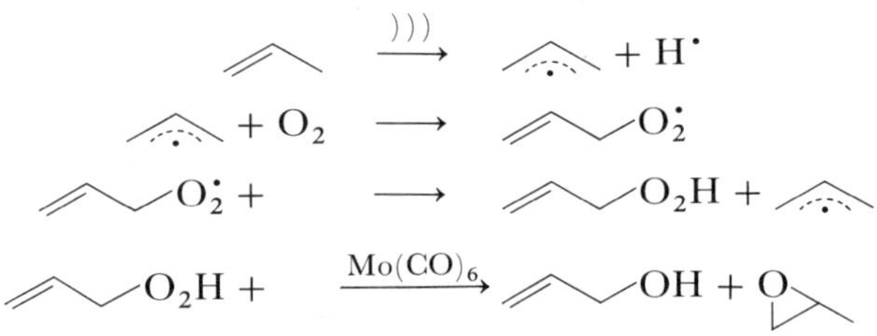

(4-48)
(4-49)
(4-50)
(4-51)

In the case of alkene isomerization, the catalyst was sonochemically activated. In the case of alkene oxidation, however, it is the substrate that is activated.

CONCLUDING REMARKS

The use of ultrasound in both homogeneous and heterogeneous reactions will see increasing study. The potential to do high-energy chemistry in condensed phases at room temperature remains an attractive feature of sonochemistry. Unique examples of sonochemical reactivity quite different from thermal or photochemical processes have been noted. There are analogies to shock wave and gas-phase pyrolyses, to "bomb" reactions, and to metal vapor chemistry, which will continue to be explored.[205]

A primary limitation of homogeneous sonochemistry remains its energy inefficiency. This may be dramatically improved, however, if a more effective means of coupling the sound field with preformed cavities can be found. The question of selectivity in and control of sonochemical reactions, as with any thermal process, remains a legitimate concern. There are, however, clearly defined means of controlling the conditions generated during cavitational collapse, which permit the variation of product distributions in a rational fashion.

The study of sonochemistry is still in its early stages. Acoustic cavitation represents a fundamental and unique means of interacting energy and matter. The combination of extraordinarily high temperatures with exceptionally short residence times provides a regime of activation unobtainable in nearly any other fashion in condensed phases. The development of sonochemistry over the next decade should have major intellectual and technological impact on a wide range of fields.

ACKNOWLEDGMENTS

The author sincerely appreciates the irreplaceable efforts of his graduate students and postdoctoral research associates. The generosity of funding from the National Science Foundation, the Research Corporation, and the donors of the Petroleum Research Fund, administered by the American Chemical Society, is gratefully acknowledged. The author greatly appreciates the receipt of a Sloan Foundation Research Fellowship and of a Research Career Development Award from the National Institute of Health.

REFERENCES

1. Richards, W. T.; Loomis, A. L. *J. Am. Chem. Soc.* **1927**, *49*, 3086.
2. Wood, R. W.; Loomis, A. L. *Phil. Mag., Ser. 7* **1927**, *4*, 417.
3. Boudjouk, P. *J. Chem. Ed.* **1986**, *63*, 427.
4. Suslick, K. S. *Adv. Organomet. Chem.* **1986**, *25*, 73.
5. Suslick, K. S. *Modern Synth. Meth.* **1986**, *4*, 1.
6. Crum, L. A. *IEEE Ultrasonics Symp.* **1982**, 1.
7. Epstein, P. S.; Plesset, M. S. *J. Chem. Phys.* **1950**, *18*, 1505.
8. Crum, L. A. *Nature (London)* **1979**, *278*, 148.
9. Winterton, R. H. S. *J. Phys. D: Appl. Phys.* **1977**, *10*, 2041.
10. Apfel, R. E. In "Methods of Experimental Physics: Ultrasound"; Edmonds, P. D., Ed.; Academic Press: New York, 1964, Vol. 19, pp. 356–411.
11. Crum, L. A.; Nordling, D. A. *J. Acoust. Soc. Am.* **1972**, *52*, 294.
12. Flynn, H. G. In "Physical Acoustics"; Mason, W. P., Ed.; Academic Press: New York, 1964, Vol. IB, p. 57.
13. Lauterborn, W. *Appl. Sci. Res.* **1982**, *38*, 165; In "Finite-Amplitude Wave Effects in Fluids, Proc. 1973 Symp."; Bjorno, L., Ed.; I.P.C. Science and Technology Press: Guildford, Surrey, UK, 1974, pp. 195–202.
14. Weissler, A.; Hine, E. J. *J. Acoust. Soc. Am.* **1962**, *34*, 130.
15. Pugin, B. *Ultrasonics* **1987**, *25*, 49.
16. Aerstin, F. G. P.; Timmerhaus, K. D.; Fogler, H. S. *American Institute of Chemical Engineers J.* **1967**, *13*, 453.
17. Dmitrieva, A. F.; Margulis, M. A. *Zh. Fiz. Khim.* **1986**, *60*, 1751.
18. Suslick, K. S.; Schubert, P. F.; Goodale, J. W. *IEEE Ultrason. Symp. Proc.*, **1981**, 612.
19. Among others: Heat Systems-Ultrasonics, 1938 New Highway, Farmingdale, NY 11735; Branson Cleaning Equipment, Parrott Dr., P.O. Box 768, Shelton, CT 06484; Branson Sonic Power, Eagle Rd., Danbury, CT 06810; Sonics & Materials, Kenosia Av., Danbury, CT 06810; Lewis Corp., 324 Christian St., Oxford, CT 06483.
20. Kurochkin, A. K.; Smorodov, E. A.; Badikov, Yu. V.; Margulis, M. A. *Zh. Fiz. Khim.*, **1986**, *60*, 893.
21. Valitov, R. B.; Kurochkin, A. K.; Margulis, M. A.; Badikov, Yu. V.; Galiakhmetov, R. N. *Zh. Fiz. Khim.*, **1986**, *60*, 889.
22. Griffing, V. *J. Chem. Phys.* **1950**, *18*, 997.
23. Griffing, V. *J. Chem. Phys.* **1952**, *20*, 939.

24. Harvey, E. N. *J. Am. Chem. Soc.* **1939**, *61*, 2392.
25. Frenkel, Y. I. *Zh. Fiz. Khim.* **1940**, *12*, 305.
26. Degrois, M.; Baldo, P. *Acustica* **1969**, *21*, 222.
27. Degrois, M.; Baldo, P. *Ultrasonics* **1974**, *12*, 25.
28. Sehgal, C. M.; Verrall, R. E. *Ultrasonics* **1982**, *20*, 37.
29. Margulis, M. A. *Zh. Fiz. Khim.* **1981**, *55*, 154.
30. Margulis, M. A. *Zh. Fiz. Khim.* **1985**, *59*, 1497.
31. Lord Raleigh, *Phil, Mag. Ser. 6* **1917**, *34*, 94.
32. Noltingk, B. E.; Neppiras, E. A. *Proc. Phys. Soc. London* **1950**, *B63*, 674; **1951**, *B64*, 1032.
33. Margulis, M. A.; Dmitrieva, A. F. *Zh. Fiz. Khim.* **1981**, *55*, 159.
34. Margulis, M. A.; Dmitrieva, A. F. *Zh. Fiz. Khim.* **1982**, *56*, 323, 875.
35. Fujikawa, S; Akamatsu, T. *J. Fluid Mech.* **1980**, *97*, 481.
36. Margulis, M. A. *Zh. Fiz. Khim.*, **1976**, *50*, 897.
37. Suslick, K. S.; Hammerton, D. A.; *IEEE Trans. Ultrason. Ferroelec., Freq. Cont.* **1986**, *33*, 143.
38. Suslick, K. S.; Gawienowski, J. W.; Schubert, P. F.; Wang, H. H.; *Ultrasonics* **1984**, *22*, 33.
39. Suslick, K. S.; Gawienowski, J. W.; Schubert, P. F.; Wang, H. H.; *J. Phys. Chem.* **1983**, *87*, 2299.
40. Tsang, W. In "Shock Waves in Chemistry"; Lifshitz, A., Ed.; Marcel Dekker: New York, 1981, p. 59–130.
41. Lewis, K. E.; Golden, D. M.; Smith, G. P. *J. Am. Chem. Soc.* **1984**, *106*, 3905.
42. Suslick, K. S.; Cline, Jr., R. E.; Hammerton, D. A. *J. Am. Chem. Soc.* **1986**, *108*, 5641.
43. Suslick, K. S.; Hammerton, D. A.; *IEEE Ultrason. Symp. Proc.* **1985**, *4*, 1116.
44. Margulis, M.; Grundel, L. M. *Zh. Fiz. Khim.* **1982**, *56*, 1445, 1941, 2592, 2987.
45. Margulis M. A. *Ultrasonics*, **1985**, *23*, 157.
46. Renaud, P. *J. Chim. Phys.* **1953**, *50*, 135.
47. Busnel, R.; Picard, D.; Bouzigues, H. *J. Chim. Phys.* **1953**, *50*, 97; Prudhomme, R.; Picard, D.; Busnel, R. *J. Chim. Phys.* **1953**, *50*, 107.
48. Sata, N.; Nakasima, K. *Bull. Chem. Soc. Jpn.* **1943**, *18*, 220.
49. Akulichev, V. A.; Sirotyuk, M. G.; Rozenberg, L. D. In "High Intensity Ultrasonic Fields"; Rozenberg, L. D., Ed.; Plenum Press: New York, 1971, pp. 203–420.
50. Kruus, P. *Ultrasonics* **1987**, *25*, 20.
51. Ibishi, M; Brown, B. *J. Acoust. Soc. Am.* **1967**, *41*, 568.
52. Weissler, A. *J. Acoust. Soc. Am.* **1953**, *25*, 651.
53. Bronskaya, L. M.; Vigderman, V. S.; Sokol'skaya, A. V.; El'Piner, I. E. *Sov. Phys. Acoust.* **1968**, *13*, 374.
54. Fitzgerald, M. E.; Griffing, V.; Sullivan, J. *J. Chem. Phys.* **1956**, *25*, 926.
55. Prudhomme, R. O. *Bull. Soc. Chim. Biol.* **1957**, *39*, 425.
56. Mead, E. L.; Sutherland, R. G.; Verrall, R. E. *Can. J. Chem.* **1975**, *54*, 1114.
57. Ceschia, M.; Iernetti, G. *Acustica* **1973**, *29*, 127.
58. El'piner, I. E. "Ultrasound: Physical, Chemical, and Biological Effects"; Sinclair, F. A., trans.; Consultants Bureau: New York, 1964.
59. Margulis, M. A. *Zh. Fiz. Khim.* **1976**, *50*, 1.
60. Schmitt, F. O.; Johnson, C. H.; Olson, R. A. *J. Am. Chem. Soc.* **1929**, *51*, 370.
61. Bresler, S. E. *Zh. Fiz. Khim.* **1940**, *14*, 309.
62. Prudhomme, R. O.; Grabar, P. *J. Chim. Phys.* **1949**, *46*, 323.
63. Lindstrom, O. *J. Acoust. Soc. Am.* **1955**, *27*, 654.
64. Henglein, A. *Naturwissenschafton* **1956**, *43*, 277; **1957**, *44*, 179.
65. Del Duca, M.; Yeager, E.; Davies, M. O.; Hovorka, F. *J. Acoust. Soc. Am.* **1958**, *30*, 301.
66. Gueguen, H. *C. R. Hebd. Seances Acad. Sci (Paris)* **1961**, *253*, 260, 647.
67. Margulis, M. A.; Mal'tsev, A. N. *Zh. Fiz. Khim.* **1968**, *42*, 1441.
68. Weissler, A.; Cooper, H. W.; Snyder, S. *J. Am. Chem. Soc.* **1950**, *72*, 1769.
69. Weissler, A. *J. Am. Chem. Soc.* **1959**, *81*, 1077.
70. Beuthe, H. *Z. Phys. Chem. A* **1933**, *163A*, 161.
71. Schultes, H.; Gohr, H. *Angew. Chem.*, **1936**, *49*, 420.
72. Virtanen, A. I.; Ellsfolk, N. *J. Am. Chem. Soc.* **1950**, *72*, 1046.
73. Frenzel, H.; Schultes, H. *Z. Phys. Chem. B* **1934**, *27*, 421.
74. Porter, C. W.; Young, L. *J. Am. Chem. Soc.* **1938**, *60*, 1497; **1941**, *63*, 3434.
75. Miyagawa, I. *J. Soc. Org. Syn. Chem.* **1942**, *7*, 167.
76. Riesz, P.; Berdahl, D.; Christman, C. L. *Environ. Health Perspect.* **1985**, *64*, 233.
77. Margulis, M. A.; Mal'tsev, A. N. *Zh. Fiz. Khim.* **1968**, *42*, 2660.
78. Margulis, M. A.; Mal'tsev, A. N. *Zh. Fiz. Khim.* **1972**, *46*, 2970.

79. Lippitt, B.; McCord, J. M.; Fridovich, I. *J. Biol. Chem.* **1972**, *247*, 4688.
80. Makino, K.; Mossoba, M.; Riesz, P. *J. Am. Chem. Soc.* **1982**, *104*, 3537.
81. Makino, K.; Mossoba, M.; Riesz, P. *J. Phys. Chem.* **1983**, *87*, 1369.
82. Makino, K.; Mossoba, M.; Riesz, P. *Radiat. Res.* **1983**, *96*, 416.
83. Carmichael, A. J.; Mossoba, M. M.; Riesz, P.; Christman, C. L. *IEEE Trans. Ultrason. Ferroelec. Freq. Control* **1986**, *UFFC-33*, 148.
84. Flynn, H. G. *J. Acoust. Soc. Am.* **1982**, *72*, 1926.
85. Christman, C. L.; Carmichael, A. J.; Mossoba, M. M.; Riesz, P. *Ultrasonics* **1987**, *25*, 31.
86. DelDuca, M.; Yeager, E.; Davies, M. O.; Hovorka, F. *J. Acoust. Soc. Am.* **1958**, *30*, 301.
87. Anbar, M.; Pecht, I. *J. Phys. Chem.* **1964**, *68*, 352.
88. Henglein, A. *Ultrasonics* **1987**, *25*, 6.
89. Fischer, C. H.; Hart, Edwin J.; Henglein, A. *J. Phys. Chem.*, **1986**, *90*, 1954.
90. Fischer, C. H.; Hart, Edwin J.; Henglein, A. *J. Phys. Chem.*, **1986**, *90*, 3059, 222.
91. Margulis, M. A. *Zh. Fiz. Khim.* , **1976**, *50*, 2536.
92. Margulis, M. A.; Didenko, Yu. T. *Zh. Fiz. Khim.*, **1984**, *58*, 1402.
93. Hart, E. J.; Henglein, A. *J. Phys. Chem.*, **1985**, *89*, 4342.
94. Hart, E. J.; Henglein, A. *J. Phys. Chem.*, **1986**, *90*, 3061.
95. Haissinsky, M.; Mangeot, A. *Nuovo Cimento* **1956**, *4*, 1086.
96. Sokol'skaya, A. V.; El'piner, I. E. *Akust. Zhur.* **1957**, *3*, 293.
97. Sokol'skaya, A. V. *J. Gen. Chem. USSR* **1978**, *48*, 1289.
98. Polotzki, I. G. *J. Gen. Chem. USSR (Engl. Transl.)* **1947**, *17*, 649.
99. Margulis, M. A.; Didenko, Yu. T.; Gorbanrenko, S. A. *Zh. Fiz. Khim.*, **1985**, *59*, 2026.
100. Hart, E. J.; Henglein, A. *J. Phys. Chem.*, **1986**, *90*, 5992.
101. Sokol'skaya, A. V. *J. Evol. Biochem. Physio.* **1975**, *11*, 446.
102. Solov'eva, L. R. *Zh. Fiz. Khim.* 1937, *9*, 77.
103. Margulis, M. A.; Didenko, Yu. T. *Zh. Fiz. Khim.*, **1984**, *58*, 1402; **1986**, *60*, 1490.
104. Margulis, M.; Didenko, Yu. T. *Zh. Fiz. Khim.*, **1980**, *54*, 1587, 1591.
105. Margulis, M. A. *Zh. Fiz. Khim.*, **1976**, *50*, 2267, 2271, 2531, 2536.
106. Weissler, A. *J. Acoust. Soc. Am.* **1960**, *32*, 283.
107. Sehgal, C. M.; Wang, S. Y. *IEEE Trans. Sonics Ultrason.* **1983**, *SU-30*, 374.
108. Sehgal, C.; Sutherland, R. G.; Verrall, R. E. *J. Phys. Chem.* **1980**, *84*, 2920.
109. Rehorek, D.; Janzen, E. G. *Z. Chem.* **1984**, *24*, 228.
110. Gueguen, H. *Ann. Chim.* **1963**, *8*, 667.
111. Margulis, M. A. *Zh. Fiz. Khim.* **1976**, *50*, 2271.
112. Margulis, M. A. *Zh. Fiz. Khim.* **1976**, *50*, 2267.
113. Mal'tsev, A. N.; Margulis, M. A. *Akust. Zh.* **1968**, *14*, 295.
114. Cauwet, G.; Coste, C. M.; Knoche, H.; Longuemard, J. P. *Bull. Soc. Chim. Fr.* **1976**, 45.
115. Margulis, M. A. *Zh. Fiz. Khim.* **1974**, *48*, 2812.
116. Wawrzyczek, W.; Tylzanowska, D. *Nature (London)* **1962**, *194*, 571.
117. Anbar, M.; Pecht, I. *J. Phys. Chem.* **1964**, *68*, 352; **1967**, *71*, 1246.
118. Elpiner, I. E.; Sokolskaya, A. V.; Margulis, M. A. *Nature (London)* **1965**, *208*, 945.
119. Starchesvskii, V. L.; Vasilina, T. V.; Grindel, L. M.; Margulis, M. A.; Mokryi, E. N. *Zh. Fiz. Khim.* **1984**, 58, 1940.
120. Valitov, R. B.; Galiakhmetov, R. N.; Kurochkin, A. K.; Margulis, M. A. *Zh. Fiz. Khim.*, **1985**, *59*, 2973.
121. Yu, T. J.; Sutherland, R. G.; Verrall, R. E. *Can. J. Chem.* **1980**, *58*, 1909.
122. Mead, E. L.; Sutherland, R. G.; Verrall, R. E. *Can. J. Chem.* **1975**, *53*, 2394.
123. Mead, E. L.; Sutherland, R. G.; Verrall, R. E. *J. Chem. Soc., Chem. Commun.* **1973**, *1973*, 414.
124. McKee, J. R.; Christman, C. L.; O'Brien, Jr., W. D.; Wang, S. Y. *Biochemistry*, **1977**, *16*, 4651.
125. Sehgal, C. M.; Wang, S. Y. *J. Am. Chem. Soc.* **1981**, *103*, 6606.
126. Wang, S. Y.; Gupta, A. B. *1977 IEEE Ultrason. Proc.* **1977**, CH1264.
127. Henglein, A.; Kormann, C. *Int. J. Radiat, Biol. Relat. Stud. Phys., Chem. Med.*, **1985**, *48*, 251.
128. Couppis, E. C.; Klinzing, G. E. *AIChE J.* **1974**, *20*, 485.
129. Lorimer, J. P.; Mason, T. J.; *J. Chem. Soc., Chem. Commun.* **1980**, *1980*, 1135.
130. Mason, T. J.; Lorimer, J. P.; Mistry, B. P. *Tetrahedron Lett.* **1982**, *23*, 5363.
131. Mason, T. J.; Lorimer, J. P.; Mistry, B. P. *Tetrahedron Lett.* **1983**, *24*, 4371.
132. Mason, T. J. *Lab. Pract.* **1984**, *33*, 13.
133. Lorimer, J. P.; Mason, T. J.; Mistry, B. P. *Ultrasonics* **1987**, *25*, 23.
134. Kling, A.; Kling, R. *Compt. Rend. Hebd Seances Acad. Sci. (Paris)* **1946**, *223*, 1131.
135. Jennings, B. H.; Townsend, S. N. *J. Phys. Chem.* **1961**, *65*, 1574.

136. Chendke, P. K.; Fogler, H. S. *J. Phys. Chem.* **1983**, *87*, 1362.
137. Tuszynski, T. M.; Graydon, W. F. *Indust. Engin. Chem. Fund.* **1968**, *7*, 396.
138. Sakai, Y.; Sadaoka, Y; Takamaru, Y. *J. Phys. Chem.* **1977**, *81*, 509.
139. Prasad, B.; Sharma, P. K. *Chim. Anal.* **1969**, *51*, 619.
140. Zechmeister, L.; Wallcave, L. *J. Am. Chem. Soc.* **1955**, *77*, 2853.
141. Zechmeister, L.; Magoon, E. F. *J. Am. Chem. Soc.* **1956**, *78*, 2149.
142. Prudhomme, R. O.; Grabar, P. *Bull. Soc. Chim. Biol.* **1958**, *29*, 122.
143. Spurlock, L. A.; Reifsneider, S. B. *J. Am. Chem. Soc.* **1970**, *92*, 6112.
144. Reifsneider, S. B.; Spurlock, L. A. *J. Am. Chem. Soc.* **1973**, *95*, 299.
145. Margulis, M. A. *Zh. Fiz. Khim.* **1976**, *50*, 2531.
146. Currell, D. L.; Zechmeister, L. *J. Am. Chem. Soc.* **1958**, *80*, 205.
147. Chen, J. W.; Chang, J. A.; Smith, G. V. *Chem. Eng. Prog. Symp. Ser.* **1966**, *67*(109), 18.
148. Weissler, A. *J. Am. Chem. Soc.* **1949**, *71*, 419.
149. Currell, D. L.; Nagy, S. S. *J. Acoust. Soc. Am.* **1968**, *44*, 1201.
150. Chen, J. W.; Kalback, W. M. *Indust. Engin. Chem. Fund.* **1967**, *6*, 175.
151. Kristol, D. S.; Klotz, H.; Parker, R. C. *Tetrahedron Lett.* **1981**, *22*, 907.
152. Needham, T. E.; Gerraughty, R. J. *J. Pharm. Sci.* **1969**, *58*, 62.
153. Moon, S.; Duchin, L.; Cooney, J. V. *Tetrahedron Lett.* **1979**, *41*, 3917.
154. Fayter, Jr., R. G.; Spurlock, L. A. *J. Acoust. Soc. Am.* **1974**, *56*, 1461.
155. Staas, W. H.; Spurlock, L. A. *J. Chem. Soc., Perkin Trans. I* **1975**, *1*, 1675.
156. Weissler, A.; Pecht, I.; Anbar, M. *Science* **1965**, *150*, 1288.
157. Gordeev, V. E.; Serbinov, A. I.; Troshin, Y. K. *Dokl. Akad. Nauk SSSR* **1967**, *172*, 383.
158. Porter, C. W.; Young, L. *J. Am. Chem. Soc.* **1938**, *60*, 1497.
159. Berkowitz, N.; Srivastava, S. C. *Can. J. Chem.* **1963**, *41*, 1787.
160. Prakash, S.; Pandey, J. D. *Tetrahedron* **1965**, *21*, 903.
161. Rosenthal, I.; Mossoba, M. M.; Riesz, P. *J. Mag. Res.* **1981**, *45*, 359.
162. Sehgal, C.; Yu, T. J.; Sutherland, R. G.; Verrall, R. E. *J. Phys. Chem.* **1982**, *86*, 2982.
163. Diedrich, G. K.; Kruus, P.; Rachlis, L. M. *Can. J. Chem.* **1972**, *50*, 1743.
164. Donaldson, D. J.; Farrignton, M. D.; Kruus, P. *J. Phys. Chem.* **1979**, *83*, 3130.
165. Suslick, K. S.; Doktycz, S. J. Unpublished work, 1987.
166. Suslick, K. S.; Schubert, P. F. *J. Am. Chem. Soc.* **1983**, *105*, 6042.
167. Henglein, A.; Fischer, C. H. *Ber. Bunsen-Ges. Phys. Chem.*, **1984**, *88*, 1196.
168. Soehnlen, F.; Perrin, D.; Masson, D.; Gaulard, M. L. *Ultrason. Intl. 1979*, **1979**, 603.
169. Kachakhidze, T. G. *Nauchn. Tr. Gruz. Politeth. Inst. Lenina* **1980**, *1980*, 45.
170. Rehorek, D.; Janzen, E. G. *J. Prakt. Chem.* **1984**, *326*, 935.
171. Starchevskii, V. L.; Mokryi, E. N.; Margulis, M. A. *Zh. Fiz. Khim.*, **1984**, *58*, 1397.
172. Vasilina, T. V.; Starchevskii, V. L.; Mokryi, E. N.; *Zh. Fiz. Khim.*, **1984**, *58*, 1926.
173. Szalay, A. *Z. Physik. Chem.* **1933**, *164A*, 231.
174. Flosdorf, E.; Chambers. L. A. *J. Am. Chem. Soc.* **1933**, *55*, 3057.
175. Szent-Gyorgyi, A. *Nature* (*London*) **1933**, *131*, 278.
176. Schmid, G.; Rommel, O. *Electrochim.* **1939**, *45*, 659.
177. Schmid, G.; Rommel, O. *Z. Physik. Chem.* **1939**, *A185*, 97.
178. Demann, W.; Asbach, H.R. *Tech. Mitt Krupp Forsch.* **1940**, *3*, 12.
179. Basedow, A. M.; Ebert, K. H. *Adv. Polym. Sci.* **1977**, *22*, 83.
180. Sheth, P. J.; Johnson, J. F. In "Polymer Stress Reactions"; Casale, A.; Porter, R. S., Eds.; Academic Press: New York, 1979, pp. 501–544.
181. Henglein, A. *Makromol. Chem.* **1955**, *15*, 188; **1956**, *18*, 37.
182. Henglein, A. *Z. Naturforsch.* **1955**, *10b*, 953.
183. Tabata, M.; Miyazawa, T.; Kobayashi, O.; Sohma, J. *Chem. Phys. Lett.* **1980**, *73*, 178.
184. Tabata, M.; Sohma, J. *Eu. Polym. J.* **1980**, *16*, 589.
185. Alexander, P.; Fox, M. *J. Polym. Sci.* **1954**, *12*, 533.
186. Schmid, G. *Phys. Z.* **1940**, *41*, 326.
187. Jellinek, H. H.; White, G. *J. Polym. Sci.* **1951**, *4*, 745; *6*, 757; *7*, 21, 33.
188. Mostafa, M. A. *J. Polym. Sci.* **1958**, *33*, 323.
189. Thomas, J. R.; De Vries, L. *J. Phys. Chem.* **1959**, *63*, 254.
190. Okuyama, M.; Hirose, T. *J. Appl. Polym. Sci.* **1963**, *7*, 591.
191. Lindstrom, O.; Lam, O. *J. Phys. Colloid Chem.* **1951**, *55*, 1139.
192. Berlin, A. A.; El'tsefon, B. S. *Khim. Nauka. Prom.* **1957**, *2*, 667.
193. Henglein, A.; Schulz, R. *Z. Naturforsch.* **1952**, *7B*, 484.

194. Fox, R.; Yeager, E.; Hovorka, F. *J. Acoust. Soc. Am.* **1960**, *32*, 1499.
195. Hatate, Y.; Keura, T. I.; et al. *J. Chem. Eng. Jpn.* **1981**, *14*, 38.
196. Kruus, P. *Ultrasonics* **1983**, *21*, 201.
197. Kruus, P.; Patraboy, T. J. *J. Phys. Chem.* **1985**, *89*, 3379.
198. Fujiwara, H.; Kunio, G. *Kobun. Ronbun.* **1982**, *39*, 481.
199. Fujiwara, H.; Goto, K. *Kobun. Ronbun.* **1984**, *41*, 133.
200. Fujiwara, H.; Kimura, K.; Mori, H.; Goto, K. *Polym. J.* (*Tokyo*) **1981**, *13*, 927.
201. Fujiwara, H.; Okazaki, K.; Goto, K. *J. Polym. Sci. Polym. Phys. Ed.* **1975**, *13*, 953.
202. O'Driscoll, K. F; Sridharah, A. U. *J. Polym. Sci. Polym. Chem. Ed.* **1984**, *11*, 1111.
203. Allen, P. E.; Downer, J. M.; Hastings, G. W.; Melvill, H. W.; Molyneux, P.; Urwin, J. R. *Nature* (*London*) **1956**, *177*, 910.
204. Malhotra, S. L.; Gauthier, J. M. *J. Macromol. Sci. Chem.* **1982**, *A18*, 783.
205. Suslick, K. S., Ed. "High Energy Processes in Organometallic Chemistry"; Symposium Series #333 Am. Chem. Soc.: Washington, D.C., 1987.
206. Suslick, K. S.; Schubert, P. F.; Goodale, J. W. *J. Am. Chem. Soc.* **1981**, *103*, 7324.
207. Carlton, H. E.; Oxley, J. H. *AIChE J.* **1965**, *11*, 79.
208. Geoffroy, G. L.; Wrighton, M. S. "Organometallic Photochemistry"; Academic Press: New York, 1979.
209. Langsam, Y.; Ronn, A. M. *Chem. Phys.* **1981**, *54*, 277.
210. Poliakoff, M.; Weitz, E. *Adv. Organomet. Chem.* **1986**, *25*, 277.
211. Suslick, K. S.; Goodale, J. W.; Wang, H. H.; Schubert, P. F. *J. Am. Chem. Soc.* **1983**, *105*, 5781.
212. Fischler, I.; Hildenbrand, K.; Koerner von Gustorf, E. *Angew. Chem. Int. Ed. Engl.* **1975**, *14*, 54.
213. Abrahamson, H. B.; Wrighton, M. S. *J. Am. Chem. Soc.* **1977**, *99*, 5510.
214. Coville, N.; Stolzenberg, A. M.; Muetterties, E. L. *J. Am. Chem. Soc.* **1983**, *105*, 2499.
215. Suslick, K. S.; Wang, H. H. Unpublished results, 1985.
216. Dickson, R. S.; Fraser, P. J. *Adv. Organomet. Chem.* **1974**, *12*, 323.
217. Suslick, K. S.; Johnson, R. E. *J. Am. Chem. Soc.* **1984**, *106*, 6856, and unpubished work.
218. Fischer, E. O.; Jira, R. *Z. Naturforsch.* **1955**, *10b*, 355.
219. Rehorek, D.; Janzen, E. G. *J. Organomet. Chem.* **1984**, *268*, L35.
220. Mitchener, J. C.; Wrighton, M. S. *J. Am. Chem. Soc.* **1981**, *103*, 975.
221. Parshall, G. W. "Homogeneous Catalysis"; Wiley-Interscience: New York, 1980.
222. Masters, C. "Homogeneous Transition-Metal Catalysis"; Chapman and Hall: New York, 1981.
223. Casey, C. P.; Cyr, C. R. *J. Am. Chem. Soc.* **1973**, *95*, 2248.
224. Graff, J. L.; Sanner, R. D.; Wrighton, M. S. *Organometallics* **1981**, *1*, 837.

5

Heterogeneous Sonochemistry

PHILIP BOUDJOUK

Department of Chemistry,
North Dakota State University

INTRODUCTION

The main focus of this chapter is the effect of ultrasonic waves on heterogeneous reactions with an emphasis on reactions with synthetic utility. That focus notwithstanding, it seems appropriate to preface the body of this chapter with a survey of the chemical effects of ultrasonic waves on heterogeneous systems and some practical applications.

Early Work in Sonochemistry

In 1927, Richards and Loomis published the first account of the chemical effects of ultrasonic waves.[1] In what was titled a "preliminary survey," the authors examined the effects of high-frequency sound waves on the physical and chemical properties of a variety of simple systems. Their observations and experimental techniques opened, for scientists and engineers of diverse interests, frontiers that are still expanding today.

Table 5-1. First Observations of the Chemical Effects of Ultrasonic Waves

System	Effect of ultrasonic waves
Nitrogen triodide	Rate of explosion accelerated
Superheated liquids	Rate of evaporation accelerated
Conversion of yellow HgI to red HgI	Rate of conversion accelerated
"Atomization" of liquids and solids	Rate of "atomization" of mercury and glass in water accelerated
Gases dissolved in water	Rate of expulsion of gases increased
Hydrolysis of methyl sulfate	Rate of hydrolysis increased
The "iodine clock" reaction	Rate of reduction of iodate by sulfite was increased
Boiling point of liquids	"Apparent" depression of boiling point.

Using quartz crystals of 6–12 mm thickness and 50–80 mm diameter held between two electrodes under oil, Loomis and his co-workers could produce high-intensity ultrasonic waves at frequencies in excess of 100,000 cps. The high intensities are the result of the ability to generate voltages in the range of 50,000 V with this configuration since the amplitude of vibration of quartz crystal increases directly with the voltage applied to it. With power levels of approximately 2 kW readily available, investigation of the effects of ultrasonic waves on chemical systems became feasible. Prior to this work the utility of ultrasonic waves was rather limited, perhaps the best known application being submarine signaling.

Using frequencies of approximately 280 kHz, Richards and Loomis examined the effects of high-intensity waves on the systems listed in Table 5-1. The impact of their paper was immediate and far ranging. Within a few years, papers from laboratories around the world appeared in the literature.

In the early 1930s, Freundlich and Söllner embarked on a series of investigations of the effects of ultrasonic waves on gels, discovering that thixotropic gels of iron and aluminum oxide could be liquefied and that the process is initiated at the liquid–air interface.[2] In a related study, Sata and Watanabe were able to produce permanently dispersed iron hydroxide from the reaction of iron chloride and ammonia in water in the absence of a peptization agent while irradiating the reaction mixture with a frequency of 450 kHz.[3]

During this period, Moriguchi conducted a number of studies in which he observed that the reactions of strong acids were enhanced by ultrasonic waves. For example, the consumption of zinc by hydrochloric acid was accelerated:[4]

$$Zn + 2\,HCl \xrightarrow{)))} ZnCl_2 + H_2$$

as was the reaction of calcium carbonate and sulfuric acid:

$$CaCO_3 + H_2SO_4 \xrightarrow{)))} CaSO_4 + H_2O + CO_2$$

and the reduction of copper sulfate by zinc to give colloidal copper:[5]

$$Zn + CuSO_4 \xrightarrow{)))} ZnSO_4 + Cu$$

He also observed that finely divided metal powders produced colloids when sonicated in water.[5]

Moriguchi was among the first to evaluate the effects of ultrasonic waves on electrochemical processes, finding that irradiation eliminated overvoltage in the electrolytic reduction of copper sulfate, presumably by removal of the diffusion barrier between the electrode and the electrolyte.[6] Using a cell in which the cathode and the electrolyte were vibrated with ultrasonic waves during the reduction of metal salts, Claus produced fine dispersions of iron, silver, and platinum.[7] However, conditions could be altered to produce high-quality films of electrodeposited metal. Mueller, for example, used ultrasonic waves during the reduction of nickel, copper, chromium, and brass and obtained bright films with finer grain texture and higher tensile strength than when they were deposited in the absence of ultrasonic irradiation.[8] By 1953 activity in this area progressed to the point where a detailed review was justified.[9]

Industrial Applications

In keeping with the intent of this chapter, only a brief review of the development of commercial uses of ultrasonic waves is given here. For the most part, chemical processes are highlighted, with an emphasis on heterogeneous systems and with potential for synthesis applications.

The earliest commercial uses of ultrasound, some of which were in place 30 years ago, were based mostly on the cleaning[10] and emulsifying actions of ultrasonic waves.[11] In this arena also, Woods and Loomis played a key role. In the same year they published their survey of the chemical effects of ultrasound,[1] they reported in a different journal that, "If two non-miscible liquids such as oil and water are simultaneously subjected to the radiation . . . , an emulsion or colloid is formed . . . ."[12] They also observed that ultrasonic waves have a coagulating effect on the organism *Paramecium*. This paper became widely cited and appears to have played a pivotal role in evoking interest in the industrial community, even though most of the research it inspired initially was limited to fundamental or "pure" aspects of ultrasonic waves. This limitation was primarily the result of the small size and thermal instability of the available crystals.

Those restraints have been progressively minimized over the years by the use of synthetic crystals and polycrystalline ceramics, such as lead zirconate–titanate,[13] as well as the development of different methods of producing ultrasonic waves, such as magnetostrictive devices and liquid and gas whistles. Today industrial applications enjoy great variety, spanning the food industry, where ultrasonic waves are used for emulsifying the ingredients in processed cheese, sauces, soups, and baby foods and for pasteurizing as well as homogenizing fruit juices; heavy manufacturing, where ultrasonic waves are used for drilling, soldering, fatigue and flaw detection, and the cleaning of complex machinery; and chemical processing, where commercial plants use ultrasound for coagulation, extraction, drying, defoaming, and crystallization.[14–18]

Cleaning and emulsification are the two most widely used applications of ultrasound. Both were developed on an industrial scale after World War II even though precedents for these techniques had been in the literature for nearly 20 years. Extended discussion of both processes is not warranted here. However, it is worth noting that for ultrasonic cleaning, low frequencies (13–90 kHz) are normally used. Important exceptions are when corrosion (by cavitation) of the part to be cleaned is a problem, and when the part has deep recesses. High frequency is required in both cases but for different reasons: in the former the frequency must be sufficiently high to prevent cavitation, and in the latter high-frequency ultrasonic waves are necessary because, unlike low-frequency waves, they can be focused to penetrate to the most hidden crevice.

These examples hold implications for the use of ultrasonic waves in synthesis: low-frequency ultrasound should be generally more useful in promoting reactions of metals because it induces cavitation, which continually cleans, and therefore activates, metal surfaces; and it may be possible to optimize frequency and the ability to focus sound beams to perform "site-specific" reactions, e.g., selective crosslinking on a surface.

Ultrasonic emulsification, or homogenization, was the subject of a patent[11] as early as 1944 but was not successfully implemented until the 1950s, when reliable, durable equipment became available to industry. Today, nearly every food-processing plant employs an ultrasonic homogenizer, usually in the form of a liquid whistle, at one or more stages of production, with a vast improvement in performance and economics over the earlier use of high-pressure mechanical stirring or shaking machines.[19] For example, liquid whistles can homogenize orange juice at the rate of 900 gallons an hour. Recently, high-intensity flow units have been used to pasteurize fruit juices at similar rates, thereby avoiding the negative effects of heat on the taste of the juices.[14,20]

Industries based on lubricants, pharmaceuticals, cosmetics, and textiles also rely heavily on ultrasonic technology, since oils can be mixed efficiently with water. In tests conducted by the U.S. Bureau of Mines, crude oil and brine were irradiated with low-frequency ultrasonic waves (20 kHz) to produce stable emulsions of 5% oil in brine. In the presence of 1% of an emulsifying agent, essentially 1:1 emulsions could be stabilized. Ultrasound induced greater and more uniform dispersion and, when used in combination with conventional recovery techniques, increased the amount of oil recovered by as much as 22%.[21] Ultrasonic waves also promote the distillation of immiscible liquids. Decker and Holz patented a process using ultrasound in lieu of emulsifying agents to distill water–chlorocarbon mixtures.[22] The improved mixing and emulsification of liquid phases by cavitation was employed successfully in a chemical reaction by Zechmeister and Wallcave in 1955, when they used ultrasound to accelerate the cleavage of halides of benzene and thiophene by aqueous suspensions of silver nitrate.[23] Silver halides and small organics (e.g., acetylene) were the main products.

Ultrasonically produced emulsions are generally more translucent and far more stable to coagulation. These favorable properties arise because ultrasonic waves help meet one of the key criteria for stable emulsions: the particle size of emulsified droplets should be as fine as possible. Typically, ultrasonically prepared emulsions have particle sizes in the 8–10 μm range compared to sizes well in excess of 20 μm for those particles in emulsions produced without ultrasound.

Dispersion of solids in liquids is also improved when carried out in a cavitating liquid. In 1938, Söllner reported the successful dispersion of mica, gypsum, and hematite in water, producing colloidal or semicolloidal solutions.[24] This technique is now widely used for the peptization of gels and gel-like substances in favor of mechanical methods and is commonly used in the preparation of drugs, dyestuffs, and preceramics. With the exceptions of the reduction in particle size of the suspended particles and the increased stability of the suspension, the properties of the ultrasonically dispersed materials were usually the same as those produced by the conventional techniques of heating, mixing, and/or vigorous shaking. There are numerous potential applications to synthesis, one of which was reported in 1959 by Margulis when he showed that ultrasonically dispersed $TiCl_3$ in polyolefins resulted in improved Ziegler-Natta catalysis.[25]

Cavitation at the interface of the immiscible phases resulting in the breakdown of the liquid–liquid and solid–liquid barriers is the primary cause of the formation of nearly all emulsions. Variation in the frequency of ultrasound appears to have no significant effect as long as cavitation occurs. Increases in intensity, however, generally improve homogenization and dispersion. The efficiency of this process holds promise for multiphase synthesis and has been used, as discussed later, to accelerate not only liquid–liquid reactions but also solid–liquid reactions without the normally required phase-transfer agent.

Coagulation can also be promoted by ultrasound. Thompson and Vilbrandt, studying the effects of ultrasound on clay particles observed that ultrasonic waves of intensity too low to produce cavitation increased the rate of coagulation, while intensity above the cavitation threshold dispersed the clay and disrupted aggregates.[26] This phenomenon has application in the chemical process industry promoting the agglomeration of particles from air. The technology is adaptable to very large scale: ultrasonic waves are used to coagulate sulfuric acid fog at the rate of 300 tons per day.[14] This approach applies to suspended solids, such as smoke, and has been used successfully in synthesis to improve the yield and quality of crystals and films generated by chemical vapor deposition.[27]

Ultrasound facilitates mass transport and has proved very useful in large-scale drying applications. In 1964, for example, Soloff demonstrated that drying coal could be accelerated using 10.9-kHz sound waves, reporting rate increases of 100%.[28] A careful study by Wilson and co-workers on the effect of ultrasonic energy on fluidized bed drying of coal revealed the following: (1) low-frequency waves (20 kHz) are more efficient; (2) initial rates of drying are greatest; (3) temperature remains constant during the time of fastest drying and then rises sharply; and (4) application of sound above the fluidized bed produces a higher drying rate than application below the bed.[29] The approach is also useful for fibrous materials: Purdy and co-workers have demonstrated that sound energy accelerates drying of textiles by a factor of three and that the rate is proportional to sound pressure.[30] They also found that ultrasound facilitates heat transfer. The technique can be applied to solid hydrates: the dehydration rates of $CuSO_4 \cdot 5\ H_2O$, $SrCl_2 \cdot 6\ H_2O$ and $BaCl_2 \cdot 2\ H_2O$ were increased 25- to 100-fold by ultrasonic waves.[31]

Mass transport of solids is increased by sound. The sublimation rate of naphthalene, for example, is approximately doubled when sonic waves are applied.[32] This technique might find application in syntheses requiring the removal of solvent for thin-film coating of reagents on solid supports as well as in aiding sublimation of solid products. The potential for combining mass transport with reaction rate acceleration is attractive, e.g., using ultrasound to first remove solvent and then to increase the rate of the reaction of the substrate. Conversely, a reaction might be accelerated first by ultrasound and then product removal could be facilitated using sound energy.

In the pharmaceutical industry, ultrasound is used routinely not only for homogenization and dispersion in the latter stages of product preparation but also for extraction and crystallization of compounds basic to the industry. The ultrasonic extraction of alkaloids and other natural products has been examined by many workers and in nearly all cases the rate is greatly accelerated by ultrasonic waves compared to that of the traditional Soxhlet techniques. However, in some cases, atropine, quinine, ergotoxin, and morphine, for example, ultrasound did degrade the extracts, presumably by fragmentation or accelerated hydrolysis.[32] The products were not characterized and these reactions are certainly worthy of further study.

Ultrasonic waves have been used extensively to increase the rate of crystallization of supercooled melts producing smaller crystals and more uniform size distributions than conventional crystallization techniques. Applications span a great variety of compounds including simple organics such as thymol,[33] benzophenone,[33] and tartaric acid,[34] as well as the normally refractory supersaturated solutions of arabinose, fructose, sorbitol, and glucose[35] and more complex organics such as progesterone and hydrocortisone.[36]

The action of ultrasonic waves on supersaturated solutions and melts is not yet completely understood, although most workers invoke a mechanism in which cavitation plays a key role in increasing the number of nucleation centers. This could be accomplished by shattering nuclei in the early stages of formation and scattering the fragments into the liquid. Consistent with this picture is that the rates of crystallization increase and the sizes of the crystals diminish with increasing ultrasonic intensity.

Cavitation of molten metals by low-frequency (10–50 kHz), high-intensity ultrasonic waves, by promoting degassing and multiplying nucleation sites, produces metals with uniform and highly refined grain structures.[37] Typically this leads to a metal with greater tensile strength. Intense cavitation also promotes efficient dispersion and has been used to prepare alloys. A small amount of lead increases the malleability of aluminum but the lead–aluminum alloy is difficult to prepare by conventional methods. Ultrasonic energy applied to a mixture of the molten metals yields a highly uniform alloy.[38] This technology has been successfully applied to synthetic problems in heterogeneous systems. Ultrasound has been employed to prepare alloys of sodium and of mercury, for example, as the first step in the synthesis of organometallic compounds.[39]

Immersing sheet aluminum in molten tin and rapidly vibrating it by transmitting ultrasonic waves plates tin on the aluminum efficiently. Presumably, the rapid vibrations of the metal, which give rise to forces of 1000–10,000 times that of gravity, permit penetration of the aluminum oxide layer by tin droplets.[40] The successful penetration of a metal oxide layer, in particular aluminum oxide, by molten tin under the influence of ultrasonic waves is the basis for the technology behind ultrasonic soldering. The action of a cavitating molten solder removes the oxide layer and permits soldering without fluxes and without laborious surface cleaning. The technique is particularly useful for joining very small pieces and ultrasonically soldered joints have been made with 0.002-in. diameter gold wire on gold plated silicon. Ultrasonic tinning baths rely on the same principle and are widely used for tinning and soldering small components.[41]

The dispersive power of ultrasonic waves has been used to produce more uniform mixtures of powdered solids, resulting in higher quality alloys[42,42a] and preceramics,[43] as well as more active supported-catalyst systems.[44] Dispersions of nonmetals in metals have also been similarly prepared. For example, ultrasound leads to a more uniform dispersion of graphite in bronze than any other technique, a critical factor in the production of self-lubricating bearings, and also permits the concentration of graphite to reach 25%.[45] A similar approach using metal powders, salicylaldehyde anils, and high-intensity ultrasonic waves leads to the formation of complexes.[46] For example, a 1:2 copper:salicylaldehyde anil complex prepared by this technique and characterized by x-ray diffraction improved the oil lubricating properties of copper–steel friction pairs.

While most industrial uses of ultrasound take advantage of changes in the physical properties produced by ultrasonic waves, improving chemical reactions of commercial value has always been of interest. For example, as early as 1937 Bachmann and Wilkins were issued a patent on the use of ultrasonic waves to promote the aging of raw, fermented, and distilled alcoholic beverages.[47] Two years later, the first of two ultrasonically improved vulcanization processes, in which the rubber was exposed simultaneously to an electric field and high-frequency ultrasound, was patented.[48] In 1957, El'piner reported that the bromination of coal samples in water to give material that was pyrolyzed more quickly and at lower temperatures occurred 160 times faster in an ultrasonic field.[49] He also reported that carbon tetrachloride further catalyzed the process and that ultrasound facilitated the extraction of germanium from the coal. Glukov showed that ultrasonic waves increase the rate of cyaniding of gold by more than an order of magnitude.[50]

The effects of ultrasonic waves on catalyzed systems have been extensively investigated and only a small fraction of the work is summarized here. Most of the work appears to fall in two broad catgories: the effects of ultrasonic waves on the activities of catalysts as a result of sonication during preparation of the catalysts, and the effects of ultrasound on the catalyzed reaction without prior ultrasonic treatment of the catalyst.

There are studies, of course, in which ultrasound was used on the overall process: for preparation of the catalyst as well as throughout the reaction. For example, Tyutyunnikov found that irradiation of solutions of copper and nickel sulfate during the precipitation with sodium carbonate and during the reduction of oils with hydrogen increased the efficiency of the catalyst by 40%.[51] How much of the increased activity is from sonication of the catalyst during its preparation as compared to during its use is not clear. Studies by Mal'tsev and others indicate that the former effect can be significant.

Using the rate of decomposition of hydrogen peroxide as an indicator, Mal'tsev and co-workers showed that the catalytic activity of platinum prepared in an ultrasonic field from chloroplatinic acid, H_2PtCl_6, and formaldehyde greatly increased the activity of that metal. Further examination showed that the specific area of the metal powder was increased by 62% and the magnetic susceptibility nearly doubled by the use of ultrasonic waves.[52] Platinum prepared in this fashion was also more efficient in accelerating the hydrogenation of hexene and the oxidation of ethanol to acetic acid. Palladium metal prepared in the same way from $PdCl_2$ was also more active than when prepared in the absence of ultrasonic waves.

Similar effects have been observed on metal oxides. Kowalska reported that irradiation of an ammonia solution of chromium nitrate led, ultimately, to a more active catalyst:

$$Cr(NO_3)_3 + NH_3(aq.) \xrightarrow{)))} Cr(OH)_3 \xrightarrow{>100^\circ C} \{Cr{-}O \text{ catalyst}\}$$

The calcined chromium hydroxide gel catalyst is approximately 30% more active toward the decomposition of hydrogen peroxide than gels prepared without sonication. Ultrasound increased the specific area of the catalyst and decreased the activation energy toward the peroxide.[53] Nickel carbonate precipitated from a basic solution of nickel nitrate while being irradiated with ultrasonic waves exhibits a 25% increase in activity toward peroxide decomposition and a 16% increase in surface area.[54] Precipitation of calcium carbonate in an ultrasonic field also increases its activity by 30%.[55] Ultrasound improves the synthesis of a high-quality iron oxide–chromium oxide catalyst for the oxidation of ammonia.[56]

The dispersive power of ultrasonic waves has proved useful in producing more active supported catalysts. In 1964, Marti and Waldmeier patented a process using pulsed ultrasonic waves claiming greater activity of the supported catalysts as a result of improved impregnation of the support by the catalyst. Reduced impregnation times and lower catalyst load requirements were additional benefits of using ultrasound.[57] These claims are supported in a later study by Matthews and co-workers on the preparation of a chromia–alumina catalyst system. Sonication of the catalyst support mixture for 1 h during the impregnation stage led to a catalyst with better dispersion of the chromium as determined by electron probe analyzer measurements.[44] There is some evidence that ultrasound may have a significant effect on the support: Skorik has observed that sonication of layered silicates increases the number of acid hydroxyl groups bound to silicon.[58]

The reaction to be catalyzed has received more attention than the preparation of the catalyst. As early as 1940, an attempt to improve the Haber process with ultrasonic equipment was tried, unsuccessfully. Pshenitsyn and co-workers reported that high-frequency ultrasound (555 kHz) at 300 atm failed to accelerate a mixed-catalyst system:[59]

$$H_2 + N_2 + Fe_3O_4 + Al_2O_3 + K_2O \xrightarrow{)))} 2\,NH_3$$

However, a decade later, Richardson obtained a patent for increasing the yields of ammonia by irradiating 200-μm iron powder suspended in a gas stream of hydrogen and nitrogen at 100–200°C and 1–2 atm pressure.[60] Yields of 29–32% per pass were reported on runs using 100 lbs of catalyst per 100 ft^3 of gas. Fastening iron wires to an ultrasonic driver and vibrating them in the longitudinal direction during exposure to the reactants also improves the process.[61] This approach is also useful for other reactions. Coating wires with iron, platinum, or nickel and coupling the wires to a transducer accelerates not only the synthesis of ammonia but also the hydrogenation of unsaturated fats, as well as the oxidations of sulfides to sulfur trioxide, carbon monoxide to carbon dioxide, methanol to formaldehyde, and ethanol to acetic acid.[62] Platinum, rhodium, and palladium black also catalyze ammonia formation and, in the presence of ultrasonic waves, the yields for each metal are 2–6.5 times greater with rhodium demonstrating the greatest overall activity.[52] Recently it was claimed that sonication accelerated the palladium-catalyzed quadricyclane–norbornadiene rearrangement by almost a factor of 10.[63]

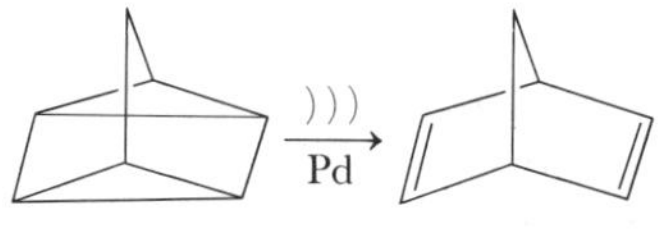

Alloyed catalysts can also benefit from sonication: Graves and co-workers claimed in their 1966 patent that the efficiency of brass used in the oxidation of isopropanol to acetone was increased from 63% to 82% using ultrasonic waves.[64]

Nonmetallic catalysts can be enhanced with ultrasonic irradiation. Isoparaffinic and olefinic hydrocarbons are more efficiently alkylated when the sulfuric acid–hydrocarbon emulsion is sonicated.[65] The reaction is faster, applies to a variety of isoparaffin–olefin mixtures and requires 5–50% less acid. The polymerization of ethylene and propylene is greatly facilitated by ultrasonic waves when the reaction is catalyzed by chromium oxide on silica–alumina as well as at least one dozen other catalysts.[66] The solid polymer product plugs up and deactivates the catalyst in the absence of ultrasonic waves.

Ultrasound also accelerates the acid-catalyzed polymerization of octamethylcyclotetrasiloxane by more than a factor of 10.[67] Ultrasonic waves overcome the chief difficulty in the polymerization of vinyl butyl ether: efficient distribution of the iron(III) chloride catalyst in the monomer.[68] Sonication accomplishes the same task for the preparation of the Ziegler-Natta catalyst, titanium trichloride dispersed in a hydrocarbon solution of triethylaluminum, for improved production of polyolefins.[25] Ultrasonic waves are also useful in the final step of the process, removal of the catalyst from the product.

Ultrasound may also occasionally inhibit a reaction. For example, Chervinskii and Malt'sev demonstrated that sonication reduced the rate of magnesium stearate–benzyl alcohol-catalyzed air oxidation of cyclohexanone.[69] The deceleration is attributed to ultrasound-induced decomposition of the stearate–alcohol complex.

The effect of ultrasonic waves on the reactions of hydrogen gas were studied by a number of workers. The ultrasonically accelerated reduction of carbon monoxide by hydrogen at low temperatures and low pressures was examined by Mayer and Marinesco in 1944.[70] Möckel found that pressure hydrogenation of olefins could be facilitated by ultrasound and presented some experimental data supporting ultrasonic acceleration of hydrogenation in open vessels.[71] Using the traditional Raney Nickel catalyst, Saracco found that ultrasound greatly accelerated the hydrogenation of olive oil dissolved in cyclohexane or cyclohexanol.[72] Destructive hydrogenation at low temperatures and pressures promoted by ultrasound was the subject of a French patent.[73] Using frequencies of 1–3 megacycles, Marinesco claimed that the autoclave that was normally required could be replaced in the hydrogenation of shale, lignite, coal, peat, and petroleum residues.[72] This technique was also applied successfully to the reduction of metal oxides and carboxylic acids by hydrogen to give the free metals and alcohols, respectively. Pchelkin and co-workers conducted a similar study on the reduction of iron ore pellets, observing that hydrogen and carbon monoxide reductions were 1.7–1.8 times faster in the presence of ultrasonic waves and that the rate of reduction of Fe_2O_3 by carbon was accelerated by 20%.[74] Similarly, ultrasound enhanced the activity of a magnesium-promoted aluminum–nickel–molybdenum catalyst system used for the hydrogenolysis of organic sulfides by increasing the dispersity and effective surface area of the catalyst.[75] Increased rates and higher yields were observed for the hydrodesulfurization of thiophene and sulfur-containing hydrocarbon raw materials.

Heterogeneous Reactions of Synthetic Interest: 1950–1980

From the beginning, the potential for commercial use propelled the expansion of the practical applications of ultrasonic waves at a rate that greatly exceeded the growth in basic research on the chemical effects of ultrasound. In recent years, however, there has been a resurgence of interest in ultrasound among chemists. Much of this renewed awareness has focused on the application of ultrasound to accelerating synthetically useful reactions. Typically, the advantages of using ultrasonic waves are faster reaction rates, lower temperature and/or pressure requirements, higher yields, and narrower product distributions. Of the systems studied, heterogeneous sonochemical reactions appear to be more synthetically useful, in general, than homogeneous systems. This brings us to the main focus of this chapter: recent successful applications of ultrasonic waves in heterogeneous reactions.

The preceding three decades saw only a modest increase in the use of ultrasound by synthetic chemists. However, while their number is small, the systems studied were important and provided precedent for some of the most recent applications. For example, in 1950, in an often overlooked paper, Renaud reported that ultrasonic waves accelerated a variety of synthetically useful reactions involving metals.[76] Noteworthy among his observations was that magnesium would react with ethyl, butyl, and phenyl bromides even in slightly aqueous diethyl ether or in di-*n*-butyl ether containing up to 50% benzene or light petroleum to produce the respective Grignard reagents:

$$\mathrm{R{-}X + Mg \xrightarrow{)))} R{-}Mg{-}X}$$

where R = C_2H_5, *n*-C_4H_9, C_6H_5; X = Cl, Br. Ultrasound could not promote the reaction in the absence of solvent and attempts to broaden the scope of the reaction to metals such as calcium, aluminum, beryllium, mercury, and zinc were unsuccessful.

Additionally, he reported that organolithium compounds could be prepared rapidly in diethyl ether from the organic halide and lithium and that organoaluminum compounds were available through the use of aluminum powder and a Grignard reagent in the presence of ultrasonic waves:

$$\mathrm{R{-}Mg{-}X + Al \xrightarrow{)))} R_3Al}$$

This modification eliminated the preparation of an Al–Mg alloy. The study also included comparisons with other metals. It was found, for example, that calcium, zinc, and beryllium did not mimic aluminum in this reaction.

In 1951, Slough and Ubbelohde used ultrasonic waves to disperse sodium in diethyl ether solutions of benzoquinoline to accelerate the formation of charge-transfer complexes:[77]

Na + 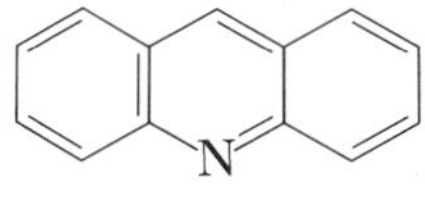 $\xrightarrow{)))}$ $Na^{+}_{1.5}[(C_{13}H_9N)]^{\dot{-}}$

Ultrasound eliminated the need for refluxing the solvent and shortened the reaction time from 48 h to 45 min. The stoichiometry of the complexes isolated was a function of the duration of the reaction. In 1959, Pratt and Helsby employed ultrasonically dispersed sodium to facilitate sodium–halide exchange reactions, e.g., in the preparation of phenylsodium from chlorobenzene:[78]

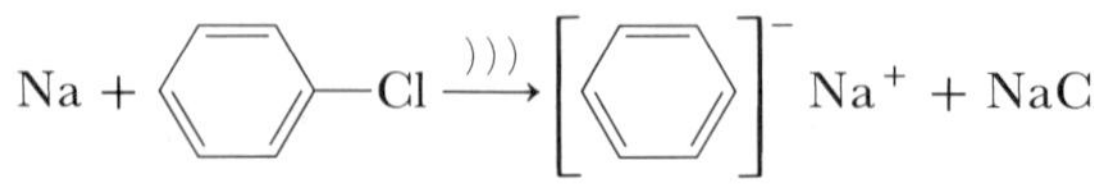

Sodium particles as small as 1 μm are easily prepared using ultrasonic waves and rate enhancements over mechanically prepared sodium are typically 10- to 20-fold.

Each of these papers contained specific suggestions about the potential utility of ultrasonic waves in synthesis so, at first glance, it is surprising that they did not strike a more responsive chord among synthetic chemists. However, there is the distinct possibility that the potential of ultrasound as a general synthetic tool was considered small even by the workers in the field. Renaud, for example, in a 1964 review on the applications of ultrasonic waves to organic chemistry, devotes little space to synthesis.[79] In retrospect, it appears that the lack of inexpensive, reliable ultrasonic equipment was a sufficient barrier to widespread acceptance of the technique.

In 1966, in a brief communication that also appears to have attracted little attention, Sjöberg described a procedure for preparing stable solutions of the synthetically useful dimsyl anion:[80]

$$(CH_3)_2SO + NaH \xrightarrow{)))} Na^+ \ {}^-[CH_2(SO)CH_3] + H_2$$

This anion, which is used routinely as a non-nucleophilic base in organic synthesis, was normally prepared by refluxing dimethyl sulfoxide and sodium hydride in tetrahydrofuran (THF), a procedure that causes some decomposition and produces an unstable solution. When ultrasound was used, in this case a frequency of 800 kHz, the reaction proceeded quickly at room temperature and the resulting solutions were far more stable than those prepared by heating. Although this paper did not receive much credit in the literature, synthetic chemists who frequently used the dimsyl anion quickly incorporated this improvement in their work.[81]

An extensive study of ultrasonic waves in synthesis was made by Fry in 1978 when he reported that ultrasonic waves from a common laboratory ultrasonic cleaner accelerated the dehalogenation of α,α'-dibromoketones by mercury, a reaction that did not occur at an appreciable rate under ordinary conditions.[82] This appears to be the first use of the simple ultrasonic bath in synthesis. The reaction was interesting from a synthetic as well as a mechanistic point of view, and Fry and co-workers proceeded to study the system thoroughly.

The initial work was on a series of alkyl-substituted α,α'-dibromoketones in carboxylic acid solvents and compared the ultrasound-promoted mercury reduction with electrochemical reduction at a mercury cathode.[82] The two methods of reduction gave the same products.

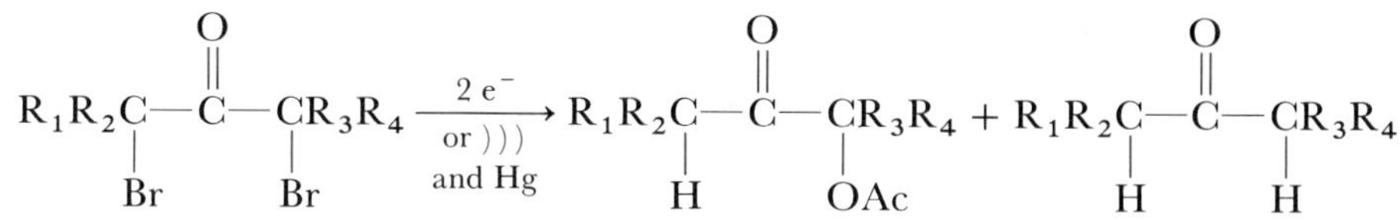

The electrochemical route was faster and gave higher yields of the two products. Even with continuous irradiation, the ultrasonically promoted mercury reduction required 1–4 days, very much less time than when ordinary stirring was used, but slower than with a high-speed laboratory homogenizer.[82]

The mechanisms of the electrochemical and ultrasonic routes are similar but not identical. For example, while increasing alkyl substitution favors acetate formation in both cases, the product distribution varies: the ratio of acetate to the simple ketone is always higher for a given dibromoketone in the ultrasound reaction than in the electrochemical reduction. Why mercury reduction should favor the tertiary product is not obvious. Nevertheless, this difference provides a simple route to unsymmetrical α-functionalized ketones.

Yields of substituted and reduced ketones improved significantly when one of the alkyl groups was replaced with an aryl group.[83] However, the ultrasound reactions were not any faster and the relative ratios of products followed the same trend as the alkyl examples. Typically, electrochemical reductions of the alkyl and cycloalkyl dibromoketones gave $>90\%$ yield of the two products, whereas ultrasound gave 30–75%.[84] Aryl substitution in the α-position has a leveling effect and yields of the electrochemical and sonochemical routes are usually above 80%.[85]

Switching to alcohol solvents improves the performance of the ultrasonically promoted mercury reduction of dibromoketones dramatically:[86]

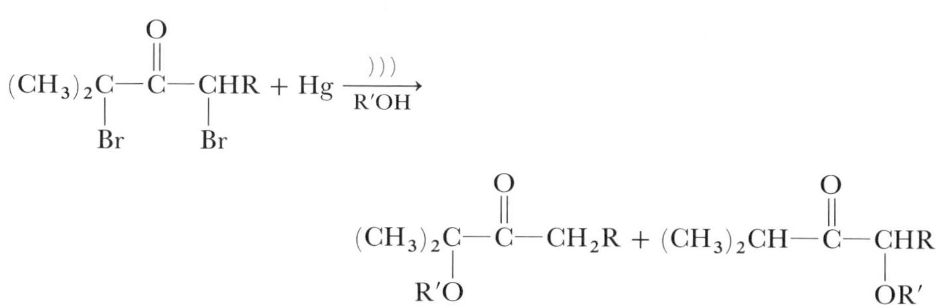

where R = CH_3, C_2H_5, C_3H_7; R′ = CH_3, C_2H_5, C_3H_7. The reaction becomes highly regioselective, exhibiting an overwhelming preference ($>90:1$ in most cases) for the tertiary ether product, and the yields increase to an average of about 70%, better than when the reaction is performed electrochemically.

When the solvent is acetone the ultrasonically dispersed mercury reduction of dibromoketones results in a convenient, one-step synthesis of 1,3-dioxolans.[87] The key step is probably the attack of the ketonic oxygen on an oxyallyl intermediate formed via rearrangement of the initial product of the insertion of mercury into the α-carbon bromine bond:

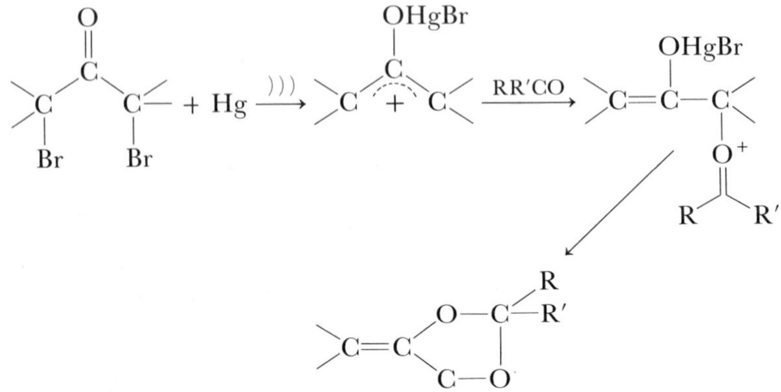

where if R = R′ = Me, the yield is 49%; if R = R′ = Et, yield is 25%, if R = Me, R′ = Et, yield is 59%; and if R = Me, R = i-Pr, yield is 47%.

Fry also compared the electrochemical and ultrasonically dispersed mercury reductions of α,α′-dibromodibenzylsulfide[88] and discovered that there was a significant difference in the two reactions: electrochemical reduction produced dibenzylsulfide, 2,3-diphenylthiirane, and the reactive intermediate, thiobenzaldehyde, whereas sonochemical reduction proceeded cleanly to *trans*-stilbene in 95% yield:

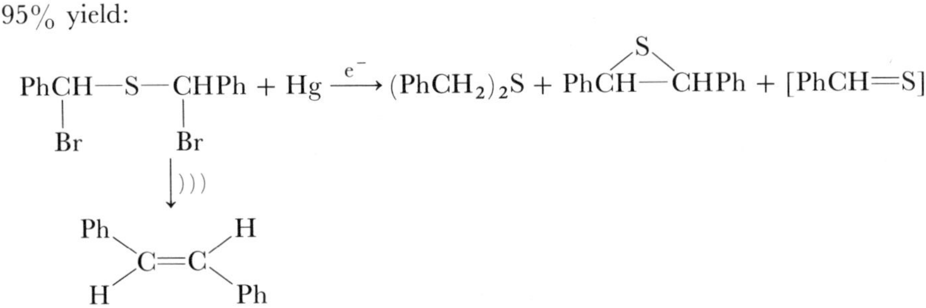

These successes provide the bridge between the largely practical and commercial applications of ultrasound, responsible for the development of ultrasonic technology, and the newly emerging field of sonochemistry with its emphasis on synthesis.

REACTIONS WITH METALS

Lithium, Sodium, and Potassium

In late 1980, Luche published a communication describing the beneficial effects of ultrasonic waves on the lithium-modified Barbier reaction:[89]

$$R{-}X + Li + R'R''C{=}O \xrightarrow{)))} RR'R''C{-}OH$$

There were several remarkable aspects of this work: higher yields of the alcohol addition products were obtained in reaction times significantly reduced from those previously reported;[90] the solvents were not meticulously dried, normally a strict requirement; and, as in the studies by Fry, these benefits were obtained using a simple ultrasonic cleaner.

The reactions were conducted at or near room temperature, which contributed to the minimization of the side reactions (Wurtz coupling, reduction, and enolization) that normally accompany and reduce the yields of the Barbier transforms. Avoiding coupling is particularly useful for the benzylation of ketones: normal reaction conditions do not give good yields of benzyllithium, and Wurtz coupling dominates to give bibenzyl. Moreover, allylic and vinylic groups add to carbonyls very efficiently under these new Barbier conditions.

The observation that ether solvents that contained some water could be used was cited as a potential advantage for industrial use and indeed it is. However, it should be noted that the solvent is in fact dried in situ by using excess lithium and organic halide. Presumably, the actual drying is accomplished by forming excess organolithium reagent, which reacts efficiently with water.

Some of these results are given in Table 5-2. These results found immediate application in cyclization reactions. Trost and Coppola demonstrated that an ultrasonically promoted intramolecular Barbier reaction gave excellent yields

Table 5-2. Ultrasound-Promoted Barbier Reaction

Organic halide	Carbonyl compound	Product	Reaction time (min)	Yield[a]
MeI	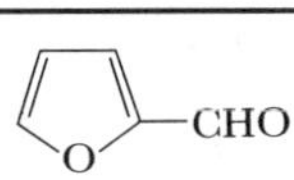	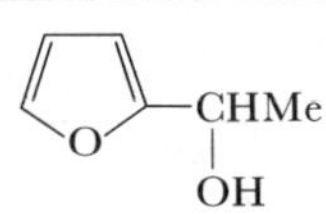	10	100 (92)
n-PrBr	PhCHO	$PhCH_2(OH)_n$—Pr	10	100
i-PrBr	*p*-$MeOC_6H_4CHO$	*p*-$MeOC_4H_4CH(OH)_n$—Pr	15	100 (96)
n-BuBr	$(n\text{-Bu})_2C{=}O$	$(n\text{-Bu})_3C$—OH	15	100 (90)
c-$C_6H_{11}Br$	$Me_2C{=}O$	$Me_2(c\text{-}C_6H_{11})C$—OH	30	80 (68)
n-BuBr	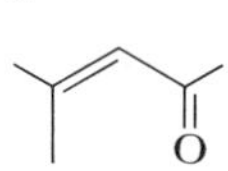	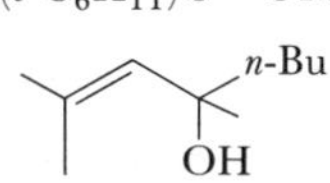	15	100 (80)
t-BuBr	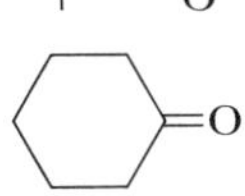	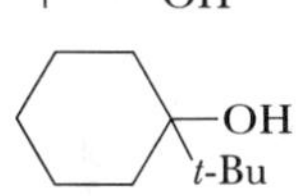	30	76
PhBr	$Me_2C{=}O$	$PhMe_2C$—OH	30	100 (95)
$CH_2{=}C(Me)Br$	$C_5H_{11}CHO$	C_5H_{11}—CH(OH)—C(Me)$=CH_2$	40	96 (71)

[a] Yields estimated by vapor phase chromatography (VPC). Isolated yields are in parentheses.

of the cyclopentanol below:[91]

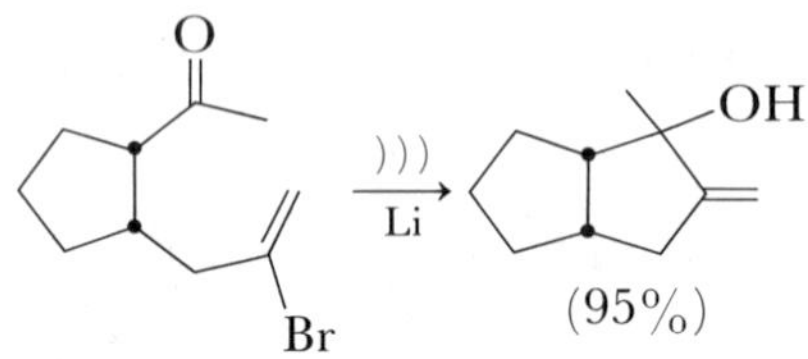

This approach succeeded where conventional methods of organometallic generation, such as using magnesium and lithium and the use of low-temperature halogen–metal exchange (t-BuLi/−120°C), failed.

The successful preparation of *n*-propyl-, *n*-butyl-, and phenyllithium in ether using lithium wire or 2% sodium lithium sand in a few minutes by substituting ultrasonic waves for high-speed stirring was also described in the Luche communication:[99]

$$\mathrm{R{-}X + Li \xrightarrow{)))} R{-}Li + Li{-}X}$$

In this reaction, when R = *n*-propyl, the yield is 90%; if R = *n*-butyl, yield is 61%; and if R = phenyl, yield is 95%. Hindered alkyl halides, such as *t*-butyl bromide and isopropyl bromide, required longer irradiation times, usually more than an hour, but this is still an improvement over a tricky procedure requiring several hours of stirring, carefully monitored temperatures, and precisely controlled addition of reagents.[92]

This approach is useful for generating vinyllithium intermediates from vinyl bromides and vinyl chlorides and is a convenient method of preparing vinyl silanes:[93]

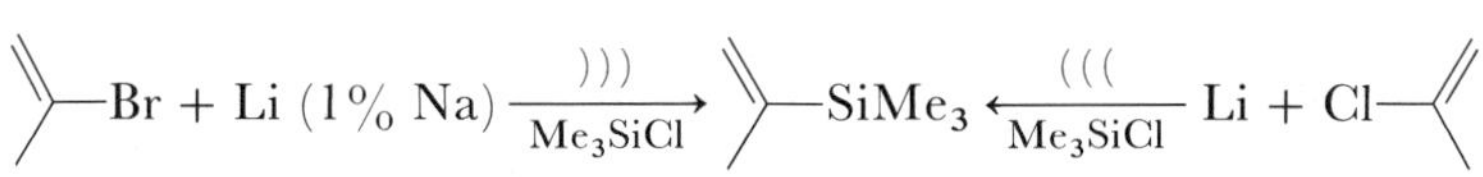

$$\text{trans-}\mathrm{C_6H_{13}{-}CH{=}CH{-}Br + Li\ (1\%\ Na) \xrightarrow[Me_3SiCl]{)))}}$$

$$\text{trans-}\mathrm{C_6H_{13}{-}CH{=}CH{-}SiMe_3}$$
$$+$$
$$\mathrm{C_6H_{13}{-}C{\equiv}C{-}SiMe_3}$$

In 1982, Luche reported an important extension of his original work in which he demonstrated that ultrasonic waves permitted a one-pot, high-yield modification of the Bouveault reaction:[94]

$$\mathrm{R{-}X + Li + Me_2NCHO \xrightarrow{)))} [R{-}CH(NMe_2)O^- Li^+] \xrightarrow{H^+} RCHO}$$

Conditions for the reaction are mild: room temperature and the reaction is complete in 15 min. It is a simple way to make aldehydes from a variety of organic halides. A representative list is given in Table 5-3.

Table 5-3. Ultrasound-Promoted Bouveault Reaction

Halide	Aldehyde	Yield (%)[a]
n-C_4H_9—X	n-C_4H_9—CHO	
X = Cl		78[b]
X = Br		88[b]
n-C_7H_{15}—Br	n-C_7H_{15}—CHO	83
Br-$(CH_2)_n$—Br	OCH—$(CH_2)_n$—CHO	
$n = 4$	$n = 4$	64
$n = 7$	$n = 7$	84
$n = 10$	$n = 10$	83
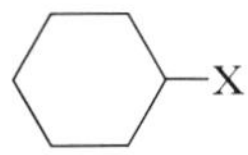	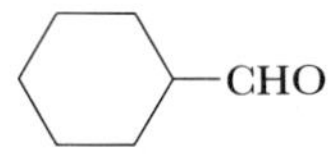	
X = Cl		70
X = Br		76
C_2H_5—C$(CH_3)_2$—X	C_2H_5—C$(CH_3)_2$—CHO	
X = Cl		84[b]
X = Br		80[b]
$C_6H_5CH_2$—Br	$C_6H_5CH_2$—CHO	85
C_6H_5—X	C_6H_5—CHO	
X = Cl		78
X = Br		86

[a] Isolated yield.
[b] Yields estimated by VPC.

Luche's group found carbonyl chemistry fertile ground for expanding the synthetic applications of ultrasonic waves. In a very short time they developed two new procedures for the synthesis of organocopper reagents and mild conditions for efficient 1,4-addition of organocuprates to α,β-unsaturated ketones.[95] A one-step method involving irradiation of a slurry of an organic bromide, lithium sand, cuprous iodide and an α-enone at 0°C in diethyl ether–THF gave good yields of the desired β-substituted ketone:

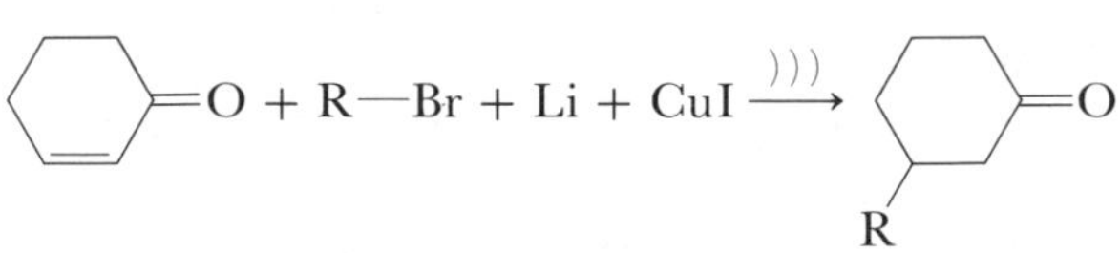

(64–91%)

where R = n-Bu, t-Bu, vinyl, phenyl. However, the product was often contaminated with compounds from side reactions of the intermediate enolate and very careful temperature control was necessary to optimize the rate of the reaction of the cuprous iodide with the organolithium reagent and to minimize attack at the acyl position to give alcohols.

To remedy this, a second procedure was developed in which the organocuprate was prepared separately by sonicating pentynylcopper, hexamethylphosphorus triamide, lithium sand, and an organic bromide in ether–THF. Upon consumption of the lithium, the α-enone was added. The yields from this modification were better than the one-step method in each case and the products were easier to purify. Neither method was proficient at β-allylation, giving yields ranging from a high of 45% in the one-step procedure to only traces of product using the separately prepared organocuprate.

Further work produced attractive alternatives to organocopper reagents. By sonicating a slurry of an aryl bromide, lithium, and zinc bromide, Luche produced high yields of diarylzinc intermediates, which, in the presence of a catalytic amount (~1%) of nickel acetylacetone, $Ni(acac)_2$, underwent 1,4-addition to conjugated ketones in 70–98% yields.[96] Presumably, the diarylzinc compounds are prepared from the reaction of the in situ-generated aryllithium with zinc bromide and not from a reaction of lithium-generated activated zinc powder and an aryl bromide. The reaction is efficient even for hindered bromides and/or hindered ketones:

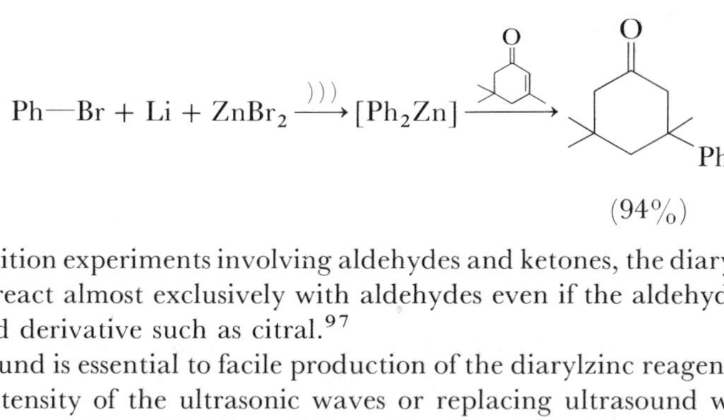

(94%)

In competition experiments involving aldehydes and ketones, the diarylzinc intermediates react almost exclusively with aldehydes even if the aldehyde is an α,α-substituted derivative such as citral.[97]

Ultrasound is essential to facile production of the diarylzinc reagents. Decreasing the intensity of the ultrasonic waves or replacing ultrasound with stirring led to slower and less efficient reactions. The diarylzinc intermediates compare favorably with the organocuprates with respect to ease of preparation, thermal stability, selectivity, and relative reactivity. Organocopper compounds normally require low temperatures to prevent decomposition and will add to the carbonyl group of β,β-disubstituted enones to give allylic alcohols.[98] It was suggested that because of their thermal stability, which permits reactions to be run at room temperature, diarylzinc intermediates add 1,4 to β,β-substituted enones in high yield.

By changing from an ultrasonic bath to an ultrasonic cell disruptor, Luche and co-workers were able to prepare dialkyl and divinyl zinc intermediates quantitatively in 10–20 min at 0°C.[99] They also switched from a diethyl ether or THF solvent system to a mixture of toluene and THF. The catalyst, $Ni(acac)_2$, is essential for good yields and is added with the enone to the organozinc intermediate. Some results are summarized in Table 5-4.

Table 5-4. Conjugate Addition to α-Enones via R_2Zn Intermediates

Organic halide	Enone	Addition product	Isolated yield (%)
CH_3I	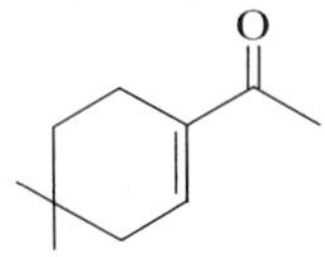	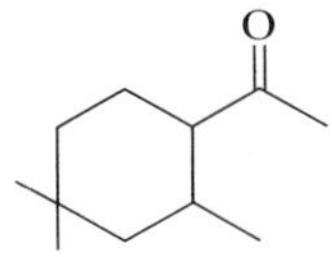	90
t-BuBr	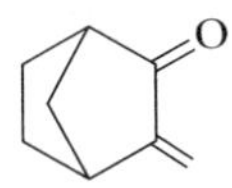	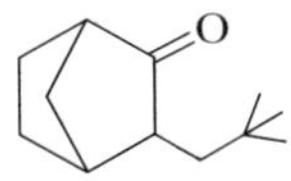	78
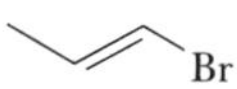	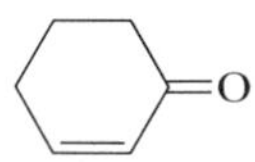	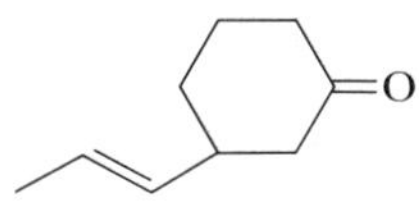	83
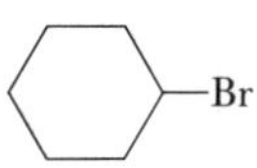	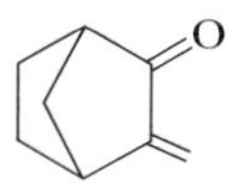	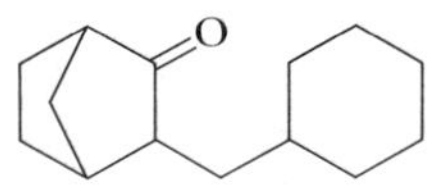	66
$PhCH_2Br$	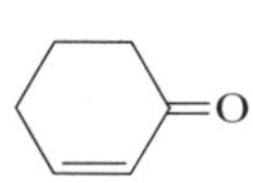	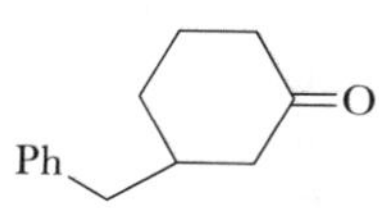	64

The change to a toluene–THF solvent system produced significant increases in yields and is worthy of note. The solvent effects in sonochemistry are complex because it must also act as an energy carrier. In this case, it appears the lower vapor pressure of the toluene–THF mixture leads to more energetic cavitation and faster reaction rates. This is in concert with the recent work of Suslick, in which he finds that the rates of some reactions of metal carbonyls are inversely related to the vapor pressure of the solvent.[100] This important aspect of sonochemistry is treated in Chapter 3.

Boudjouk was able to apply ultrasonic waves successfully to the reductive coupling of organometallic chlorides:[101]

$$R_3M{-}Cl + Li \xrightarrow{)))} R_3M{-}MR_3 + LiCl$$

where M = Si, Sn. The results are summarized in Table 5-5.

These results, although not spectacular and certainly not optimized, pointed the way to an improved method of preparing compounds with metal–metal bonds. Some advantages were obvious: the reaction times were shortened by several hours for each disilane; room temperature could be maintained throughout the reaction; vigorous stirring was unnecessary because cavitation kept the lithium wire clean; and, the shorter reaction times at lower temperatures reduced the quantities of the usual silane and siloxane byproducts, the latter often very difficult to separate from the disilane.

Table 5-5. Coupling of Chlorosilanes with Lithium Wire

Chlorosilane	Disilane	Solvent	Time (h)	Isolated yield (%)
Ph_3SiCl	$Ph_3SiSiPh_3$	C_6H_6:THF (2.5:1)	10	73
Ph_2HSiCl	$Ph_2HSiSiHPh_2$	THF	10	68
Et_3SiCl	$Et_3SiSiEt_3$	THF	60	15
Me_3SiCl	$Me_3SiSiMe_3$	THF	24	9

Application of the technique to organotin chlorides was straightforward, giving the ditin compound in less than 2 h:[101]

$$R_3SnCl + Li \xrightarrow{)))} R_3Sn—SnR_3$$

where if R = Me, yield was 60%; and if R = n-Bu, yield was 94%.

In the absence of ultrasonic waves the reaction time more than doubled to 4 h and resulted in significantly lower yields, provided the reaction was vigorously stirred. Little or no reaction was observed without stirring.

Parallel studies with organic halides, i.e., the Wurtz and Wurtz-Fittig reactions, also showed rate and yield enhancements:[102]

$$R—X + Li \xrightarrow{)))} R—R + LiCl$$

Table 5-6 summarizes the results.

Boudjouk and co-workers subsequently discovered that replacement of lithium wire with lithium dispersion or lithium powder further accentuated the effects of ultrasonic waves.[103] For example, for nearly every halide in Tables 5-5 and

Table 5-6. Coupling of Organic Halides with Lithium

RX	Product	Time (h)	Isolated yield (%)
C_6H_5Cl	$C_6H_5—C_6H_5$	12	70
C_6H_5Br	$C_6H_5—C_6H_5$	10	70
C_6H_5I	$C_6H_5—C_6H_5$	12	73
p-$CH_3C_6H_4Br$	$(p\text{-}CH_3C_6H_4)_2$	10	52
p-$CH_3C_6H_4I$	$(p\text{-}CH_3C_6H_4)_2$	10	42
m-$CH_3C_6H_4Br$	$(m\text{-}CH_3C_6H_4)_2$	10	36
C_6H_5COCl	$C_6H_5—C(=O)—C(=O)—C_6H_5$	17	73
$CH_3CH_2CH_2CH_2Cl$	$CH_3(CH_2)_6CH_3$	17	72

5-6, 85% or better yields of the listed product can be obtained in less than 1 h. Precise work has yet to be done but, generally, reaction rates are 2 to 10 times longer in the absence of ultrasonic waves. In almost all cases, the reactions do not go to completion without the aid of ultrasound because the lithium becomes coated with lithium halide and other impurities. In fact, ultrasound is very useful for restarting reactions that have stopped because of passivation of the metal. The reactions are sufficiently simple and reproducible to be easily adapted for undergraduate synthesis experiments.[104]

Ultrasound permits the use of lithium metal rather than alkyllithium in the generation of cyclopropylidenes from *gem*-dibromocyclopropanes leading to an improved synthesis of allenes or C—H insertion products:[105]

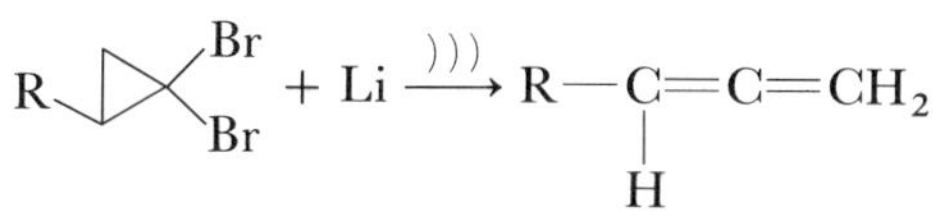

where when R = n-C_5H_{11}, the yield is 81%, and when R = C_6H_5, yield is 76%;

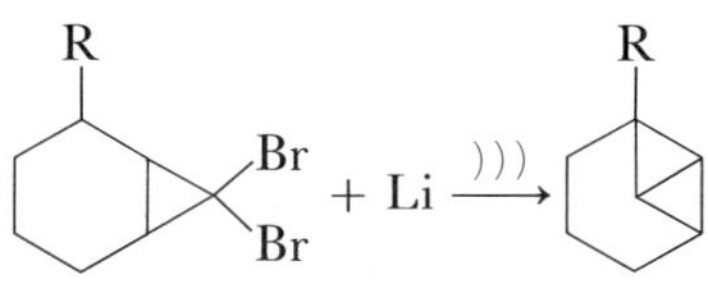

where when R = H, yield is 46%, and when R = CH_3, yield is 52%. The induction period was eliminated using ultrasound and the reactions were complete in less than 20 min. Magnesium produced similar results. For the less reactive dichlorides, however, magnesium was inert even at reflux temperatures and lithium required over 5 h. However, if sodium sand is employed in conjunction with ultrasonic waves, the reaction is complete in 5 min, giving the same products in similar yields as the dibromo analogs with lithium and magnesium.

Dehalogenation of a variety of α,ω-diiodides in the gas phase using sodium or potassium vapor is also promoted by ultrasonic waves.[106] The procedure is useful for preparing small amounts of highly strained olefins and ring compounds and is particularly useful in generating reactive compounds for isolation and characterization by low-temperature matrix techniques. Thus, the very reactive bridgehead olefin, adamantene,

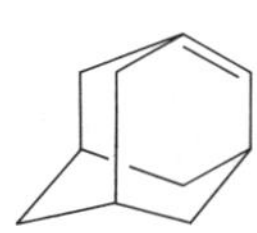

could be prepared with only traces of radical products. Similarly, 2.2.1-propellane, a highly strained cage system,

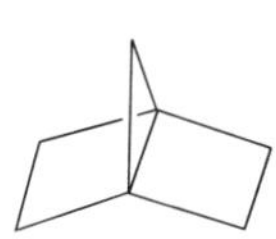

Table 5-7. Ultrasonically Accelerated Lithiation Reactions

Substrate	Organolithium	Time (min)	Electrophile	Product	Yield (%)
PhC≡CPh	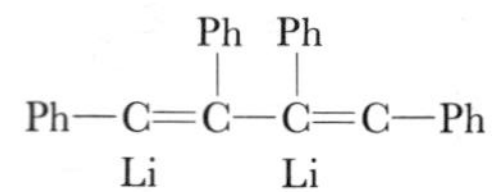	10	$MeHSiCl_2$	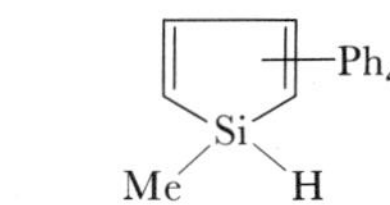	68
Ph—C≡C—H	Ph—C≡C—Li	5	MeI	Ph—C≡C—Me	95
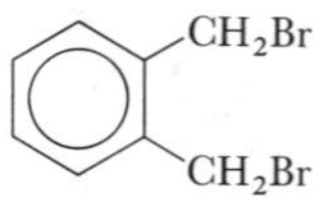	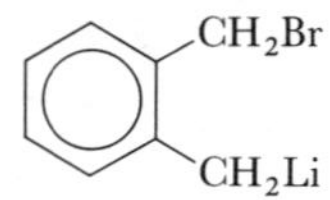	60–90	—	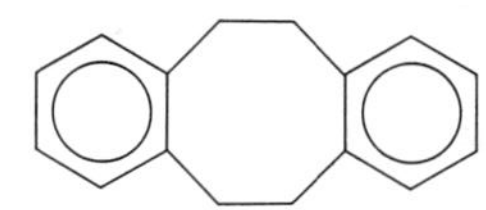	80
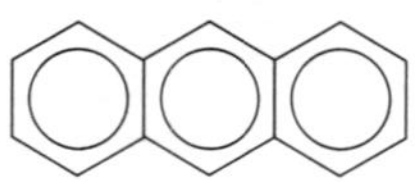	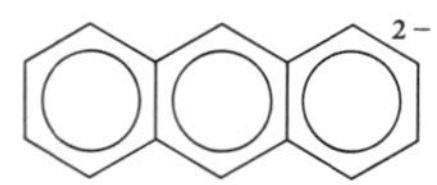	30–45	H_2O	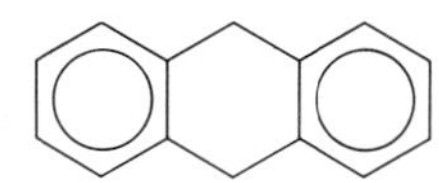	90
	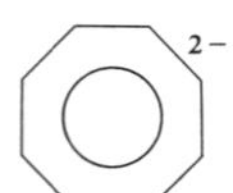	10	Me_3SiCl	c—$C_8H_8(SiMe_3)_2$	82
	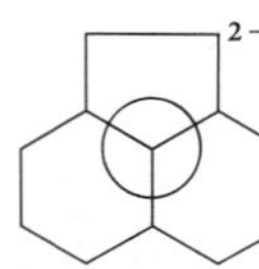	60–90	MeOH	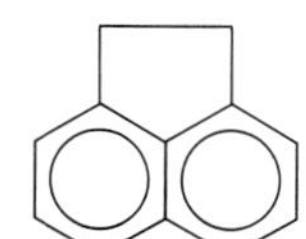	90

was prepared with fewer products from competing hydrogen transfer reactions when ultrasound was used. The side reactions are from radicals whose intermolecular reactions are minimized by maintaining a high concentration of metal vapor and efficient mixing of gases. Ultrasonic waves provide a clean surface for maximum evaporation of the metal and improve the mixing of gases. With ultrasound, furnace temperatures could be lowered 40°C.

Boudjouk and co-workers recently reported studies of the reactions of lithium with a variety of organic substrates, observing rate accelerations for reductive coupling of diphenylacetylene, liberation of an acetylenic hydrogen, lithium halide exchange, and reductions of some aromatics to dianions.[107] With lithium powder or lithium dispersion the reactions are complete in less than 90 min at room temperature. In the absence of ultrasound the reactions require 3–9 h and do not always go to completion. The production of the dilithiotetraphenylbutadiene from lithium and diphenylacetylene, for example, gave only low yields at room temperature without ultrasound, requiring heating and vigorous stirring to approach the yields obtained with sonication.

Each of the reactions generates a synthetically useful organolithium intermediate as evidenced by the efficient quenching with appropriate electrophiles. The anions and dianions are produced in yields higher than indicated by the yields of isolated product. For lithiation of the derivatives of anthracene, cyclooctatetraene, and acenaphthene, for example, nmr spectroscopy shows quantitative formation of the respective dianions. Table 5-7 summarizes the results. Alkali metal reduction of aromatics under the influence of ultrasonic waves was also investigated by Yanagida,[108] who prepared lithium and sodium derivatives of naphthalene as well as sodium derivatives of anthracene and biphenyl. No yields were given.

Reactions of difunctional organosilanes with lithium are of interest because of the novel cyclic compounds produced and the potential for generating new reactive intermediates. Ultrasound proved useful in promoting these reactions. For example, when lithium wire was suspended in a solution of THF, dimethyldichlorosilane, and a catalytic amount of anthracene (to act as an electron carrier) and irradiated with ultrasonic waves from a cleaning bath, dodecamethylcyclohexasilane was obtained in yields averaging about 85% in about 2 h:

$$Me_2SiCl_2 + Li \xrightarrow{)))} (Me_2Si)_6 + (Me_2Si)_5$$

$$Ph_2SiCl_2 + Li \xrightarrow{)))} (Ph_2Si)_4 + (Ph_2Si)_5$$

Slightly better results were obtained for the production of octaphenylcyclotetrasilasilane from diphenyldichlorosilane. With lithium dispersion and ultrasound the reaction time can be shortened to less than 20 min. The five-membered ring is a minor product in both reactions.[101]

The size of the cyclopolysilane formed is inversely related to the size of the substituent, and the next logical step was to increase the size of the organic group, perhaps progressing to the point where cyclization could be prevented. West's novel compound tetramesityldisilene,[109] the first example of a stable species with a silicon–silicon double bonds, was the target. This species was generated using dimesityldichlorosilane, lithium, and ultrasonic waves.[110] The disilene (and/or its precursor formed during the reaction) is very reactive toward lithium, however, and we have found it very difficult to obtain consistent results. Most often, hexamesitylcyclotrisilane is isolated in very good yields:[111]

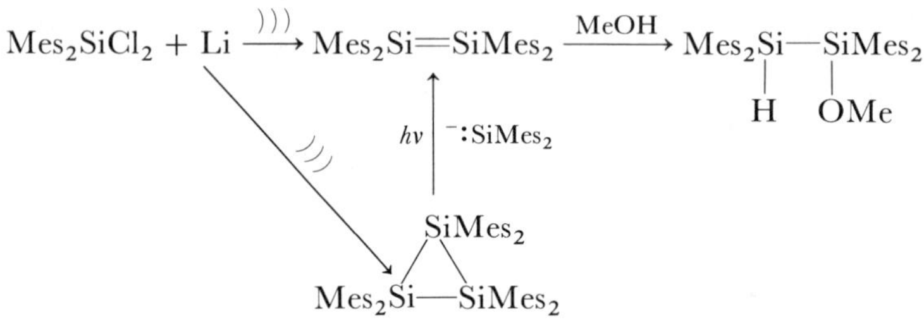

where Mes = 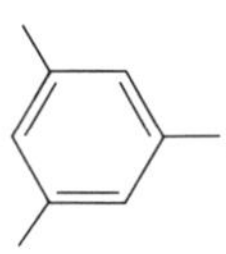. This stable cyclotrisilane is valuable in its own right, being one of only a few cyclotrisilanes, and is, upon photolysis, a potentially useful source of tetramesityldisilene. Delicate hexaalkylcyclotrisilanes were later prepared by Masamune using ultrasonic techniques.[112]

An unexpected result was obtained when di-*t*-butyldichlorodisilane and lithium were irradiated with ultrasonic waves:

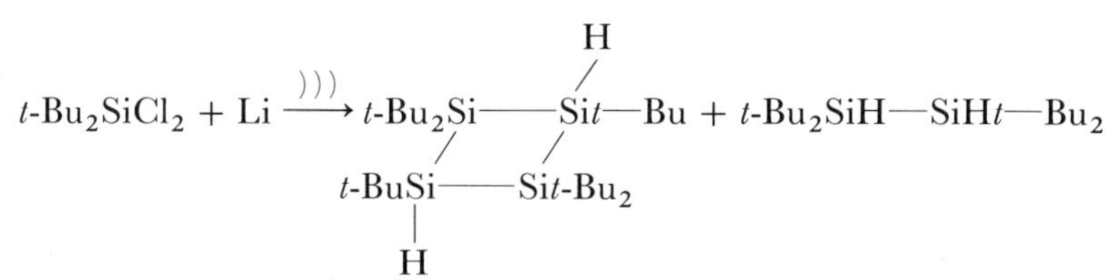

the novel four-membered ring, *trans*-1,1,2,3,3,4-hexa-*t*-butylcyclotetrasilane, was produced in 20% yield in addition to the main product, tetra-*t*-butyldisilane.[113] The reaction mechanism is clearly complex and in need of further study. At this stage our evidence suggests that, for dichlorosilanes with two large organic groups, reactions with lithium generate the reactive intermediates, silylenes, by α-elimination.[114] This was determined by a series of trapping reactions:[114,115]

$$RRSiCl_2 + Li \xrightarrow{)))} RR'Si\colon \xrightarrow{Et_3Si-H} Et_3Si-SiRR'-H$$

Substituent	Yield (%)	Substituent	Yield (%)
R = R′ = *t*-Bu	60	R = *t*-Bu, R′ = Ph	3
R = *t*-Bu, R′ = Mes	38	R = R′ = Ph	0
R = R′ = Mes	18	R = R′ = Me	0

Table 5-8. Lithium Wire Reduction: Ultrasound vs. Stirring

	Reduction time	
Metal halide	Ultrasound (min)	Stirring (h)
$NiCl_2$	<40	18
$FeCl_2$	<40	24
$PdCl_2$	<40	26
$CoCl_2$	<40	19
$PbCl_2$	<40	20

It appears that steric hindrance is necessary to permit α-elimination to occur at a rate competitive with intermolecular processes. Thus, ultrasound provides, in the case of dichlorosilanes, a mild (30–60 min at room temperature or below) and efficient route to novel cyclics and hindered silylenes. In the absence of ultrasonic waves several hours of stirring or higher temperatures are usually required. Higher temperatures normally produce more complex product mixtures.

Alkali metal reductions of metal halides to form activated metal powders are accelerated by ultrasonic waves.[116] Rieke and co-workers have demonstrated that reactive metal powders have a wide variety of synthetic applications.[117] Typical procedures call for refluxing mixtures of potassium, sodium, or lithium bulk metal or dispersion and a halide of the desired metal powder in a polar solvent like THF for 2–12 h:

$$n\,\mathrm{A} + \mathrm{MX}_n \longrightarrow \mathrm{M}^* + n\,\mathrm{AX}$$

where X = Cl, Br, I; A = K, Na, Li.

Ultrasonic waves accelerate this reaction so that, for a wide variety of metal halides, the reduction is complete in less than 40 min at room temperature.[118] With sonication, lithium wire and lithium powder reduce metal halides to metal powder 10–50 times faster than stirring alone at room temperature and 3–10 times faster than potassium metal in refluxing THF. These comparisons are summarized in Tables 5-8, 5-9, and 5-10.

Table 5-9. Lithium Powder Reduction: Ultrasound vs. Stirring

	Reduction time	
Metal halide	Ultrasound (min)	Stirring (h)
$NiCl_2$	<40	14
NiI_2	<40	12
$CuBr_2$	<40	13
CuI	<40	12

Table 5-10. Comparison of Potassium and Lithium Reduction of Metal Halides

Metal powder	Metal halide	Reducing agent	Conditions[a]	Reduction time
Zn	$ZnBr_2$	K	A	4 h
Zn	$ZnBr_2$	Li	B	<40 min
Mg	$MgCl_2$	K	A	1–2 h
Mg	$MgCl_2$	Li	B	<40 min
Cr	$CrCl_3 \cdot 3$ THF	K	A	2 h
Cr	$CrCl_3$	Li	B	<40 min
Cu	CuI	K	A	8 h
Cu	$CuBr_2$	Li	B	<40 min

[a] A = reduction in refluxing THF[117]; B = reduction with lithium powder and ultrasound at room temperature in THF.

Using the metal-induced coupling of benzylbromide to bibenzyl as the benchmark, it was found that nickel and copper powders produced by lithium reduction (accelerated by ultrasound or by stirring) gave high yields of bibenzyl.[118] Suslick has observed that some transition-metal powders produced by sodium reduction of their halides can be converted to carbonyl derivatives under the influence of ultrasonic waves.[119] The reactions of the "less active" metals are discussed later in Transition Metals.

Kimmel and Becker developed a high-yield synthesis of sodium and potassium salts of 3-aminopropylamide, a reagent useful for the isomerization of internal acetylenes to terminal acetylenes and in the synthesis of pheromones and fatty acids, by sonicating a mixture of the metal, 1,3-propanediamine, ammonia, and a trace of ferric nitrate for 10 min at 90°C.[120] Chou and co-workers found that ultrasound improved the reductive cleavage of the phenyl–phosphorous bond in triphenylphosphine by lithium to give good yields of the diphenylphosphide anion:[121]

$$Ph_3P + Li \xrightarrow{)))} [Ph_2P]^-Li^+ + Ph^-\,Li^+$$

Extension of this reaction to diphosphines led to a high-yield synthesis of diphosphide dianions under mild conditions.[122] The dianions can be treated with electrophiles, making a variety of diphosphines:

$$Ph_2P(CH_2)_nPPh_2 + Li \xrightarrow{)))} \underset{Li^+}{PhP^-}\!-\!(CH_2)_n\!-\!\underset{Li^+}{{}^-PPh} \xrightarrow{R-X} \underset{R}{\underset{|}{PhP}}\!-\!(CH_2)_n\!-\!\underset{R}{\underset{|}{PPh}}$$

where $n = 2$–6, R—X = MeI, c-C_6H_{11}. With ultrasound the reaction is complete in a few minutes using either lithium at 0°C or potassium at −78°C, and the product is freer of contaminants than when conventional conditions (potassium, 0°C, 20 min) are employed. The shorter reaction times minimize reactions of the phosphide anions with the solvent and with t-butyl chloride, which is used to quench the phenyllithium generated from cleaving the P—Ph bond.

Luche[123] used ultrasonic waves to disperse potassium in toluene or xylene in a few minutes at 10°C in a fashion similar to Slough[77] and Pratt.[78] This "colloidal" potassium proved to be useful in several condensation reactions:

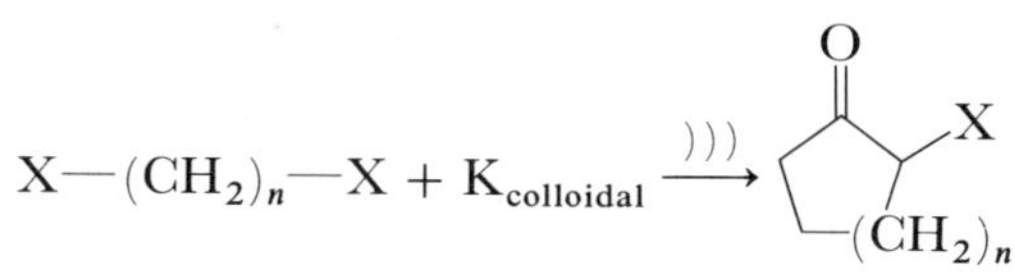

(75–83%)

where $n = 4, 5$; X = COOEt, CN.

Reaction of triethylphosphonoacetate with colloidal potassium leads to efficient production of an ylid intermediate, which reacts with cyclohexanone to give ethylcyclohexylidene acetate in 81% yield:

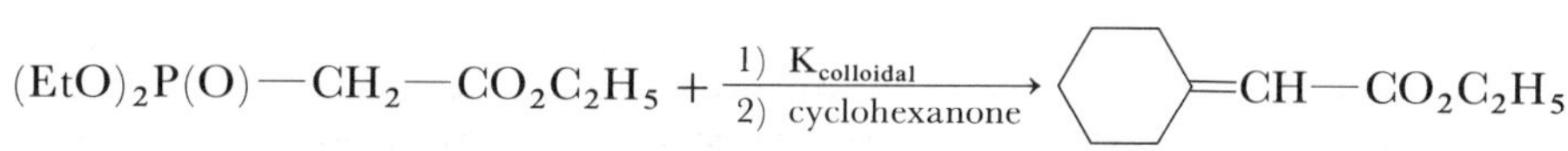

These reactions were completed in a few minutes.

Chou and You used ultrasonically dispersed potassium in toluene to open cyclic sulfones in good yield under mild conditions:[124]

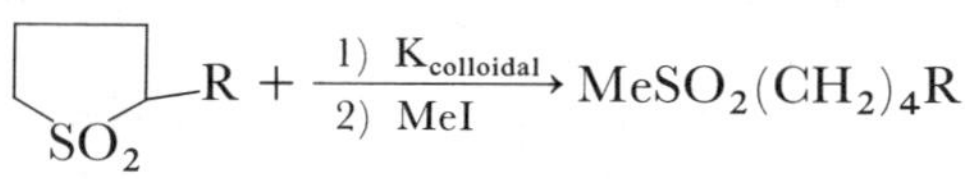

where R = CH_3, *i*-C_3H_7, C_6H_{11}, CH_2Ph and yield is 74–94%.

Zinc, Magnesium, and Aluminum

Reactions involving zinc have been very responsive to treatment with ultrasonic waves. The Simmons-Smith reaction, for example, which transfers methylene through a reaction of a zinc–copper couple with methylene iodide, is improved when run in the presence of ultrasonic waves. Ultrasound eliminates the need for copper when zinc dust is used. Thus good yields of cyclopropylbenzene and cycloprop[*a*]acenaphthene are obtained from styrene and acenaphthene, respectively.[116]

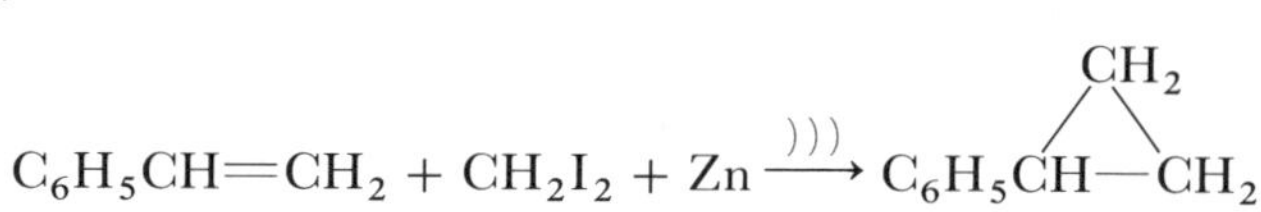

(60%)

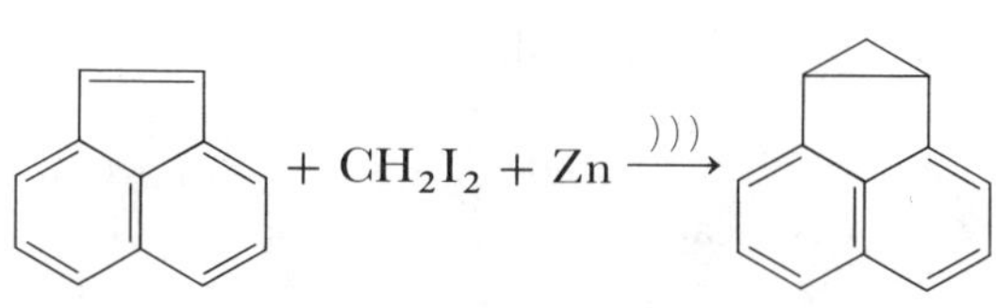

(50%)

Table 5-11. Cyclopropanations of Alkenes Using Dibromomethane

Alkene	Product	Conditions[a]	Time (h)	Isolated yield (%)
$CH_3(CH_2)_2CH{=}CH_2$	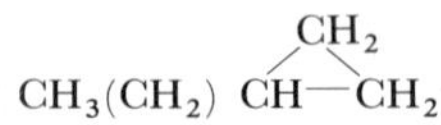	A	4	30
$CH_3(CH_2)_5CH{=}CH_2$	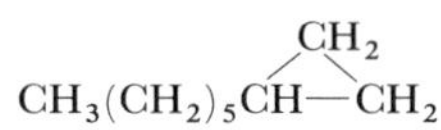	A	4	28
$CH_3CH{=}CHCH_2OH$	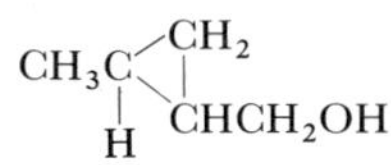	B	2	57
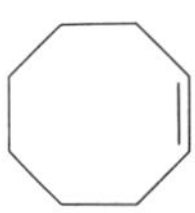	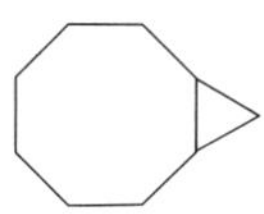	A	4	72
		B	2.5	64

[a] Alkene: CH_2Br_2:Zn:CuCl, 0.2:0.4:0.8:0.08; A = all reagents present initially and sonication throughout reaction. B = alkene added after vigorous reaction of Zn and CH_2Br_2 was under way.

Ultrasound so activates the system that it is sometimes difficult to control the reaction.[116] However, Repic and Vogt found that the use of mossy zinc provides sufficient inhibition to permit a smooth reaction.[125] Thus, methyl oleate, methyl linoleate, β-pinene, and camphene, for example, were converted to their cyclopropane derivatives in 74–97% isolated yields (compared to literature yields of 12–75% by conventional methods) in 2–6 h using mossy zinc, diiodomethane, and ultrasonic waves in refluxing dimethoxyethane. The reaction has been sealed up to hundred gram quantities.[126]

Friedrich, Domek and Pong developed a technique based on ultrasound that permits the use of the much less expensive dibromomethane in ether and a conveniently prepared zinc dust–cuprous chloride reagent.[127] Following an induction period of 1 h and 3 h of sonication, moderate yields of cyclopropane products were isolated from olefins. Higher yields were obtained using freshly prepared zinc powder.[117] Table 5-11 summarizes the results.

The methylenation of carbonyl compounds under very mild conditions was carried out by Yamashita and co-workers using methylene iodide, ordinary zinc dust and THF solvent:[128]

$$R_2C{=}O + Zn + CH_2I_2 \xrightarrow{)))} R_2C{=}CH_2$$

The reaction may be carried out at room temperature, and methylene bromide may be substituted for methylene iodide. Methylene bromide requires significantly longer reaction times and gives slightly lower yields. The results are summarized in Table 5-12.

The reactions of zinc with activated organic halides are also accelerated by ultrasonic waves. Zinc dust reacts with α,α'-dibromo-*o*-xylene in dioxane[129] to give good yields of the reactive intermediate *o*-xylylene, which can be trapped

Table 5-12. Ultrasound-Promoted Methylenation of Carbonyl Compounds[a]

Carbonyl compounds	Time (min)	Products	Yield[b] (%)
Benzaldehyde	60	Styrene	82
p-Methylbenzaldehyde	120	*p*-Methylstyrene	73
p-Methoxybenzaldehyde	90	*p*-Methoxystyrene	50
p-Chlorobenzaldehyde	15	*p*-Chlorostyrene	69
1-Naphthaldehyde	120	1-Naphthylethene	54
Cinnamaldehyde	60	Phenylbutadiene	37
Heptanal	20	1-Octene	71
Cyclohexanone	60	Methylenecyclohexane	18
Acetophenone	300	α-Methylstyrene	8

[a] Typical experiment: Zn 24 mmol, diiodomethane 10 mmol, carbonyl compound 5 mmol, THF 20 mL, room temperature.
[b] Yields estimated by gas–liquid chromatography.

by dienophiles:

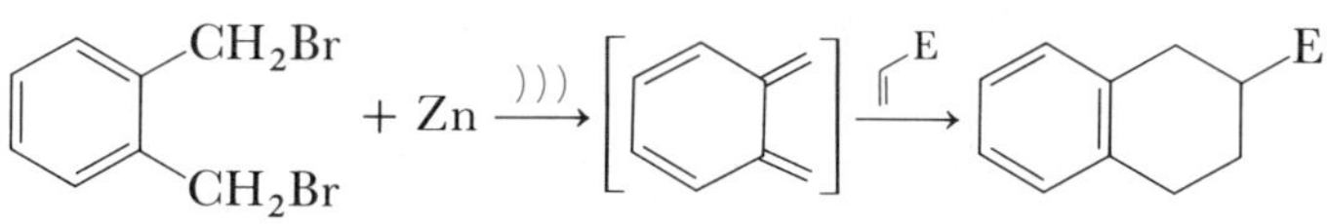

where  = activated olefin. The cycloaddition reaction is efficient, leading to a useful synthesis of functionalized tetrahydronaphthalenes. Table 5-13 summarizes the results of some trapping reactions.

Table 5-13. Reactions of *o*-Xylylene with Dienophiles[a]

Dienophile	Product	Time (h)	Isolated Yield (%)
$CH_2{=}CHCO_2CH_3$	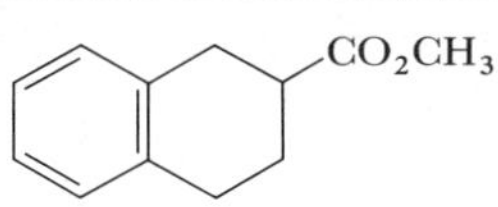	12	67
$CH_2{=}CHCOCH_3$	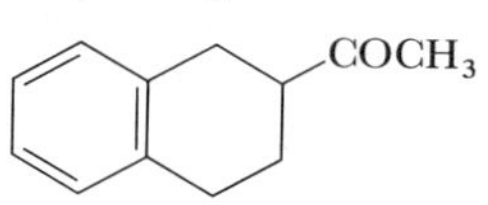	12	87
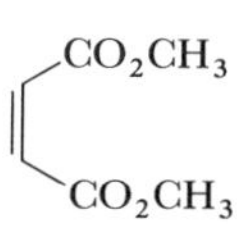	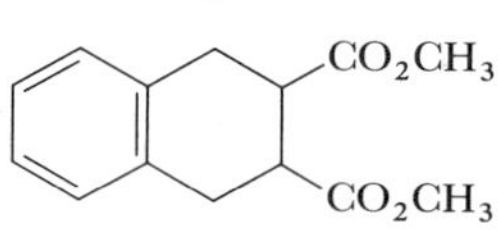	12	70
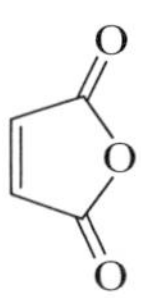	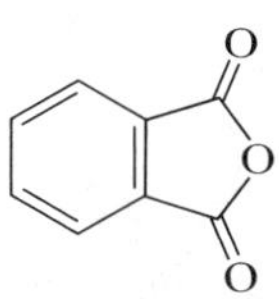	15	89

[a] Typical experiment: 10 mmol dibromide, 11 mmol dienophile, 23 mmol Zn in 10 mL dioxane at 20–25°C.

These results contrast sharply with those obtained when lithium reacts with this dibromide. The intermediate in the lithium reaction is ionic and can be trapped with an appropriate electrophile, such as trimethylchlorosilane, but fails to react with the dienophiles in Table 5-13 to give cycloaddition products.[107] The zinc-generated intermediate is not trapped by trimethylchlorosilane and does not dimerize efficiently, the latter an expected result from *o*-xylylene since dimerization would be a 4 + 4 thermally forbidden cycloaddition. Scheme 5-1 compares the two fates of α,α-dibromo-*o*-xylene.

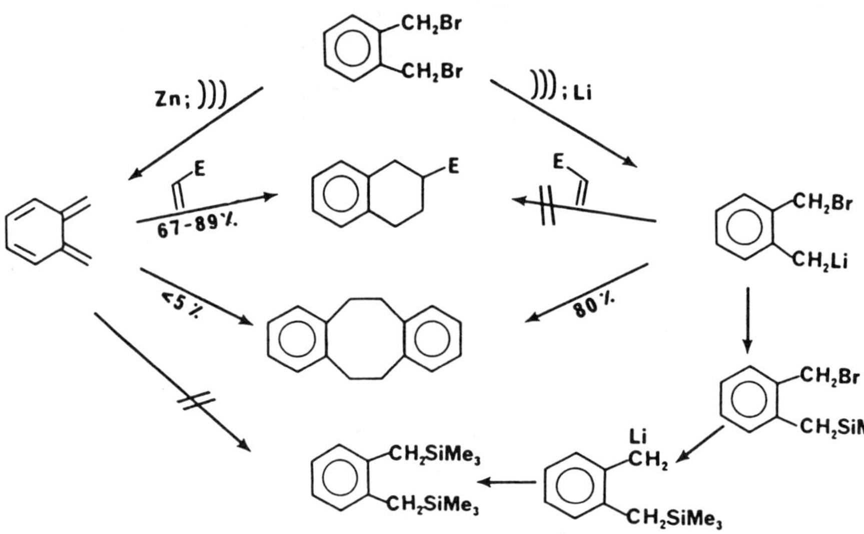

Using this technique, Chew and Ferrier prepared a series of optically pure functionalized hexahydroanthracenes and hexahydronaphthacenes.[130] These workers found that zinc-generated *o*-xylylene reacted with a D-glucose enone to give the cycloadduct in good yield:

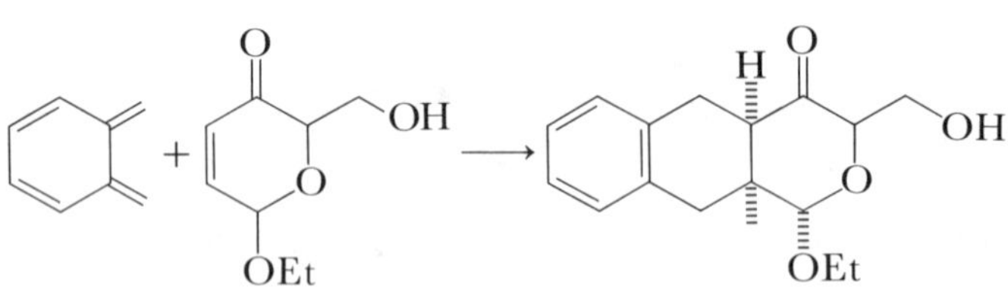

The naphthalene analog, generated from 2,3-bis(bromomethyl)naphthalene and zinc, performs similar reactions but in lower yield. This approach is an improvement on methods relying on benzocyclobutene derivatives as a source of *o*-xylylene since they usually require several days of heating at temperature above 100°C.[131]

The Reformatsky reaction is the most generally applicable procedure for converting aldehydes and ketones to β-hydroxy esters and, consequently, has been the subject of intensive study.[132] Boudjouk found that ultrasonic waves not only greatly accelerated this reaction but also dramatically improved the yields compared to conventional methods.[133] The reaction times and yields are comparable to those obtained using freshly prepared zinc powder.[134]

$$RR'C{=}O + BrCH_2CO_2Et + Zn \xrightarrow{)))} RR'C(OH)CH_2CO_2Et$$

Substituent	Yield (%)	Substituent	Yield (%)
$R = C_3H_7$; $R' = H$	90	$R = R' = C_3H_7$	100
$R = C_7H_{15}$; $R' = H$	100	$R = C_6H_5$; $R' = CH_3$	90
$R = C_6H_5$; $R' = H$	100	$R = R' = C_6H_5$	82

The reactions were run at 25–30°C and were complete in less than 30 min with the exception of the more hindered benzophenone, which took 3 h. When the zinc is activated by washing with acid, iodine or potassium iodide is also needed, as in conventional methods,[132] to promote the reaction and inhibit side reactions. However, if zinc powder is prepared by lithium reduction of zinc with sonication and the reaction is continuously sonicated, no promoters are necessary.[118]

Bose et al. applied this methodology successfully to the synthesis of β-lactams using Schiff bases in place of carbonyl compounds.[135] Normally yields of 25–50% are obtained in refluxing toluene[136] or somewhat higher yields (50–70%) using pure zinc foil.[137] However, employing the ultrasound Reformatsky conditions described above, yields as high as 95% were realized:

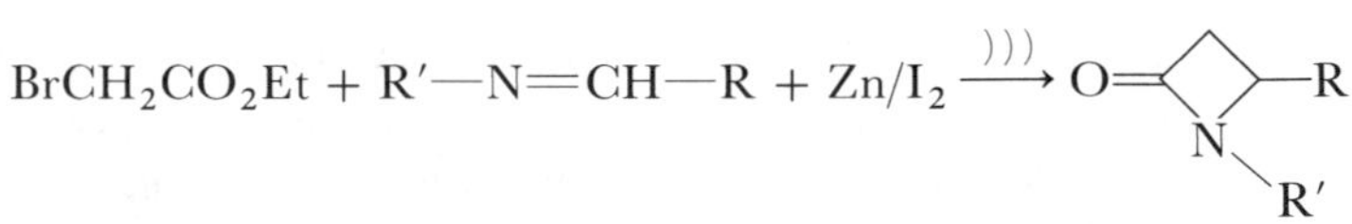

Substituent	Yield (%)	Substituent	Yield (%)
$R = p\text{-}MeC_6H_4$, $R' = p\text{-}MeOC_6H_4$	95	$R = C_6H_5$, $R' = C_6H_5$	70
$R = p\text{-}ClC_6H_4$ $R' = p\text{-}MeOC_6H_4$	77	$R = C_6H_5$, $R' = p\text{-}MeOC_6H_4$	82

Brennan and Hussain used this approach for the removal of bromine from 6α-bromopenicillanate esters.[138] Tributyltin hydride reduction or palladium-catalyzed hydrogenolysis can perform the task but both reagents are expensive and produce side products. Ultrasound Reformatsky conditions circumvent these difficulties, leading to moderate to good yields of penicillanate esters:

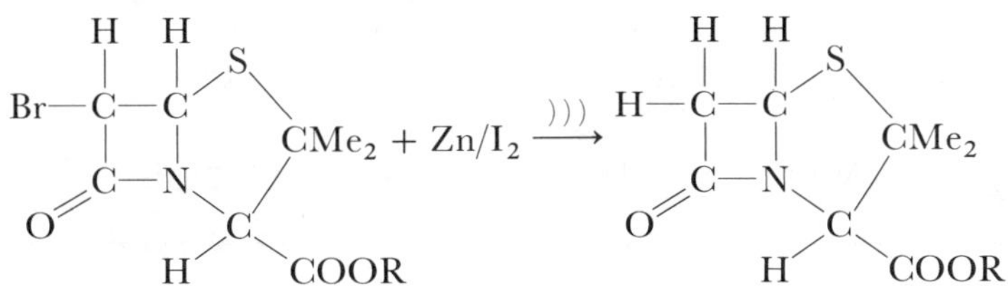

Yamashita and co-workers found that ultrasonic waves activated a nickel chloride–zinc–hexamethylphosphoric triamide medium sufficiently to dehalogenate the relatively inert halobenzenes at 60°C in 0.5–4 h, 20–50% faster than without ultrasound:[139]

$$\mathrm{R{-}X + NiCl_2 + Zn} \xrightarrow{)))} \mathrm{R{-}H} \quad (81\text{–}100\%)$$

where R = phenyl, naphthyl, cyclohexyl, heptyl, and X = chlorine, bromine, iodine. The reaction, believed to operate by an oxidative addition mechanism, requires 1 or 2 equiv of nickel chloride. It is improved by sodium iodide and gives much higher yields for aryl systems with electron-donating groups than for those with electron-withdrawing groups. This appears to be a particularly useful method for reducing aryl chlorides.

The reactive allylic halides, on the other hand, can be treated with zinc in the presence of aldehydes or ketones to prepare homoallylic alcohols. Luche and Petrier were able to conduct this reaction in aqueous media:[140]

$$\mathrm{R{-}CH{=}CH{-}CH_2{-}X + R'R''C{=}O + Zn} \xrightarrow{)))} \mathrm{R{-}CH{=}CH{-}CH_2{-}C(OH){-}R'R''}$$

This surprising reaction is simply executed: a mixture of the carbonyl compound and an allyl halide is sonicated in a 5:1 NH_4Cl (aq.):THF solvent system for 30–60 min at 0–20°C in an ultrasonic cleaner. Table 5-14 provides some examples.

The mechanism for this reaction has not been worked out but it is clear that the typical organozinc intermediates with zinc–carbon sigma bonds can be ruled out because they are known to react violently with water. While radicals may be involved, a simple free-radical process seems unlikely because the reaction proved to be indifferent to radical promoters as well as radical scavengers. The suggestion by Molle and Bauer[141] that a free-radical pair process on the surface of the metal is important in the classical Barbier reaction may also be relevant in this case.

Expansion of this methodolgy led to the discovery that tin also promoted allylation of carbonyls and that for tin and zinc intermediates, aldehydes were

Table 5-14. Allylations in Aqueous Media

Carbonyl compound	Organic halide	Product	Isolated yield (%)
PhCHO	$CH_2{=}CHCH_2Br$	$PhCH(OH)CH_2CH{=}CH_2$	95
Me_2CHCHO	$Me_2C{=}CHCH_2Cl$	$Me_2CHCH(OH)CMe_2CH{=}CH_2$	84
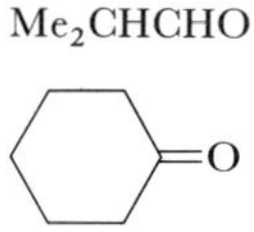	$MeCH{=}CHCH_2Br$	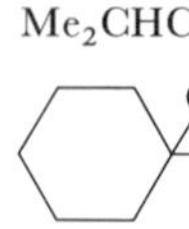 OH, $-CH(Me)CH{=}CH_2$	98

more easily converted to homoallylic alcohols than were ketones.[142] With the addition of copper to the zinc to form a zinc–copper couple the scope of the reaction was broadened to include alkyl halides.[143] However, the mechanism changes and conjugate addition results. Thus, sonication in an ultrasonic cleaner of a mixture of cuprous iodide and zinc powder in any of several aqueous solvents produces a black suspension that promotes 1,4-addition in good yields. The aqueous solvents rule out classical organometallic intermediates, and the observation that cumene completely supresses the reaction suggests a radical pathway:

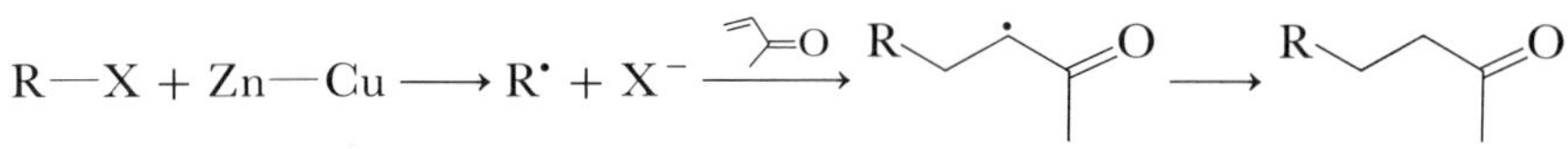

This method offers considerable potential for synthesis because it permits the use of difunctional substrates. For example, 3-iodochloropropane and 2-iodoethanol react cleanly at the iodo position with no interference from the other group:

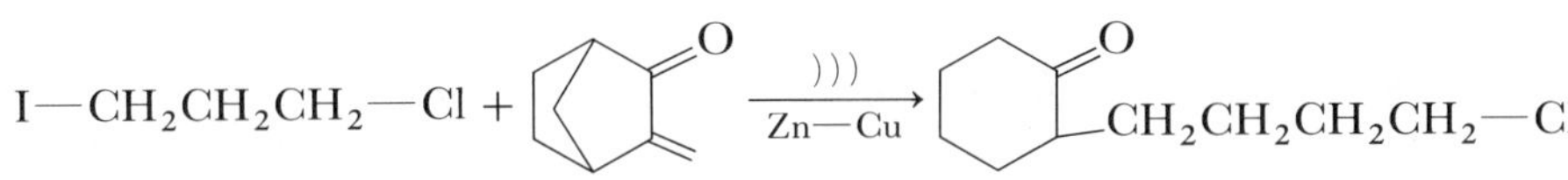

(85%)

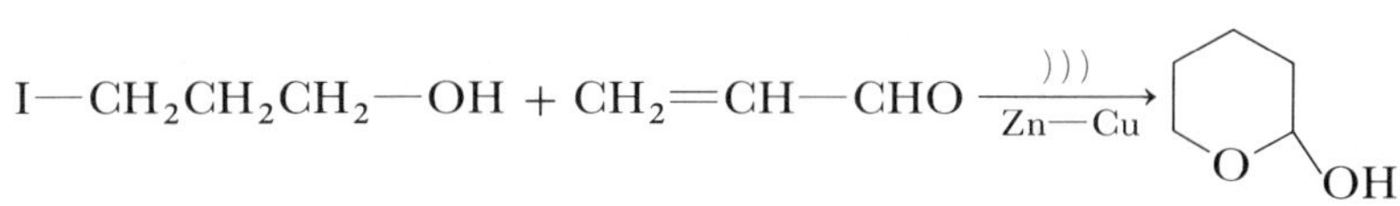

(70%)

Normant developed a new procedure using ultrasound that promoted the regioselective additions of ester and phosphonate derivatives of allyl bromide to alkynes.[144] The resulting adducts are useful precursors to various six- or seven-membered carbo- or heterocycles:

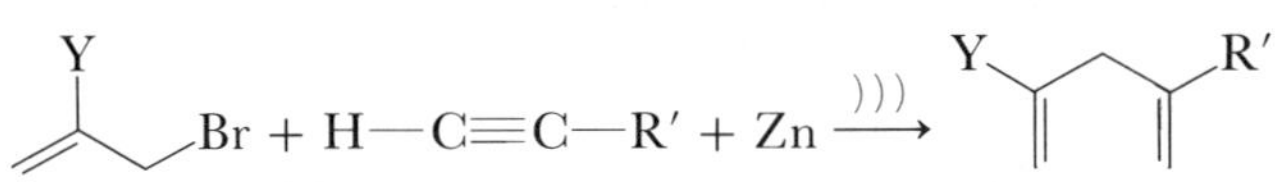

Some examples are given in Table 5-15.

Table 5-15. Regioselective Additions of Substituted Allyl Bromides to Alkynes

Y	R′	Isolated yield of adduct (%)
$—CO_2t$-Bu	—*n*-Bu	70
	$—CH_2Ot$-Bu	64
	$—CH_2SPh$	55
$—PO(OMe)_2$	$—CH_2OSiMe_3$	72
	$—CH_2CH_2OSiMe_3$	66

Table 5-16. Synthesis of α-Trifluoromethylcarbinols

RCHO or RCOR	Product	Reaction time (h)	Isolated yield (%)
PhCHO	$PhCH(CF_3)OH$	0.5	72
$C_5H_{11}CHO$	$C_5H_{11}CH(CF_3)OH$	0.5	61
MeCH=CHCHO	$MeCH{=}CHCH(CF_3)OH$	0.5	62
PhCH=CHCHO	$PhCH{=}CHCH(CF_3)OH$	0.5	68
PhCOMe	$PhC(CF_3)(Me)OH$	1	55
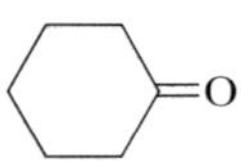	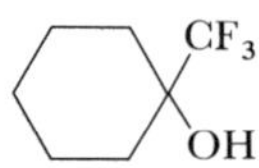	1.5	48
$CH_2{=}CHCH_2COMe$	$CH_2{=}CHCH_2C(CF_3)(Me)OH$	1	45

An interesting modification of the ultrasonically accelerated Barbier reaction to organofluorine chemistry was made by Ishikawa and co-workers. Using trifluoromethyl iodide, zinc, and a variety of aldehydes and ketones dissolved in dimethylformamide solvent and irradiating with low-intensity ultrasonic waves from a cleaning bath, good yields of α-trifluoromethylcarbinols were obtained.[145] Without ultrasound, the reaction did not proceed at all. The results are summarized in Table 5-16.

A series of papers, all dealing with applications of ultrasonic waves to the synthesis of perfluoroalkyl compounds, (R_f = perfluoroalkyl group), quickly emanated from this group. Following the success with carbonyls, Ishikawa developed a convenient method for cross-coupling perfluoroalkyl iodides with allyl, vinyl, and aryl halides.[146]

$$R_f{-}I + R'{-}X + Zn \xrightarrow[\text{“Pd”}]{)))} R_f{-}R'$$

By adding a palladium catalyst to the reagents, which include a perfluoroalkyl iodide, an organic halide, zinc powder, and THF as solvent, good yields of the cross-coupled product could be obtained after 1 h of sonication. Without ultrasound the reaction does not proceed at all.

The organic halides required different catalysts: allyl bromides worked best with palladium acetate, vinylic bromides with tetrakistriphenylphosphinepalladium, and aryl iodides required bistriphenylphosphinepalladium dichloride. Table 5-17 summarizes their results.

Extending the carbon chain of fluorocarbons and increasing their functionality was accomplished by hydroperfluoroalkylation of alkynes under very mild conditions.[147] This procedure involved the formation of a zinc–copper couple from zinc powder and cuprous iodide in the presence of a perfluoroalkyl iodide and a terminal alkyne. The method applies to aryl- and alkyl-substituted alkynes and to propargyl alcohol. These reactions, listed in Table 5-18, require about 2 h of sonication and do not proceed in the absence of ultrasonic waves.

Table 5-17. Palladium-Catalyzed Cross Coupling

R_f	Substrate	Product	Catalyst	Isolated yield (%)
CF_3	$PhCH{=}CHCH_2Br$	$Ph(CF_3)CHCH{=}CH_2$	$Pd(OAc)_2$	63
CF_3	$PhCH{=}CHBr$	$PhCH{=}CHCF_3$	$Pd(PPh_3)_4$	65
CF_3	$4\text{-}MeC_6H_4CH{=}CHBr$	$4\text{-}MeC_6H_4CH{=}CHCF_3$	$Pd(PPh_3)_4$	67
$i\text{-}C_3F_7$	$PhCH{=}CHCH_2Br$	$Ph(i\text{-}C_3F_7)CHCH{=}CH_2$	$Pd(OAc)_2$	78
$i\text{-}C_3F_7$	$PhCH{=}CHBr$	$PhCH{=}CH(i\text{-}C_3F_7)$	$Pd(PPh_3)_4$	72
$n\text{-}C_4F_9$	$MeCH{=}CHCH_2Br$	$Me(n\text{-}C_4F_9)CHCH{=}CH_2$	$Pd(OAc)_2$	68
$n\text{-}C_4F_9$	$PhCH{=}CHBr$	$PhCH{=}CH(n\text{-}C_4F_9)$	$Pd(PPh_3)_4$	62
CF_3	PhI	$PhCF_3$	$PdCl_2(PPh_3)_2$	82
$n\text{-}C_3F_7$	PhI	$Ph(n\text{-}C_3F_7)$	$PdCl_2(PPh_3)_2$	78
$i\text{-}C_3F_7$	PhI	$Ph(i\text{-}C_3F_7)$	$PdCl_2(PPh_3)_2$	87
$i\text{-}C_3F_7$	$4\text{-}MeC_6H_4I$	$4\text{-}MeC_6H_4(i\text{-}C_3F_7)$	$PdCl_2(PPh_3)_2$	81
$i\text{-}C_3F_7$	$2\text{-}MeC_6H_4I$	$2\text{-}MeC_6H_4(i\text{-}C_3F_7)$	$PdCl_2(PPh_3)_2$	59

A mechanism involving a copper–zinc complex was proposed:

$$R_fI + Zn \xrightarrow{)))} [R_fZnI] \xrightarrow{)))} [(R_f)_2CuZnI] \xrightarrow{HC\equiv CR} R_fCH{=}CHR$$

Functionalizing perfluoroalkyl groups in good yields was also readily accomplished using zinc and carbon dioxide:[148]

$$R_f{-}I + Zn \xrightarrow[2)\ CO_2,\ H+]{1)\ )))} R_f{-}COOH$$

where if $R_f = i\text{-}C_3F_7$, yield is 48%; if $R_f = n\text{-}C_4F_9$, yield is 61%; if $R_f = n\text{-}C_6F_{13}$, yield is 77%; and if $R_f = n\text{-}C_8F_{16}$, yield is 72%. Ultrasound accelerates the reduction of the perfluoroalkyliodide to the radical anion, $[R_fI]^{\overline{\cdot}}$, which reacts

Table 5-18. Perfluoroalkylation of Alkynes

R_f	Substrate	Product	Isolated yield (%)	E/Z ratio
CF_3	$HC\equiv CPh$	$CF_3CH{=}CHPh$	65	28/72
CF_3	$HC\equiv CBu$	$CF_3CH{=}CHBu$	71	24/76
CF_3	$HC\equiv CCH_2OH$	$CF_3CH{=}CHCH_2OH$	61	32/68
$i\text{-}C_3F_7$	$HC\equiv CPh$	$i\text{-}C_3F_7CH{=}CHPh$	52	31/69
$i\text{-}C_3F_7$	$HC\equiv CCH_2OH$	$i\text{-}C_3F_7CH{=}CHCH_2OH$	65	33/67
$n\text{-}C_3F_7$	$HC\equiv CCH_2OH$	$n\text{-}C_3F_7CH{=}CHCH_2OH$	66	30/70
$n\text{-}C_4F_9$	$HC\equiv CCH_2OH$	$i\text{-}C_4F_9CH{=}CHCH_2OH$	74	27/73

Table 5-19. Chiral Perfluoroalkyl Ketones from Enamines

R_fX	R_1	R_2	Isolated yield (%)	Optical purity
CF_3I	C_2H_5	CH_3	38	62
CF_3Br	C_2H_5	CH_3	46	76
C_2F_5I	C_2H_5	CH_3	51	54
n-C_3F_7I	C_2H_5	CH_3	48	59
n-C_4F_9Br	C_2H_5	CH_3	58	64
n-$C_8F_{17}I$	C_2H_5	CH_3	56	72

with carbon dioxide efficiently in dimethylformamide. However, in solvents in which carbon dioxide has much lower solubility, such as THF and acetonitrile, no carboxylic acids were isolated. The homocoupled product, $R_f—R_f$, was produced instead.

With convenient routes for attaching perfluoroalkyl groups to other carbon moieties available, attention was drawn to the problem of asymmetric induction at selected sites. Using chiral enamines, a perfluoroalkyl iodide, zinc powder, and a catalytic amount of dicyclopentadienyltitaniumdichloride, Ishikawa was able to prepare optically active ketones in moderate yields with optical purities ranging from 59 to 76%:[149]

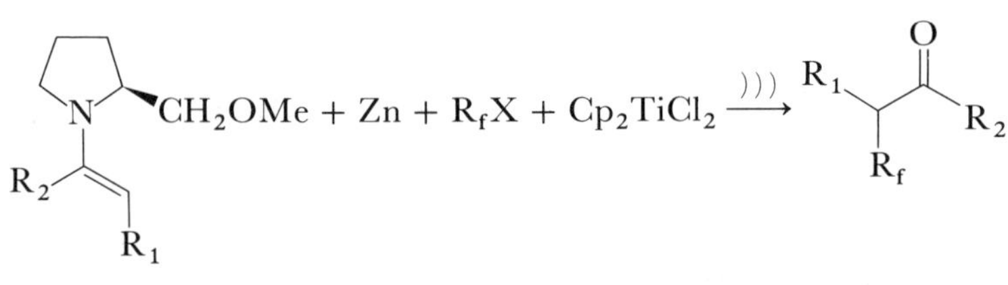

Table 5-19 contains pertinent data for one of two systems studied.

Solladie-Cavallo obtained asymmetric induction performing the zinc-mediated Barbier reaction between perfluoroalkyl iodides and a chiral chromium complex:[150]

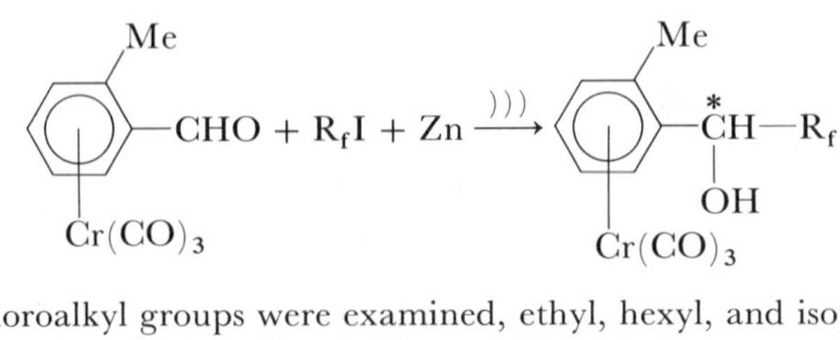

Three perfluoroalkyl groups were examined, ethyl, hexyl, and isopropyl, giving product yields of 50–85% in dimethylformamide with the percentage of asymmetric induction ranging from 30 to 66%.

Reactions of chlorosilanes with carbonyl compounds in the presence of zinc benefit from irradiation with ultrasonic waves. α-Dicarbonyls are converted to bissiloxyalkenes in very good yields in less than 30 min:[151]

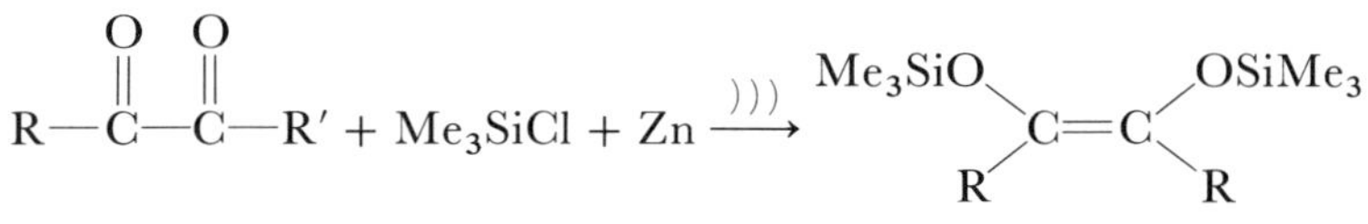

(60–90%)

In the absence of ultrasonic waves not even 2 or 3 h of stirring produce the same yields. 1,4-Benzoquinone, 1,4-naphthoquinone, 9,10-anthraquinone are readily converted to the bissiloxyarene derivates under the same conditions.

The reaction of simple ketones and aldehydes is even more interesting because, with the appropriate stoichiometry, the carbonyl cannot only be silylated but can be persuaded to couple to another carbonyl group to form bissiloxyalkanes:[152]

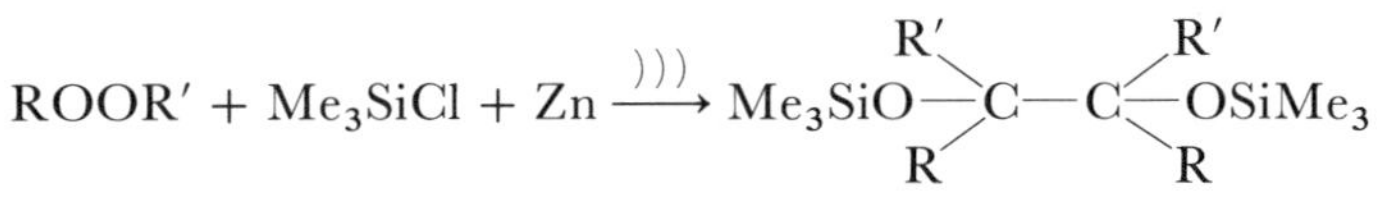

Some of the results are summarized in Table 5-20.

Benzaldehyde gave modest yields of the coupled product and 15% of *trans*-stilbene. By using large excesses of zinc and trimethylchlorosilane, the stilbene yield was increased to 36%. So far, only 1-indanone has produced high yields, 83%, of the unsaturated dimer:[152]

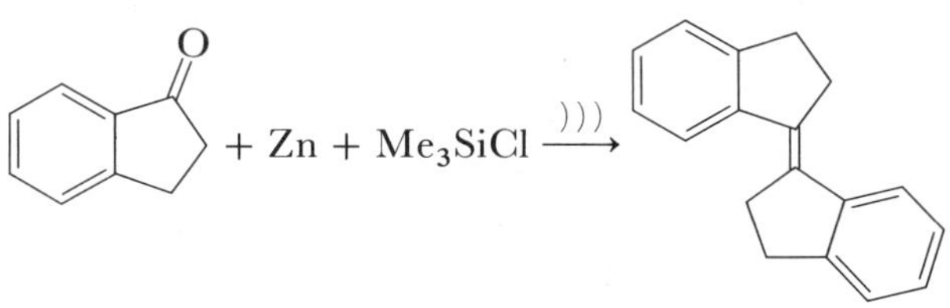

Table 5-20. Reductive Coupling of Carbonyls with Zinc and Trimethylchlorosilane[a]

R	R′	t (h)	Yield (%))))	Yield (%) Stir
p-MeO—Ph	Me	8	74	54
p-F—Ph	Me	6	63	48
Ph	Ph	6	84	65
p-Me—Ph	Me	2	76	63
Ph	Me	2	65	55
Ph	H	4	56	42

[a] Carbonyl: Me_3SiCl:Zn, 1:1:5.

Most of the carbonyls tested couple and undergo a pinacol-type rearrangement via a bissiloxyalkane intermediate:

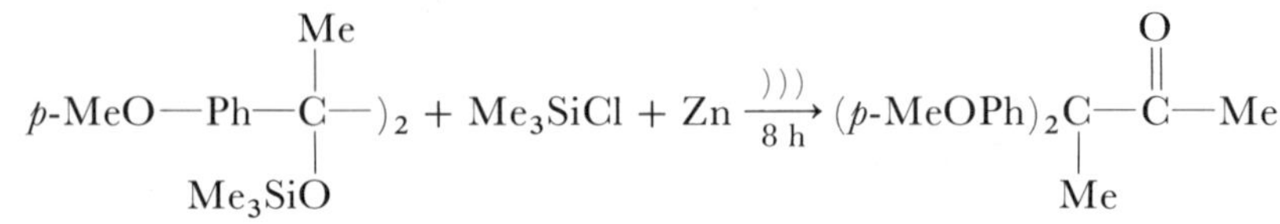

(100%)

The synthesis of biscyclooctadienyl ruthenium complexes was improved using ultrasonic waves to accelerate the reduction of ruthenium trichloride hydrate in the presence of the cyclooctadiene ligand.[153] For example, η^6-1,3,5-(cyclooctatriene)(η^4-1,5-cyclooctadiene)-ruthenium(0) was prepared in 70–93% yields in 2 h at 70°C. Comparisons to nonultrasonic conditions were not reported.

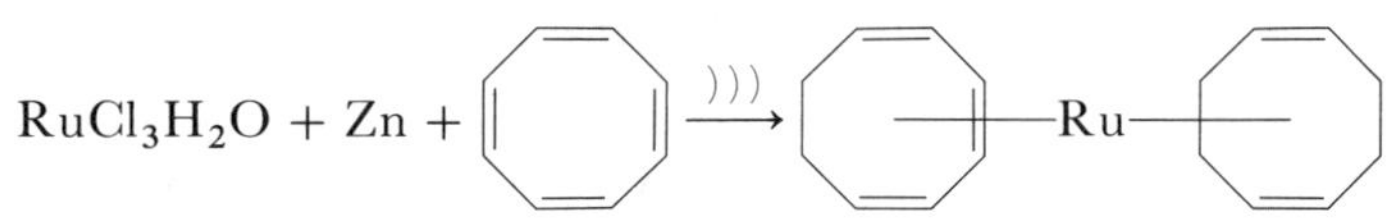

Magnesium, which has received less attention than zinc because it is less diverse in synthesis, is also activated by ultrasonic waves. As mentioned earlier, Renaud observed in 1950 that the Grignard reaction was accelerated by ultrasonic waves.[76] In 1983, Sprich and Lewandos suggested that ultrasonic waves from a low-intensity source, such as an ultrasonic cleaner, shortened the induction periods for Grignard reactions primarily through removal of adsorbed water and alcohol from magnesium surfaces and not via latticce defects and dislocations.[154] Sixty- to 100-fold reductions in reaction time were observed, but yields of the Grignard reactions with and without ultrasound were essentially identical.

Brown used ultrasound to improve a new procedure for the synthesis of organoboranes from Grignard reagents and boron trihalides in ether:[155]

$$3\ RX \xrightarrow[\text{Mg, } BF_3\cdot OEt_2]{)))} R_3B$$

The Grignard route makes easily accessible those organoboranes not achievable by hydroboration methodology. The yields are virtually quantitative and scaleup is readily accomplished. The results are summarized in Table 5-21.

Eaborn et al. found that ultrasound improved the yield of the highly hindered tris(phenyldimethylsilyl)methane from 15 to 22% and reduced the reaction time from 48 to 6 h:[156]

$$PhMe_2SiCl + CCl_4 + Mg \xrightarrow{)))} (PhMe_2Si)_3CH$$

Bönneman developed a novel approach to the synthesis of a variety of metallocenes and π-allyl complexes using magnesium powder, anthracene, and ultrasound to generate a highly reactive reducing catalyst composed of a solvated

Table 5-21. Preparation of Trialkylboranes via a Modified Organometallic Route

Alkyl Halide	Product	Reaction Time (h))))	Reaction Time (h) stir	Yield (%))))	Yield (%) stir
n-C_3H_7Br	$(n$-$C_3H_7)_3B$	0.25	2	100	98
n-C_4H_9Br	$(n$-$C_4H_9)_3B$	0.25	2	100	99
s-$C_5H_{11}Br$	$(s$-$C_5H_{11})_3B$	0.5	36	96	97
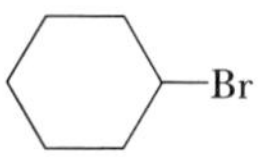	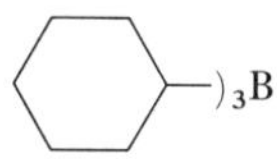	0.5	24	99	90
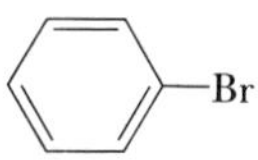	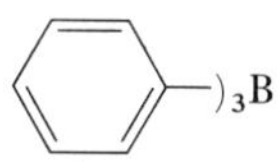	0.25	3	97	94
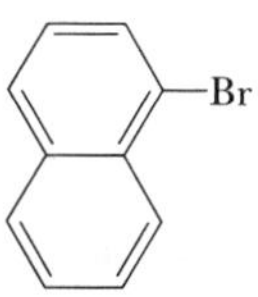	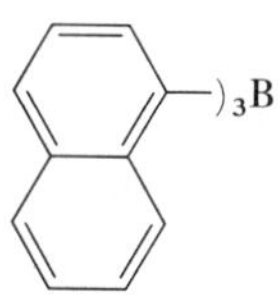	0.5	24	93	91
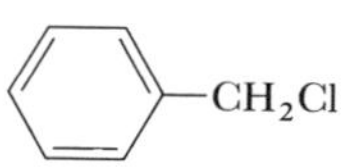	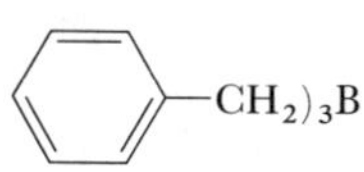	0.5	24	99	96
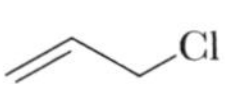	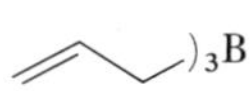	0.25	3	94	94
n-$C_7H_{15}I$	$(n$-$C_7H_{15})_3B$	0.25	2	90	88

magnesium–anthracene complex:[157]

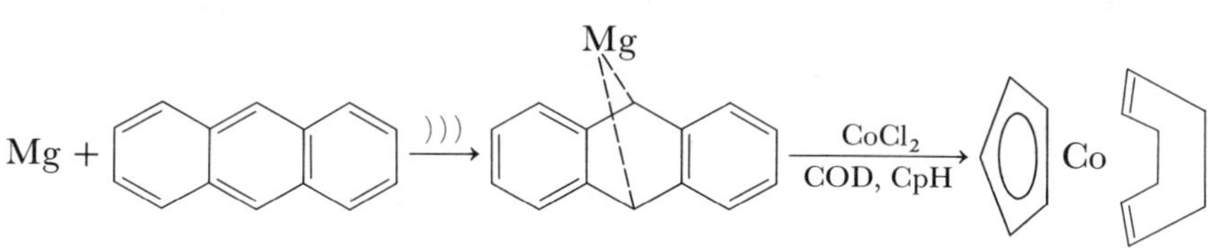

where COD = 1,5-cyclooctadiene and CpH = cyclopentadiene. The ultrasonic activation of aluminum toward alkyl halides to form alkylaluminum sesquihalides in good yields was discovered by two groups.[158,159,160] Ultrasound reduced the induction period (by half in some cases) and significantly increased the yield of the sesquihalide compared to room temperature stirring.[158,159] Lin used these improvements for a convenient sesquihalide synthesis of trialkylboranes:[160]

$$3\,R{-}X + 2\,Al \xrightarrow{)))} R_3Al_2X_3 \xrightarrow{(EtO)_3B} R_3B$$

where when R = Me, X = I; and when R = Et, X = Br.

Transition Metals

The use of ultrasonic waves with transition metals has been directed largely toward activating catalysts and early work in the field has been summarized above. In recent years, only a few new applications have been published.

A simple laboratory procedure for hydrogenating olefins was developed using formic acid as the source of hydrogen and an ultrasonic cleaning bath.[161] Nearly quantitative yields of hydrogenated product were obtained in 1 h or less, a significant improvement over earlier efforts to use formic acid and its derivatives:[162–166]

$$-\overset{|}{\underset{}{C}}=\overset{|}{C}- + HCOOH \xrightarrow[Pd/C]{)))} H\overset{|}{\underset{|}{C}}-\overset{|}{\underset{|}{C}}H + CO_2$$

The method requires only the simplest equipment and minimizes the need for hydrogen gas in the laboratory, a consideration for the occasional user and for undergraduate lab classes. The results are summarized in Table 5-22.

Using a high-intensity horn fitted to a steel reaction vessel as shown in Fig. 5-1, hydrogen gas could be used to hydrogenate quantitatively gram quantities of the olefins in Table 5-22 in less than 1 hour at 15–25° and less than 2 atm of

Table 5-22. Hydrogenation of Olefins by Formic Acid Catalyzed by Pd/C[a]

Olefin	Product	Isolated yield (%)
$C_7H_{15}CH{=}CH_2$	C_9H_{20}	100
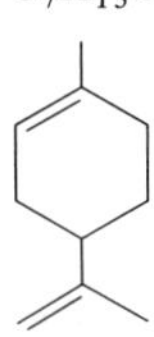	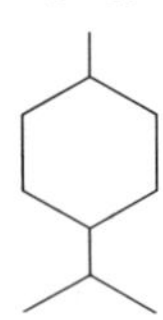	95
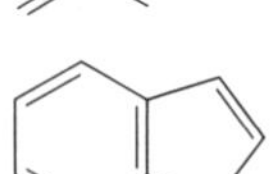	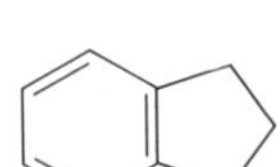	100
trans-PhCH=CHPh	$PhCH_2CH_2Ph$	100
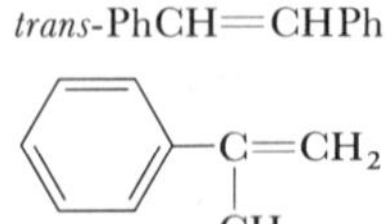	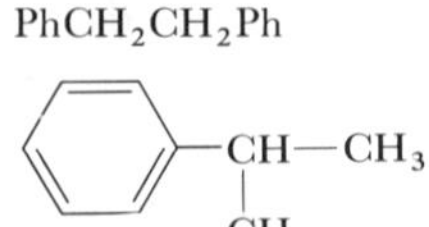	100
trans-PhC(H)=C(H)—C(=O)—CH_3	$PhCH_2CH_2C(=O)CH_3$	100
n-$C_4H_9OCH{=}CH_2$	*n*-$C_4H_9OC_2H_5$	93
Methyl 9-octadecenoate	Methyl 9-octadecanoate	90

[a] Typical olefin:formic acid:catalyst mole ratios were 1:4:0.08.

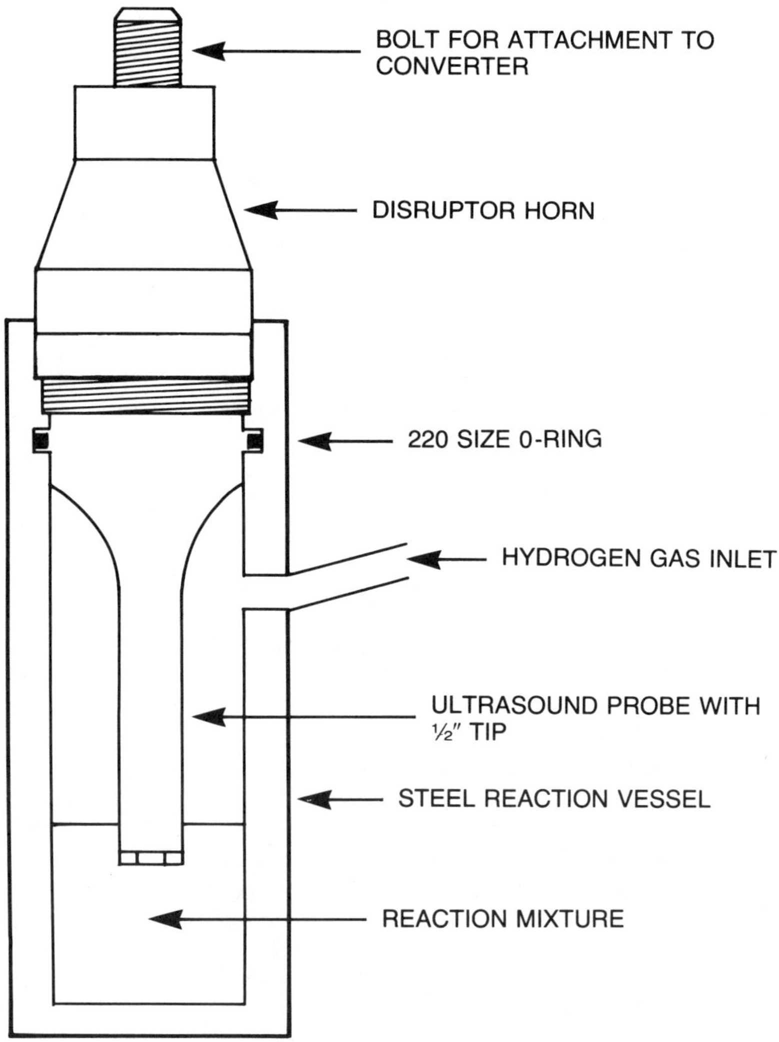

Figure 5-1. Apparatus for pressurized sonication.[167]

pressure.[167] There is precedent for this in the work of Marinesco,[70] Möckel,[71] and Saracco,[72] all of whom used equipment and conditions not commonly used for synthesis.

Recently, high-intensity ultrasonic waves were used to activate commercially available nickel powder to hydrogenate olefins,[168,169] with rate increases of 10^5. The origin of the enhanced reactivity was shown by scanning electron microscopy and Auger electron spectroscopy to be removal of the passivating oxide layer normally found on Ni surfaces. It appears that the mechanism of this surface cleaning comes from interparticle collisions.

Table 5-23. Ultrasonically Accelerated Hydrosilation Reactions[a]

Alkene	Silane	Time (h)	Product	Isolated yield (%)
1-Hexene	$HSiCl_3$	1	n-$C_6H_{13}SiCl_3$	90
	$HSiMeCl_2$	1	n-$C_6H_{13}SiMeCl_2$	95
	$HSiEt_3$	2	n-$C_6H_{13}SiEt_3$	74
4-Methyl-1-pentene	$HSiCl_3$	1	$(CH_3)_2CHCH_2CH_2CH_2SiCl_3$	94
	$HSiMeCl_2$	1	$(CH_3)_2CHCH_2CH_2CH_2SiMeCl_2$	96
	$HSi(OEt)_3$	1	$(CH_3)_2CHCH_2CH_2CH_2Si(OEt)_3$	93
Styrene	$HSiCl_3$	1.5	$PhCH_2CH_2SiCl_3$	94
	$HSiMeCl_2$	1.5	$PhCH_2CH_2SiMeCl_2$	94

[a] Typical alkene:silane:Pt mole ratios were $0.05:0.1:5 \times 10^{-6}$.

Ultrasound accelerates the hydrosilation of olefins:[170]

$$-\overset{|}{C}=\overset{|}{C}- + R_3Si-H \xrightarrow[\text{Pt/C}]{)))} R_3Si-\underset{|}{\overset{|}{C}}-\underset{|}{\overset{|}{C}}-H$$

Commercially available platinum on carbon gives high yields of addition product at 30°C and atmospheric pressure. In the absence of ultrasonic waves, even 10–48 h of stirring produced less than 10% conversion. Table 5-23 summarizes the results.

Nickel powder prepared from nickel iodide and lithium and sonicated in the presence of a hydrosilane and an olefin also catalyzes hydrosilation at 30°C and atmospheric pressure.[152]

Han has reported that ultrasound accelerates the reduction of a variety of nitroaromatics using catalytic quantities of iron on carbon and hydrazine:[171]

$$Ar-NO_2 + H_2NNH_2 \xrightarrow[\text{Fe/C}]{)))} Ar-NH_2 \quad (87\text{–}95\%)$$

where Ar = H, p-MeO, o-MeO, p-NH_2, o-NO_2, p-NO_2.

Studies of the effects of ultrasonic waves on reactions in which transition metals are involved in stoichiometric quantities have only recently begun to attract attention. Suslick and Johnson reported that sonochemical activation of transition-metal powders permitted the synthesis of transition-metal carbonyl anions at low temperature and low pressures of carbon monoxide.[172] The yields are comparable to those obtained under much more vigorous conditions (e.g. 200°C, 300 atm). The results are summarized in Table 5-24.

Typically, a 20% excess of sodium sand is added to a THF solution of a transition metal halide and the system is flushed with carbon monoxide. The slurry is then irradiated for 1000 min at 10°C. An atmosphere of carbon monoxide is maintained throughout. The reaction scale is small, 1 mmol, but scale up is feasible because large-scale ultrasound technology is well developed. This technique may

Table 5-24. Sonochemical Synthesis of Transition-Metal Carbonyl Anions

Metal carbonyl anions	Yields[a] (%) 4.4 atm CO	1 atm CO
$[W_2(CO)_{10}]^{2-}$ from WCl_6	47	40
$Mo_2(CO)_{10}]^{2-}$ from $MoCl_5$	54	39
$[Cr_2(CO)_{10}]^{2-}$ from $CrCl_3$	7	3
$[Ta(CO)_6]^-$ from $TaCl_5$	6	
$[Nb(CO)_6]^-$ from $NaCl_5$	51	23
$[V(CO)_6]^-$ from $VCl_3(THF)_3$	35	23

[a] Yields are based on metal halide and have been calculated spectrophotometrically.

be particularly useful for the synthesis of carbon-13 enriched metal carbonyl anions.

The possible mechanisms of this ultrasonic activation include (1) the creation of extremely reactive dispersions of the metals on a sodium chloride matrix, (2) improved mass transport between bulk solution and the reactive surfaces, and (3) direct trapping with carbon monoxide of reactive metallic species formed during the reduction of the metal halide. The first and second mechanisms are not operable because separate reduction of the metal halide to the metal dispersion under argon followed by exposure to carbon monoxide gave no metal carbonyl anions. Spectroscopic studies of intermediates formed during the reaction support a stepwise reduction process wherein metal species are formed upon partial reduction of the sodium surface and trapped by the carbon monoxide.

The activation of freshly generated metal powders by ultrasonic waves was observed for the coupling of benzyl halides.[118]

$$PhCH_2{-}Br + Ni^*(Cu^*) \xrightarrow{)))} PhCH_2CH_2Ph$$

Commercial powders generated no product after 8.5 h of refluxing. The optimum conditions, i.e., highest yields in the shortest time at the lowest temperature, are when ultrasound is used during generation of the activated powder and throughout the coupling reaction. Using this procedure, 80–100% yields of bibenzyl can be obtained from either copper or nickel in approximately 1 h.[118]

Inouye and co-workers reported that ultrasound sometimes promoted the reaction of palladium black with allylic halides to form π-allylic palladium complexes:[173]

$$H_2C{=}CHCH_2{-}X + Pd \xrightarrow{)))}$$ 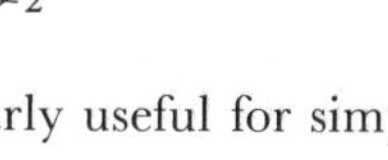

where X = Br, I. The technique is believed to be particularly useful for simple allylic complexes.

Table 5-25. Ultrasonically Accelerated Reductions of Aromatic Halids

Ar—X	Ar—H	Isolated yield (%)	Time (h)[a]	Comparison[174] (yield; time; temp.; solvent)[b]
o-Bromotoluene	Toluene	98	5	21%; 24h; 25°C, THF
p-Bromotoluene	Toluene	97	5	20%; 24h; 25°C, THF
m-Bromotoluene	Toluene	81	6	
o-Iodotoluene	Toluene	95	5	92%; 24h; 100°C; diglyme
o-Bromoanisole	Anisole	98	4	58%; 24h; 100°C; diglyme
p-Bromoanisole	Anisole	70	7	35%; 24h; 100°C; diglyme
p-Iodoanisole	Anisole	73	7	91%; 24h; 100°C; diglyme
1-Bromonaphthalene	Naphthalene	99	6	99%; 6h; 65°C; THF[c]
1-Iodonaphthalene	Naphthalene	99	6	72%; 24h; 65°C; THF
1-Chloro-3-bromobenzene	Chlorobenzene	98	4	95%; 24h; 25°C; THF[d]
1-Chloro-4-bromobenzene	Chlorobenzene	98	4	

[a] Typical conditions: dimethoxyethane solvent using a 1:1 molar ratio of $LiAlH_4$:ArX at 35°C.
[b] $LiAlH_4$:ArX was 1:1.
[c] $LiAlH_4$:Ar—X was 4:1.
[d] $LiAlH_4$:Ar—X^4 was 1.5:1.

REACTIONS WITH NONMETALS

Metal Hydrides

There is a growing literature of synthetically important reactions involving nonmetals that are accelerated by ultrasonic waves. Lithium aluminum hydride, which reduces aryl halides only reluctantly under reflux conditions,[174] is an efficient reagent when sonicated at room temperature.[175]

$$\text{Ar—X} + \text{LiAlH}_4 \xrightarrow{)))} \text{Ar—H}$$

Table 5-25 summarizes the effects of ultrasound on a variety of aryl halide reductions and compares some of the results to those obtained under optimized thermal conditions.

The activity of lithium aluminum hydride is sufficiently enhanced by ultrasonic waves that this reagent can reduce halo, alkoxy, and amino derivatives of silicon, germanium, and tin in nonpolar solvents, such as hexane and benzene.[176] Normally, donor solvents, such as diethyl ether or tetrahydrofuran, are required. In the absence of ultrasonic waves none of the derivatives was reduced. Table 5-26 summarizes the results.

Potassium hydride reacts very efficiently with hydrosilanes to give silyl anions in the presence of ultrasonic waves.[177] Sonication permits the generation of phenyl- and vinyl-substituted anions in nearly quantitative yield allowing spectroscopic characterization of novel species:

$$\text{RMe}_2\text{Si—H} + \text{KH} \xrightarrow{)))} \text{RMe}_2\text{Si:}^- \text{K}^+ + \text{H}_2$$
$$(95\text{–}100\%)$$

Table 5-26. Ultrasound-Promoted Reduction of Derivatives of Group 14 with $LiAlH_4$

Substrate[a]	Solvent	Reaction time (h)	Temp. (°C)	Hydride	Yield (%) (by GLC)
Me_3SiCl	Hexane	3	40	Me_3SiH	80[b]
Ph_2SiHCl	Hexane	2	40	Ph_2SiH_2	>95
Et_3GeCl	Pentane	4.5	40	Et_3GeH	>95
Me_3SnCl	Cyclohexane	2.5	25	Me_3SnH	>95
Me_3SiNEt_2	Hexane	3	40	Me_3SiH	70[b]
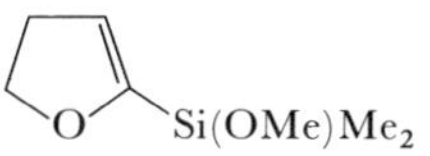	Pentane	3	25	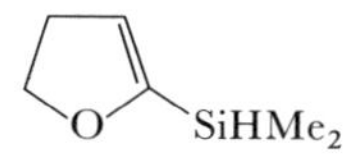	100
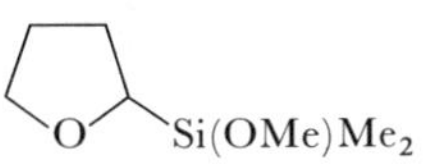	Cyclohexane-d_{12}	2	25	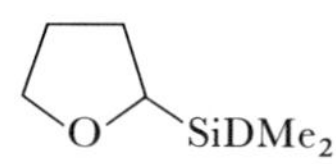	100[c]

[a] 0.2 M solutions of substrate; $LiAlH_4$:substrate molar ratio was 3:1.
[b] Isolated yield.
[c] $LiAlD_4$ was used as a reducing agent.

where R = phenyl, vinyl. This was useful in the determination of the chemical shifts of the potentially aromatic silyl anions:[177]

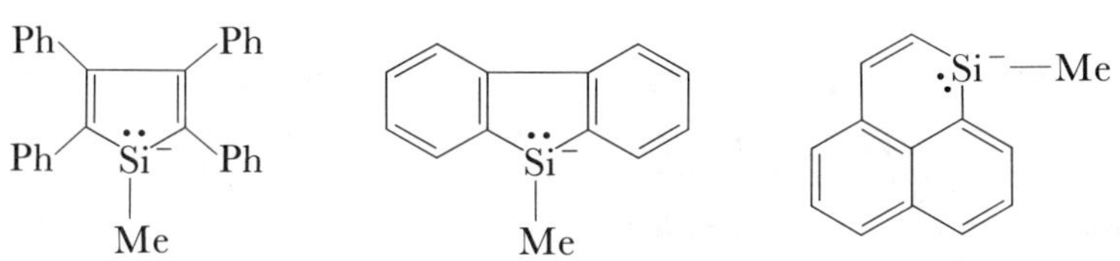

The absence of side products lead to unequivocal asignments and the conclusion that, although the carbon analogs have ring current and a high degree of resonance stabilization, silicon does not participate in significant π-bonding and these anions are not aromatic.

Brown and Racherla discovered that ultrasound dramatically enhanced the rates of many slow heterogeneous hydroboration reactions and has a modest accelerating effect on slow homogeneous reactions.[178] For example, the preparation of tricyclohexylborane involves the formation of insoluble dicyclohexylborane and slow hydroboration of the third cyclohexene, a reaction that requires 24 h at 25°C for completion:

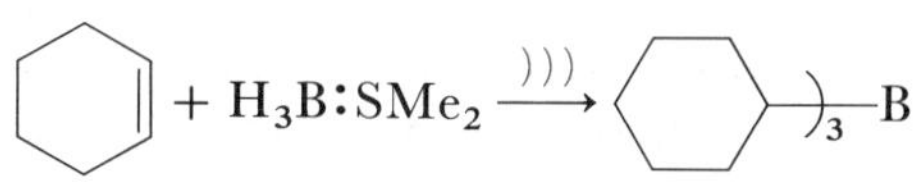

(99%)

With ultrasound the reaction is complete in 1 h.

Table 5-27. Oxidation by Solid Potassium Permanganate under Ultrasonic Irradiation[a]

Run	Substrate	Product	Solvent	Time (h)	Yield (%)	
					)))	Mech. agit.
1	2-Octanol	2-Octanone	Hexane	5	92.8	2.6
2	Cyclohexanol	Cyclohexanone	Hexane	5	53.2	4.2
3	Cyclododecanol	Cyclododecanone	Benzene	24	85.7	15.7
4	$PhCH{=}CHCH_2OH$	$PhCH{=}CHCHO$	Benzene	3	82.8	4.5
5	1-Octene-3-ol	1-Octene-3-one	Benzene	24	43.6	0.8
6	2-Decanol	2-Decanone	Benzene	40	89.4	
7	4-Decanol	4-Decanone	Benzene	40	89.7	
8	PhCH(OH)Ph	PhCOPh	Benzene	5	98.9	
9	Cyclododecanol	Cyclododecanone	Benzene	32	84.1	
10	Cholestanol	Cholestanone	Benzene	20	65.3	
11	$PhCH{=}CHCH_2OH$	$PhCH{=}CHCHO$	Benzene	14	74.7	
12	$PhCH_2OH$	PhCHO	Benzene	1.5	29.7[b]	
13	1-Octanol	Octanoic acid	Hexane	14	80.5	

[a] Runs 1–5, calibrated yields by gas chromatography; runs 6–13, isolated yields on preparative scale. Run 12 at 20°C; all others at 50°C.
[b] Unreacted benzyl alcohol (47.4%) was recovered.

Metal Oxides

Ando and co-workers found that ultrasonic irradiation facilitates heterogeneous oxidation of secondary alcohols to ketones using potassium permanganate in non-polar solvents.[179] Ultrasound eliminates the need for traces of water to promote the reaction and gives yields that are much greater than those obtained from mechanical agitation. Some primary alcohols are difficult to stop at the aldehyde and are converted to the acid. Table 5-27 summarizes the results.

It was also observed that ultrasound eliminated the normally observed solvent dependence of the reaction. For example, the yield of 2-octanone after 24 h of mechanical agitation was 7.9, 36.8, and 78.5% in hexane, benzene, and acetone, respectively, whereas the yield after 5 h of sonication was 92.8, 87.2, and 82.5%, respectively.

The same group also investigated the alumina–water-promoted cyanide substitution of alkyl halides finding that ultrasound doubled the yield of the reactions:[180]

$$RCH_2Br + KCN + Al_2O_3 \xrightarrow{)))} RCH_2CN$$

Ultrasonic pretreatment of the reaction mixture diminished the usual ether and alcohol byproducts from hydrolysis. Ultrasound produced better results (higher yields, shorter reaction times) than those obtained using phase-transfer catalysts under conventional conditions. Further studies showed that yields of aliphatic nucleophilic substitution of halides by acetate and thiocyanate in the presence of silica gel were also increased by ultrasound.[181]

During the course of these studies, it was observed that the reaction of benzyl bromide and potassium cyanide in toluene solvent in the presence of alumina gave different results when the reagents were mechanically agitated from when they were irradiated with ultrasonic waves.[181] Mechanical agitation led to *o*- and *p*-benzyltoluene:

$$PhCH_2Br + KCN + Al_2O_3 \xrightarrow[PhMe]{\text{mech. agit.}} o\text{-}, p\text{-}PhCH_2\text{—}PhMe \quad (75\%)$$

a product of Friedel-Crafts alkylation, whereas ultrasonic irradiation produced benzyl cyanide in 71% yield. Thus, ultrasound completely switched the reaction course. Ultrasound also accelerates the Friedel Crafts reaction if there is no cyanide present, leading the authors to conclude that ultrasound assists the blocking of the sites on alumina that are responsible for alkylation and increases the nucleophilicity of the cyanide ion at the alumina surface.[182]

Basic alumina catalyzes the aldol condensation of ketones[183] and, in the presence of ultrasonic waves, was reported to enhance this reaction still further.[184] This report has been challenged,[185] however, and the effect of ultrasonic waves on this reaction is unclear.

Varma and Kabalka found that sonication improved the rates of basic alumina-catalyzed condensations of *o*-hydroxybenzaldehydes with *β*-nitrostyrenes:[186]

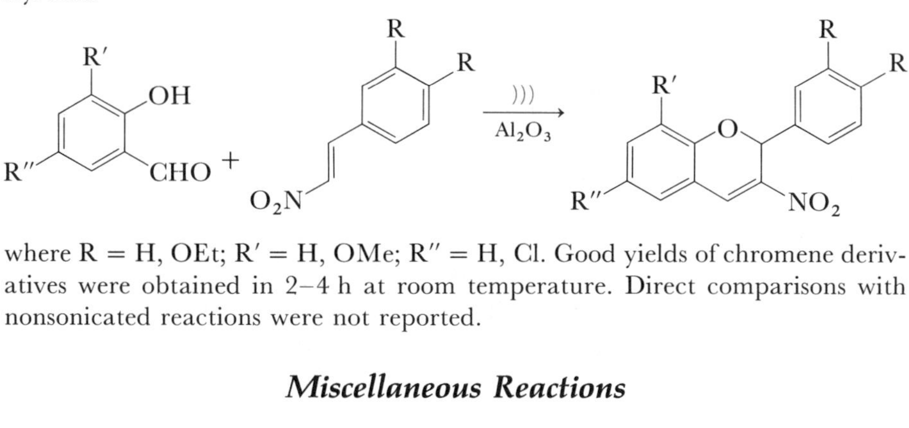

where R = H, OEt; R′ = H, OMe; R″ = H, Cl. Good yields of chromene derivatives were obtained in 2–4 h at room temperature. Direct comparisons with nonsonicated reactions were not reported.

Miscellaneous Reactions

Raucher and Klein devloped a convenient procedure for a high yield synthesis of thioamides from amides and P_4S_{10} using ultrasonic waves at room temperature:[187]

$$R\text{—}\overset{\overset{\displaystyle O}{\|}}{C}\text{—}ONR'_2 + P_4S_{10} \xrightarrow[THF]{)))} R\text{—}\overset{\overset{\displaystyle S}{\|}}{C}\text{—}ONR'_2$$

where R = Ph, Me; R'_2 = Me, Me; H, Me; H, Ph. Ultrasound eliminates the need for large excess of the sulfide, lowers the temperature requirement by 40–80°C, and shortens the reaction time from 1 or 2 d to 1 or 2 h.

Table 5-28. Dichlorocarbene Addition to Alkenes[a]

Olefin	Product	Time (h)	Isolated yield (%)
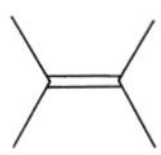	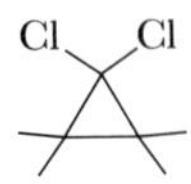	0.7	74
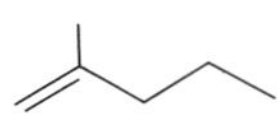	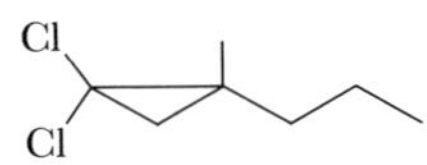	1.8	80
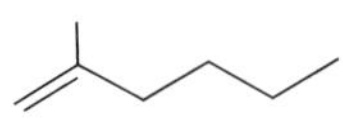	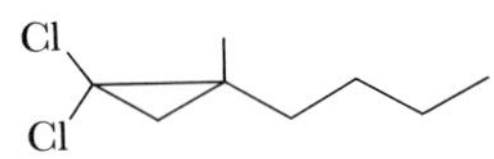	1.0	96
	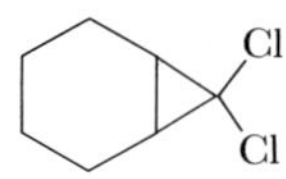	3.0	93
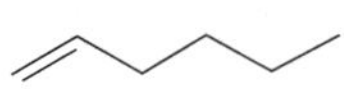	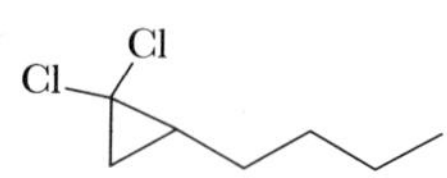	3.0	62
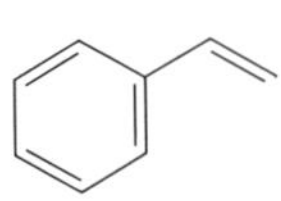	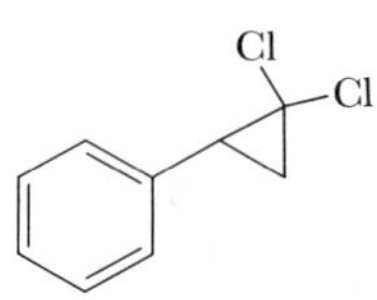	1.0	96
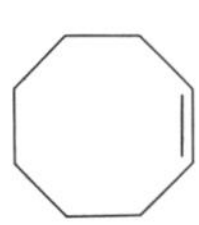	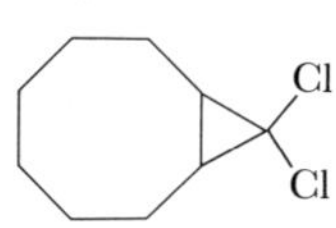	5.0	99
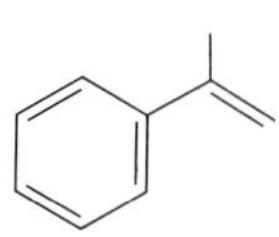	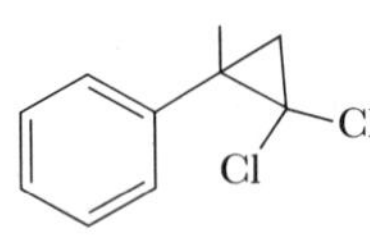	1.5	97
	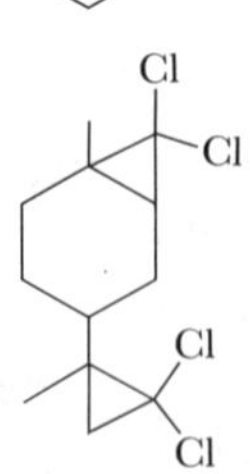	3.0	95

[a] NaOH (powdered) : olefin : $CHCl_3$ ratio was 20 mmol : 20 mmol : 20 mL.

The reaction between solid sodium hydroxide and chloroform to produce dichlorocarbene normally requires a phase-transfer catalyst[188] but Regen and Singh found that ultrasound circumvents that need and produces dichlorocarbene in excellent yields in the absence of phase-transfer catalysts:[189]

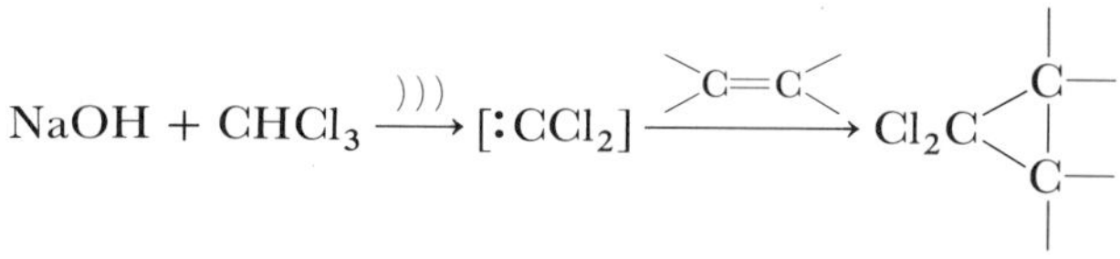

The highest yields of dichlorocyclopropanes were obtained when ultrasound and stirring were combined. The results are summarized in Table 5-28.

Some reactions involving immiscible liquids have been accelerated by ultrasonic waves. Lash demonstrated that ultrasound promoted the oxidation of alkenes by aqueous potassium permanganate, the Baeyer test, eliminating the need for a phase-transfer catalyst.[190] Moon et al. noted that the basic hydrolysis of aromatic carbonxylic esters was almost 10 times more efficient when sonicated.[191] In contrast, Kristol found that the rate of alkaline hydrolysis of nitrophenyl esters in a one-phase system did not exceed 15% when irradiated with ultrasonic waves.[192]

Two- and three-phase systems that are promoted by ultrasound have been subjected to ultrasonic waves and significant rate and yield improvements have been observed. Davidson and co-workers applied ultrasound to the *N*-alkylation of a variety of amines by alkyl halides using alkylammonium halides or polyethylene glycol methyl ether (PEGME) as a phase-transfer catalyst:[193]

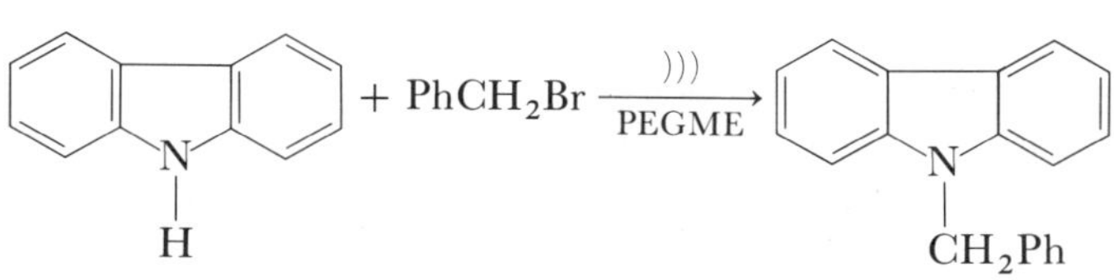

The alkylation of carbazole was monitored showing that 8 h of stirring were required to achieve 50% reaction whereas only 30 min were required with ultrasound. Table 5-29 summarizes their results.

Ezquerra and Alvarez-Builla conducted similar experiments with isoquinoline derivatives to achieve alkylation at carbon:[194]

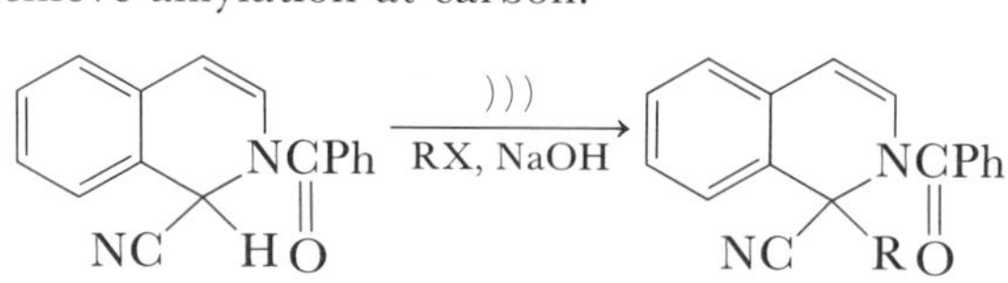

Their results are given in Table 5-30.

Sonication of phase-transfer reactions are not always faster than stirred systems. Reeves and McClusky reported that for the two-phase displacement of halide

Table 5-29. Ultrasound-Promoted Phase-Transfer Catalyzed *N*-Alkylation of Amines[a,b]

Amine	Alkylating agent	Catalyst	Conditions	Time (h)	% Yield of *N*-alkylated Product
Indole	Iodomethane	PEGME	Stirred at 20°	5	60
	Iodomethane	PEGME	Ultrasound	0.5	65
	Benzyl bromide	PEGME	Stirred at 20°	8	80
	Benzyl bromide	PEGME	Ultrasound	2	95
	1-Bromododecane	PEGME	Stirred at room temp.	72	48
	1-Bromododecane	PEGME	Ultrasound	1.3	60
	1-Bromododecane	$[(n\text{-}C_6H_{13})_4N]^+Cl^-$	Ultrasound	1	92
	1-Bromododecane	$[(n\text{-}C_4H_9)_4N]^+[NO_3]^-$	Stirred at 20°	3	19
	1-Bromododecane	$[(n\text{-}C_4H_9)_4N]^+[NO_3]^-$	Ultrasound	1.3	90
	1-Bromododecane	$[(n\text{-}C_4H_9)_4N]^+I^-$	Ultrasound	0.8	73
Carbazole	Benzyl bromide	PEGME	Stirred at room temp.	8	50
	Benzyl bromide	PEGME	Ultrasound	1	96
Diphenylamine	Benzyl bromide	PEGME	Heated at reflux	48	70
	Benzyl bromide	PEGME	Ultrasound	1	98

[a] Yield determined by gas–liquid chromolography; no reaction observed without catalyst
[b] Typical conditions: indole:halide:KOH (powder):PEGME:toluene, 20 mmol:20 mmol:80 mmol:20 mmol:50 mL

Table 5-30. Ultrasonically Accelerated Alkylation of 2-Benzoyl-1,2-dihydroisoquinoline-1-carbonitrile

R	X	Method[a]	% Yield (isolated) Stir[b]	)))[c]
$PhCH_2$	Cl	A	50	60
$o\text{-}ClC_6H_4CH_2$	Cl	A	80	88
$p\text{-}ClC_6H_4CH_2$	Cl	A	26	50
$p\text{-}NO_2C_6H_4CH_2$	Br	B	40	50
$p\text{-}NO_2C_6H_4CH_2$	Cl	B	46	76
$2,4\text{-}(NO_2)_2C_6H_3$	Cl	B	64	66

[a] Method A: isoquinoline:R—X:$[Et_3NCH_2Ph]^+$ Cl^-:NaOH(50% aq.), 5.7 mmol:8.2 mmol:13 mg:2.5 mL. Method B: isoquinoline:R—X:$[Me_3NC_{16}H_{33}]^+Br^-$:toluene:NaOH (50% aq.), 5.7 mmol:6 mmol:33 mg:20 mL:3.3 mL.
[b] 2 h.
[c] 20 min.

by thiocyanate in the presence of a variety of phase-transfer catalyst, stirring is at least as effective and sometimes more so than using ultrasonic waves.[195] Surprisingly, these authors found that even unstirred reactions proceeded at rates competitive with stirred and sonicated reactions.

MECHANISTIC CONSIDERATIONS

Metal Surfaces

Cavitation-induced cleaning and erosion of metal surfaces and improved mass transport via acoustic streaming are the major factors contributing to rate enhancements of reactions involving metals in liquids. The removal of inhibiting impurities and the pitting of the surface increase the number of reactive sites available to reagents and the accelerated flow of liquid from acoustic streaming counters the local depletion of reagents at the active sites.

Surface Cleaning

Surface contaminants can be classified according to their ability to withstand cavitational effects, their binding strength to the surface to be cleaned, and their mode of interaction with the solvent.[196] For metals in the forms normally used in chemical reactions (i.e., powders, dispersions, chips, and wires) the common surface inhibitors are dust, grease films, oxide films, and products of reactions with incidental agents such as sulfur, chloride, and other halogens. Most surface films are broken down in a liquid by cavitation and acoustic streaming. In addition, vibration of the particle causing stresses in the coating can lead to film fracture and removal.

There are three types of bubbles in a sound field: transient cavitation bubbles with large-amplitude pulsations that generate powerful shockwaves in the liquid; stable cavitation bubbles characterized by large size, long lifetimes, and small vibrations that give rise to small pressure pulses; and surface attached, large, gas-filled bubbles.

Transient cavitation bubbles are primarily responsible for the cleaning action and this can be accomplished by three mechanisms: peeling, jet cleaning, and emulsification. Peeling occurs when cavitation bubbles adhere to the interface of surface and film and, through their pulsating motion, separate them. Jet cleaning occurs as a pulsating cavitation bubble moves over the contaminant film clearing "paths" that trace its journey. This cleaning action is the result of the impact of liquid jets that are generated during the division of large unstable bubbles into smaller ones on the surface.

Emulsification, important for viscous films, can involve the cooperative action of gas-filled bubbles and cavitation bubbles. Gas-filled bubbles within the film, oscillating because of ultrasound and mobilized by acoustic streaming, entrain some of the film. Simultaneously, cavitation bubbles spray solvent on the film, enveloping the pulsing gas bubble. The pulsing action of the gas bubble disrupts and scatters the liquid(s) on its surface, leading to highly dispersed emulsions.

These are primarily mechanical means of cleaning a surface. In some cases chemical reactions promoted by ultrasonic waves must be considered as an important cleaning mechanism. For example, metal oxide films may be susceptible to aqueous solvents depending on the pH of the system. For organic solvents, free-radical side reactions may promote the removal of oxide, sulfide, or halide impurities. Oxidation or reduction by electron transfer between intermediates formed during the course of a chemical reaction and surface impurities is also feasible for some of the reactions discussed in this chapter.

Surface cleaning by acoustic streaming is most important when the contaminant film is soluble in the solvent. This mechanism differs sharply from mechanical agitation because it rapidly exchanges the solvent at the surface, reducing the thickness of the boundary layer. Higher frequencies lead to greater flow rates because of increased absorption of acoustic energy and are recommended for loosely bound contaminants. More tightly held inhibitors require cavitation, thus lower frequencies are more efficient.

Cavitation Erosion

Cavitation erosion is predominantly mechanical in nature and has two possible sources: the shockwave of isolated imploding cavities attached to the surface and the jet impact generated by clusters of cavities collapsing near the surface. The relative importance of these effects has not been determined unequivocally. The most popular theory is that erosion of metals results from asymmetric collapse of single bubbles attached to, or very near, the surface, which produces shockwaves and liquid jets of sufficient force to deform the surface of the metal.[197,198]

As early as 1944, Kornfeld and Suvorov proposed that cavitation damage might arise from liquid jets produced by a collapsing bubble attached to a surface.[199] The first experimental evidence for this phenomenon came from Naudé and Ellis in 1961.[200] In 1966, Benjamin and Ellis concluded that the nonspherical collapse of single cavities led to impact jets of sufficient energy to cause damage on a solid surface and that spherical collapse of cavities occurred only at distances so far from the solid that damage was minimal.[201]

Calculations by Plesset and Chapman support these conclusions.[202] The estimate that little or no effect will be registered on the solid surface if the distance between the solid and the collapsing bubble is greater than $b/R_0 = 1.5$, where b is the distance from the center of the bubble of radius R_0 to the surface wall. At $b/R_0 = 1$, i.e., when the cavity touches the surface, the energy from collapse of the cavity is transferred into kinetic energy of the liquid jet and this energy is directed toward the wall. Figure 5-2 compares the two cases. Liquid jet velocities of 130 m s^{-1} leading to impact pressures of 2 kbar were calculated with this model. The pressure is applied to an area smaller than the cross section of the jet for a time period that does not exceed the ratio of the radius of the jet to the sound velocity. For a 1 mm cavity this is about 10^{-8} s.

Brunton obtained photographs of the collapse of single attached bubbles produced by a vibration device demonstrating the effects of asymmetrical collapse.[203] Photographic data (Fig. 5-3) led to the calculation of jet velocities of approximately 500 m s^{-1} and shock pressures of about 5 kbar for involuting bubbles, or nearly twice that when the bubble is attached and the jet impacts the surface directly. Following collapse the bubble begins to expand again, causing a flow of liquid toward the jet. The hydrodynamic pressure resulting from this flow, while smaller than the shockwave, has a longer lifetime and can be significant, especially for detached bubbles. For collapsing bubbles near a crack in the surface the expansion phase can lead to significant damage because the bubble can be wedged into the crack, enlarging it and eventually causing fragmentation. Brunton sees three causes of erosion: impact of the intruding wall of the atttached bubble; shockwave damage from the implosion; and flow against the surface during expansion of the bubble. Of these, the first is generally viewed as the most important.

Single bubbles detached from a surface can also cause erosion. Using laser-generated bubbles and high-speed photography, Lauterborn and Bolle[204] recorded the growth and collapse of bubbles near a solid boundary. As shown in Fig. 5-4, the shape of a bubble is distorted by a nearby surface as it collapses toward implosion. Instead of a spherical collapse a vanishing bubble near a solid develops a peak or tip directed toward the surface that eventually "touches down," generating a liquid jet at the surface.

The asymmetry in the collapsing bubble is the result of the distortion of the ultrasonic field surrounding the bubble near the solid. The diminished pressure between the surface and the oscillating bubble prevents movement of the part of the bubble closest to the surface, leading to a bulge toward the surface. Once started, the process is irreversible and the tip grows rapidly toward the surface.

a

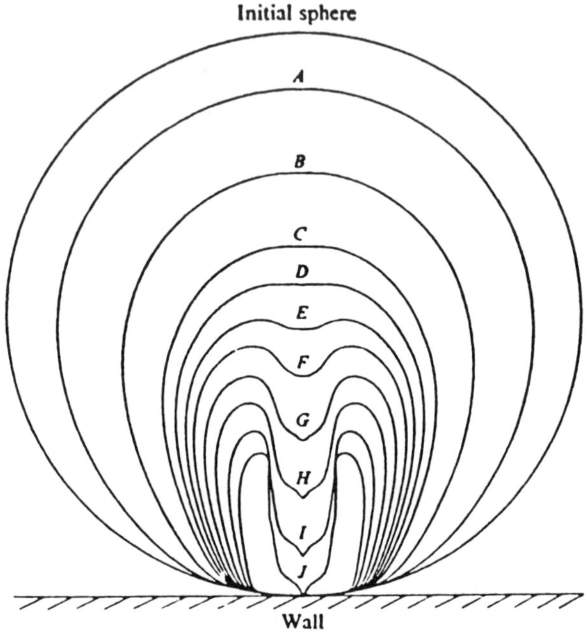

Figure 5-2. Calculated bubble surfaces for (a) attached and (b) separated bubbles.[202]

Lauterborn and Bolle[204] achieved nearly quantitative agreement with Plesset and Chapman[202] in their comparison of the experimentally observed stages of bubble collapse and the calculated shapes of the bubble during implosion. Velocities of liquid jets at the tip exceeded 100 m s^{-1}. These velocities are also in accord with theory[202] and are attributed to the high radius of curvature at the tip. In addition to the jet impact produced by the collapsing cavity, a shockwave of 1–10 kbar is also generated.

As the distance between the wall and the collapsing bubbles increases the bubbles become more spherical and their collapse becomes more symmetrical.[205] This leads to an increasing fraction of energy emitted in the form of a spherical wave, which is partly responsible for reduced impact at larger distances from the wall. This was verified by Lauterborn, who showed, using holographic techniques, that collapsing spherical bubbles generate spherical shockwaves.[206]

Gaining acceptance is the theory that a cluster of bubbles collapsing in concert at a distance from the surface may be an important mechanism for energy transfer to a solid.[197,198] Vyas and Preece have measured pressures up to 9 kbar from the simultaneous collapse of a cavity cloud generated by an ultrasonic horn.[207] A series of concerted collapses of large numbers of bubbles is more likely responsible for the formation of long slip lines (>1 cm),[208] the growth of large craters,[209] and the depth of work hardening[210] in some metals and alloys than are isolated implosions of single bubbles.

b

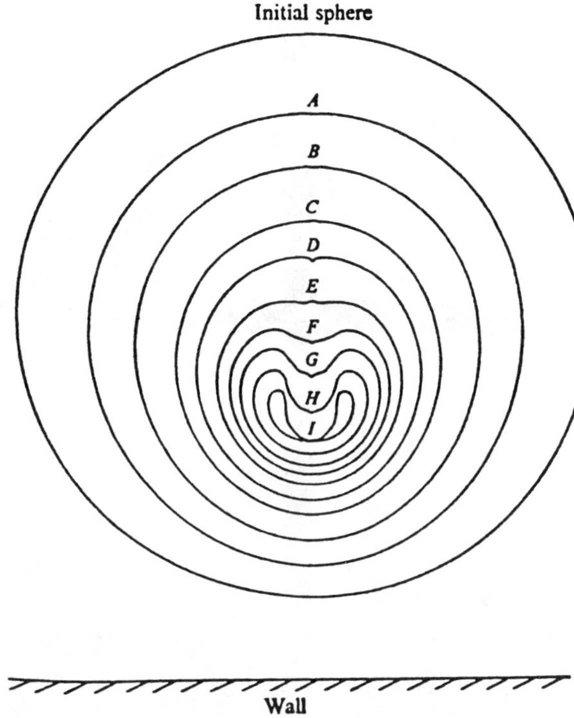

Figure 5-2. *(Continued)*

Nonmetallic Solids

Far less research has been done on nonmetallic solids than on metals primarily because of the importance of cavitational erosion of industrial machinery. Most of the interest in the effects of ultrasound on nonmetallic solids has been confined to polymers, so there are fewer studies that are as pertinent to the ultrasonic enhancement of nonmetallic reactions as there are to ultrasonically accelerated reactions involving metals.

The major factors responsible for the increase in the reaction rates of nonmetallic solids are probably: (1) increase in surface area of the solid reagent from cavitation-induced particle size reduction and (2) improved mass transport of liquid reagent to the solid surface via acoustic streaming.

The fracturing of nonmetallic solids in a cavitating liquid is well documented and is an important factor in ultrasonic machining technology.[211] The reduction of particles from average sizes of approximately 90 μm to about 20 μm is considered the main cause of the decrease in the machining rate. Since particle breakdown is rapid (cutting power can be lost in 5–10 s) ultrasonic machining requires constant replacement of the abrasive.

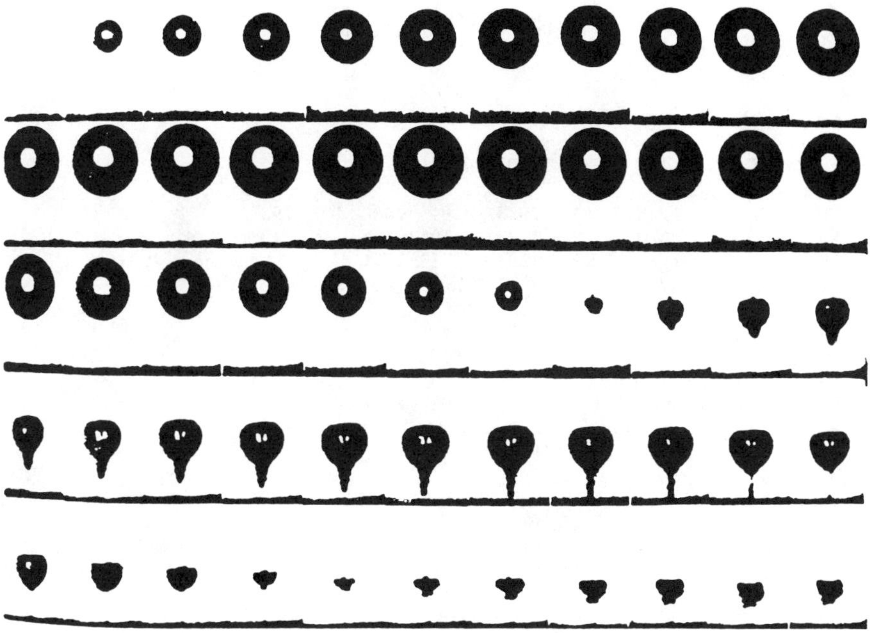

Figure 5-4. The asymmetrical collapse of a detached air bubble near a solid boundary. (From Lauterborn and Bolle.[200])

Augmenting the effects of increased surface area should be the improved "refresh rate" at the surface of the solid from acoustic streaming. As mentioned earlier, this effect is different from mechanical mixing because the flow rates at the surface are increased by the motion of the particle caused by its absorption of sound energy.

ACKNOWLEDGMENTS

The author is indebted to his graduate students and postdoctoral associates, with whom he has been a privileged coauthor. He also wishes to express his grateful appreciation to the Air Force Office of Scientific Research for their support of his program in sonochemistry.

Figure 5-3. The asymmetrical collapse of an attached air bubble in water. (From Brunton.[199])

REFERENCES

1. Richards, W. T., Loomis, A. L., *J. Am. Chem. Soc.* **1927**, *49*, 3086.
2. Freundlich, H.; Rogovski, F.; Söllner, K., *Z. Physik. Chem.* 1932, *A160*, 469; *Chem. Abstr.* **1932**, *26* 5243: *Kolloid-Beihefte* **1933**, *37*, 223; *Chem. Abstr.* **1933**, *27*, 3379.
3. Sata, N.; Watanabe, S., *Kolloid-Z.*, **1935**, *73*, 50; *Chem. Abstr.* **1936**, *30*, 13.
4. Moriguchi, N. *J. Chem. Soc. Jpn.* **1933**, *54*, 949; *Chem. Abstr.* **1934**, *28*, 398.[7]
5. Moriguchi, N. *J. Chem. Soc. Jpn.* **1933**, *54*, 1047; *Chem. Abstr.* **1934**, *28*, 955.[4]
6. Moriguchi, N. *J. Chem. Soc. Jpn.* **1934**, *55*, 749; **1934**, *55*, 751; *Chem. Abstr.* **1934**, *28*, 71,204.
7. Claus, B., *Z. Tech. Physik*, **1935**, *16*, 80; *Chem. Abstr.* **1935**, *29*, 3576.[4]
8. Mueller, F.; Kuss, H., *Helv. Chem. Acta* **1950**, *33*, 217.
9. Barnartt, S., *Quart. Rev.* (*London*) **1953**, *7*, 84.
10. Heimberger, N., German Pat. No. 733,470, 1943.
11. Siemens-Schuckertwerke Akt.-Ges.; Swiss Pat. No. 294,390, 1944.
12. Wood, R. W.; Loomis, A. L., *Phil. Mag.*, **1927**, *4*, 417.
13. Jaffe, B.; Roth, R. S.; Marzullo, S. *J. Res. Natl. Bur. Stds.* **1955**, *55*, 5.
14. Brown, B.; Goodman, J. E. "High Intensity Ultrasonics. Industrial Applications", Van Nostrand: Princeton, NJ, 1965.
15. El'piner, I. E. "Ultrasound: Physical, Chemical and Biological Effects"; Sinclair, F. A., transl.; Consultants Bureau: New York, 1964.
16. Blandamer, M. J. "Introduction to Chemical Ultrasonics"; Academic Press: New York, 1973.
17. Ensminger, D. "Ultrasonics, The Low- and High-Intensity Applications"; Marcel Dekker: New York, 1973.
18. Cracknell, A. P. "Ultrasonics"; Wykeham Publications, Ltd.: London, 1980.
19. Janovsky, W.; Pohlmann, R. *Z. Angew. Phys.*, **1948**, *1*, 22.
20. Segala, L.; Crosby, L.; Puskas, W.; Loffler, A. I.; Evans, D. A., Abstracts of Papers, 181st National Meeting of the American Chemical Society, New York, NY; American Chemical Society; Washington, DC, 1981, MICR 7.
21. Komar, C. A.; Moore, H. A. W. *Chem. Eng. Sym. Ser.* **1971**, *67*(109), 117.
22. Decker, R.; Holz, H. German Pat. No. 803,538, 1951.
23. Zechmeister, L.; Wallcave, L. *J. Am. Chem. Soc.* **1955**, *77*, 2853.
24. Söllner, K. *Trans. Faraday Soc.* **1938**, *34*, 1170.
25. Margulis, M. A. Russian Pat. 631,372, 1959.
26. Thompson, D.; Vilbrandt, F. C. *Ind. Eng. Chem.* **1954**, *46*, 1172.
27. Sugiyama, K.; Kubara, K., Itoh, H. *Thin Solid Films* **1984**, *112*, 257.
28. Soloff, R. S. *J. Acoust. Soc. Am.* **1964**, *35*(4), 961.
29. Wilson, J. S.; Moore, A. S.; Bowie, W. S. *Chem. Eng. Prog. Symp. Ser.* **1971**, *67*(109), 68.
30. Purdy, K. R.; Simmons, G. W.; Hribar, A. E.; Griggs, E. I. *Chem. Eng. Prog. Symp. Ser.* **1971**, *67*(109), 55.
31. Bachelet, M.; Huguet, J. *Compt. Rend. Hebd. Seances Acad. Sci.* (*Paris*) *Ser. C* **1966**, *262*, 1308.
32. Skauen, D. M. *J. Pharm. Sci.* **1967**, *56*, 1373.
33. Kapustin, A. P. *Zh. Techn. Fiz.* **1952**, *22*, 765; *Chem. Abstr.* **1956**, *50*, 6123c.
34. Glavatskii, D. E.; Kortner, A. V. *Nauchn. Zap. Odessk. Polite khn. Inst.* **1962**, *41*, 22; *Chem. Abstr.* **1963**, *59*, 2232g.
35. Vanhook, A.; MacInnes, M. B.; *Sugar*, J., **1953**, *16*, 20.
36. Cohn, R. M.; Skauen, D. M. *J. Pharm. Sci.* **1964**, *53*, 1040.
37. Schmidt, G.; Roll, A. *Z. Electrochem.* **1940**, *46*, 653.
38. Masing, G.; Ritzau, G. *Z. Metallk.* **1936**, *28*, 293, *Chem. Abstr.* **1937**, *31*, 1740.[4]
39. Han, B.-H.; Boudjouk, P. Unpublished results.
40. Thiemann, A. E. *Automobiltech Z.* **1942**, *45*, 668; *Chem. Fentre.* **1943**, *II*, 367; *Chem. Abstr.* **1944**, *38*, 4554.
41. Neppiras, E. A., *Iron Age*, **1954**, *173*, 97, 24/vi; 7.
42. Severanko, V. P.; Lozhechnikov, E. B.; Bock, M. A., *Vesti Akad. Navak-Belarus SSR, Ser. Fiz. Tekh Navuk.* **1970**, 115.

42a. Agranat, B. A.; Konterovich, L. E.; Novik, N. I., *Primer Ultrazunka Met. Protsessakr.* **1972**, 142; *Chem. Abstr.* **1973**, *78*, 61,289h.

43. Radiamant, A. G. Swiss Pat. 234,822, 1945; *Chem. Abstr.* **1950**, *44*, P5791g.
44. Mathus, I.; Bakhshi, N. N.; Mathews, J. W. *Can J. Chem. Eng.* **1972**, *50*, 344; *Chem. Abstr.* **1972**, *77*, 79,913w.

45. Heidemann, E. A., J. Acoust. Soc. Am., 1954, *26* 831.
46. Kuzharov, A. S.; Suchkov, V. V.; Vlasenko, L. A.; *Zh. Fiz. Khim.* **1974**, *53*, 2064; *Chem. Abstr.* **1979**, *91*, 197,452g: *Zh. Fiz. Khim.* **1984**, *58*, 894; *Chem. Abstr*, **1984**, *100*, 216,410q.
47. Bachmann, J. A.; Wilkens, R., U.S. Pat. 2,086,891, 1937.
48. Böhme, J., Walther, Ger. Pat. 735,283, 1943; *Chem. Abstr.* **1944**, *38*, 2850[5]: Fr. Pat. 848,467, **1939**; *Chem. Abstr.* **1941**, *35*, 6152.[6]
49. Losev, B. I.; Mel'nikova, A. N.; El'Piner, I. E. *Izv. Akad. Nauk SSSR Otdel. Tekh. Nauk* **1957**, 90; *Chem. Abstr.* **1958**, *52*, 5790i.
50. Glukov, V. S.; Pronischev, A. V. *Tr. Vses. Nauch.-Issled. Inst. Zolota Redk. Metal.* **1972**, No. 32, 369; *Chem. Abstr.* **1974**, *81*, 109,024.
51. Tyutyunnikov, B. N.; Novitskaya, I. I. *Masloboino-Zhorovaya Prom.* **1959**, *25*, 13; *Chem. Abstr.* **1960**, *54*, 11,520a.
52. Mal'tsev, A. N.; Solov'eva, I. V., *Zh. Fiz. Khim.* **1970**, *44*, 1092; *Chem. Abstr.* **1970**, *73*, 39,022r.
53. Kowalska, E.; Miszczyszn, M. *Rocz. Chem.* **1972**, *46*, 233; *Chem. Abstr.* **1972**, *76*, 145,322k.
54. Bronislaw, Z.; Lomnicka, A. *Zesz. Nauk. Univ. Jagiellon Pr. Chem.* **1975**, *20*, 109; *Chem. Abstr.* **1976**, *85*, 25,765w.
55. Witekowa, S.; Forbotko, W. *Soc. Sci. Lodz Acta Chim.* **1974**, *18*, 99,403; *Chem. Abstr.* **1975**, *82*, 7935k.
56. Romenskii, A. V.; Popok, I. V.; Loboiko, A. Ya. *Atroshchenko Khim. Technol.* **1985**, *1*, 21; *Chem. Abstr.* **1985**, *102*, 120,614z.
57. Marti, W.; Waldmeir, K. Swiss Pat. 376,886, 1964; *Chem. Abstr.* **1964**, *61*, P10,088.
58. Skorik, Yu. I.; Gileva, K. G.; Kukharokaya, E. V.; Fedoseev, A. D. *Izv. Akad. Nauk SSSR Otd. Khim. Nauk* **1963**, 932; *Chem. Abstr.* **1963**, *59*, 4879g.
59. Pshenitsyn, N. N.; Sidorov, N. V.; Sternina, D. G. *J. Appl. Chem. (USSR)* **1940**, *13*, 76; *Chem. Abstr.* **1940**, *34*, 8189.
60. Richardson, C. N. U.S. Pat. 2,500,008, 1950; *Chem. Abstr.* **1950**, *44*, P4643.
61. Jones, J. B. U.S. Pat. 3,245,892, 1966; *Chem. Abstr.* **1966**, *65*, P82f.
62. Aeroprojects, Inc. Br. Pat. 991,759, 1965; *Chem. Abstr.* **1965**, *63*, P4994a.
63. Taoda, H.; Yumoto, T.; Hayakawa, K.; Kawase, K. *Jpn. Kokai Tokkyo Koho* Jpn. 60, 200, 832 [85, 200, 832]; *Chem. Abstr.* **1986**, *104*, 68,495m.
64. Graves, C. A.; Steiner, D. F.; Hindler, F. C. U.S. Pat. 3,231,513, 1966; *Chem. Abstr.* **1966**, *64*, P10,453a.
65. Hughes, E. C.; Veatch, F. U.S. Pat. 2,800,444, 1957; *Chem. Abstr.* **1957**, *51*, 17,151g.
66. Mertes, T. S. U.S. Pat. 2,899,414, 1959; *Chem. Abstr.* **1959**, *53*, 23,093.
67. Kogan, E. V.; Smirnov, N. I. *Zh. Priklad. Khim.* **1962**, *35*, 1382; *Chem. Abstr.* **1962**, *57*, 10,025g.
68. Salnikov, V. V.; Pan'shina, Z. K.; Druyan, E. A. *Zh. Prik. Khim.* **1962**, *35*, 214; *Chem. Abstr.* **1962**, *57*, 1053g.
69. Chervinskii, K. A.; Mal'tsev, V. N. *Ukr. Khim. Zh.* **1966**, *32*, 69; *Chem. Abstr.* **1966**, *64*, 19,356h.
70. Mayer, H.; Marinesco, N. Fr. Pat. 893,663, 1944; *Chem. Abstr.* **1953**, *47*, 4064i.
71. Möckel, P. *Ing.-Tech.* **1952**, *24*, 1534; *Chem. Abstr.* **1952**, *46*, 5412f.
72. Saracco, G. *Chim. Ind.* **1963**, *43*, 1394; *Chem. Abstr.* **1964**, *60*, 12,244h. Saracco, G.; Arzano, F. *Chim. Ind.* **1968**, *50*, 314; *Chem. Abstr.* **1968**, *69*, 3906k.
73. Marinesco, N., Fr. Pat. 973,715, 1951; *Chem. Abstr. 47*, 2461e.
74. Pchelkin, S. A.; Mardosevich, V. A.; Kudryoutsev, V. S. "Nov. Metody. Issled Protsessov Vosstanov Chern. Met."; Dokl. Simp. **1971**, Publ. 1974, 93, Rostovtsev, S. T., Ed., "Nauka": Moscow, USSR *Chem. Abstr.* **1975**, *82*, 158,974x.
75. Skorobogat'ko, L. I.; Gai, N. V.; Abdukadyrov, A. A.; Sultanov, A. S.; Makhmudov, A. K. USSR Deposited Doc. 1979 VINITI 2158; *Chem. Abstr.* **1980**, *93*, 186,070w.
76. Renaud, P. *Bull Soc. Chim. Fr.* **1950**, 1044.
77. Slough, W.; Ubbelohde, A. R. *J. Chem. Soc.* **1957**, 918.
78. Pratt, M. W. T.; Helsby, R. *Nature (London)* **1959**, *184*, 1694.
79. Renaud, P.; Gillier, H. *Bull. Soc. Chim. Fr.* **1964**, 2381.
80. Sjöberg, K. *Tetrahedron Lett.* **1966**, 6383.
81. Ochrymowycz, L. Private communication, 1984.
82. Fry, A. J.; Herr, D. *Tetrahedron Lett.* **1978**, 1721.
83. Fry, A. J.; Bujanauskas, J. P. *J. Org. Chem.* **1978**, *43*, 3157.
84. Fry, A. J.; Donaldson, W. A.; Ginsburg, G. S. *J. Org. Chem.* **1979**, *44*, 349.
85. Fry, A. J.; Ginsburg, G. S. *J. Am. Chem. Soc.*, **1979**, *101*, 3927.
86. Fry, A. J.; Hong, S.-S. S. *J. Org. Chem.* **1981**, *46*, 1962.
87. Fry, A. J.; Ginsburg, G. S.; Parente, R. A., *J. Chem. Soc., Chem. Commun.* **1978**, 1040.
88. Fry, A. J.; Ankner, K.; Handa, V. *Tetrahedron Lett.* **1981**, *22*, 1791.

89. Luche, J.-L.; Damiano, J.-C. *J. Am. Chem. Soc.* **1980**, *102*, 7927.
90. Pearce, P. J.; Richards, D. H.; Scilly, N. F. *J. Chem. Soc. Perkin Trans. I* **1972**, 1655.
91. Trost, B. M.; Coppola, B. P. *J. Am. Chem. Soc.* **1982**, *104*, 6879.
92. Stiles, M.; Meyer, R. P. *J. Am. Chem. Soc.* **1959**, *81*, 1501.
93. Hudrlik, P. F. Private communication, 1984.
94. Petrier, C.; Gemal, A. L.; Luche, J.-L. *Tetrahedron Lett.* **1982**, *23*, 3361.
95. Luche, J.-L.; Petrier, C.; Gemal, A. L.; Zikra, N. *J. Org. Chem.* **1982**, *47*, 3806.
96. Luche, J.-L.; Petrier, C.; Lansard, J. P.; Greene, A. E. *J. Org. Chem.* **1983**, *48*, 3837.
97. de Sousa Barboza, J. C.; Petrier, C., Luche, J.-L. *Tetrahedron Lett.* **1985**, *26*, 829.
98. Petrier, C.; de Souza Barbosa, J. C.; Dupuy, C.; Luche, J.-L. *J. Org. Chem.* **1985**, *50*, 5761.
99. Petrier, C.; Luche, J.-L.; Dupuy, C. *Tetrahedron Lett.* **1984**, *25*, 3463.
100. Suslick, K. S.; Gawienowski, J. W.; Schubert, P. F.; Wang, H. H. *J. Phys. Chem.* **1983**, *87*, 2299.
101. Boudjouk, P.; Han, B.-H. *Tetrahedron Lett.* **1981**, *22*, 3813.
102. Han, B.-H.; Boudjouk, P. *Tetrahedron Lett.* **1981**, *22*, 2757.
103. Boudjouk, P., in "Science of Ceramic Chemical Processing", Hench, L. L.; Ulrich, D. R., Eds., J. Wiley & Sons Publ., New York, NY; 1986, Chap. 39.
104. Lash, T. D.; Berry, D. *J. Chem. Ed.* **1985**, *62*, 85.
105. Xu, L.; Tao, F.; Yu, T. *Tetrahedron Lett.* **1985**, *26*, 4231.
106. Otteson, D.; Michl, J. *J. Org. Chem.* **1984**, *49*, 866.
107. Boudjouk, P.; Sooriyakumaran, R.; Han, B.-H. *J. Org. Chem.* **1986**, *51*, 2818.
108. Azuma, T.; Yanagida, S.; Sakurai, H.; Sasa, S.; Yoshino, K. *Synth. Commun.* **1982**, *12*, 137.
109. West, R.; Fink, M. J.; Michl., J. *Science* **1981**, *214*, 1343.
110. Boudjouk, P.; Han, B.-H.; Anderson, K. R. *J. Am. Chem. Soc.* **1982**, *104*, 4992.
111. Masamune, S.; Murakami, S.; Tobita, H. *Organometallics* **1982**, *2*, 1464.
112. Masamune, S.; Tobita, H.; Murakami, S. *J. Am. Chem. Soc.* **1983**, *105*, 8324.
113. Sooriyakumaran, R. Ph.D. Thesis, North Dakota State University, Fargo, N.D. 1985.
114. Anderson, K. R. M. S. Thesis, North Dakota State University, Fargo, N.D. 1986.
115. Boudjouk, P.; Han. B.-H.; Sooriyakumaran, R.; Ohrbom, W.; Thompson, D.; Anderson, k.; Samaraweera, U.; Falvey, D.; Abstracts of Papers, 186th National Meeting of the American Chemical Society, Washington, D.C.; American Chemical Society; Washington, D.C., 1983, ORGN 267.
116. Boudjouk, P.; Han, B.-H. Abstracts of Papers, 183rd National Meeting of the American Chemical Society, Las Vegas, NV; American Chemical Society: Washington, DC, 1982, ORGN 190.
117. Rieke, R. D.; Burns, T. P.; Wehmeyer, R. M.; Kahn, B. E. in "High Energy Processes in Organometallic Chemistry", Suslick, K. S., Ed., American Chemical Society Symposium Series No. 333; American Chemical Society; Washington, D.C., Chap. 14; 1987.
118. Boudjouk, P.; Thompson, D. P.; Ohrbom, W. H.; Han, B.-H. *Organometallics* **1986**, *5*, 1257.
119. Suslick, K. S.; Johnson, R. E. *J. Am. Chem. Soc.* **1984**, *10*, 6856.
120. Kimmel, T.; Becker, D. *J. Org. Chem.* **1984**, *49*, 2494.
121. Chou, T; Yuan, J. J.; Tsao, C.-H. *J. Chem. Res. Synop.* **1985**, 18.
122. Chou, T.; Tsao, C.-H.; Hung, S. C. *J. Org. Chem.* **1985**, *50*, 4329.
123. Luche, J.-L.; Petrier, C.; Dupuy, C. *Tetrahedron Lett.* **1984**, *25*, 753.
124. Chou, T.; You, M.-L. *Tetrahedron Lett.* **1982**, *6*, 4495.
125. Repic, O.; Vogt, S. *Tetrahedron Lett.* **1982**, *23*, 2729.
126. Repic, O. *Org. Prep. Proc. Int.* **1984**, *6*, 25.
127. Friedrich, E. C.; Domek, J. M.; Pong, R. Y. *J. Org. Chem.* **1985**, *50*, 4640.
128. Yamashita, J.; Inoue, Y.; Kondo, T.; Harukichi, H. *Bull. Chem. Soc.* Jpn. **1984**, *57*, 2335.
129. Han, B.-H.; Boudjouk, P. *J. Org. Chem.* **1982**, *47*, 751.
130. Chew, S.; Ferrier, R. J. *J. Chem. Soc. Chem. Commun.* **1984**, 911.
131. Franck, R. W.; John, T. U.; Olejniczak, K.; Blount, J. F. *J. Am. Chem. Soc.* **1982**, *104*, 1106.
132. Rathke, M. W. *Org. React.* **1975**, *22*, 423.
133. Boudjouk, P.; Han, B.-H. *J. Org. Chem.* **1982**, *47*, 5030.
134. Rieke, R. D.; Uhm, S. J. *Synthesis* **1975**, 452.
135. Bose, A. J.; Gupta, K., Manhas, M. S., *J. Chem. Soc. Commun.* **1984**, 86.
136. Gilman, H.; Speeter, M. *J. Am. Chem. Soc.* **1943**, *65*, 2255.
137. Mohan, S.; Sethi, P. S.; Kapoor, A. L. *J. Indian Chem. Soc.* **1971**, *48*, 68.
138. Brennan, J.; Hussain, F. H. S. *Synthesis* **1985**, 749.
139. Yamashita, J.; Inoue, Y.; Kondo, T.; Hoshimoto, H., *Bull. Chem. Soc.* Jpn. **1985**, *58*, 2709.
140. Petrier, C.; Luche, J.-L. *J. Org. Chem.* **1984**, *50*, 910.
141. Molle, G.; Bauer, P. *J. Am. Chem. Soc.* **1982**, *104*, 3481.

142. Petrier, C.; Einhorn, J.; Luche, J.-L. *Tetrahedron Lett.* **1985**, *26*, 1449.
143. Petrier, C.; Dupuy, C.; Luche, J.-L. *Tetrahedron Lett.* **1986**, *27*, 3149.
144. Knochel, P.; Normant, J. F. *Tetrahedron Lett.* **1984**, *25*, 1475.
145. Kitazume, T.; Ishikawa, N. *Chem. Lett.* **1981**, 1679.
146. Kitazume, T.; Ishikawa, N. *Chem. Lett.* **1982**, 137.
147. Kitazume, T.; Ishikawa, N. *Chem. Lett.* **1982**, 1453.
148. Ishikawa, N.; Takahashi, M.; Sato, T.; Kitazume, T. J. *Fluorine Chem.* **1983**, *22*, 585.
149. Kitazume, T.; Ishikawa, N. *J. Am. Chem. Soc.* **1985**, *107*, 5186.
150. Solladie-Cavallo, A.; Farkhani, D.; Fritz, S.; Lazrak, T.; Suffert, J. *Tetrahedron Lett.* **1984**, *25*, 4117.
151. Boudjouk, P.; So, J.-H., *Synth. Commun.* **1986**, *16*, 775.
152. Boudjouk, P. in "High Energy Processes in Organometallic Chemistry", Suslick, K. S., Ed., American Chemical Society Symposium Series No. 333; American Chemical Society; Washington, D.C., Chap. 13; 1987.
153. Itoh, K.; Nagashima, H.; Ohshima, T.; Oshima, N.; Nishiyama, H., J. *Organomet. Chem.* 1984, *272*, 179.
154. Sprich, J. D.; Lewandos, G. S. *Inorg. Chem. Acta* **1983**, *76*, L241.
155. Brown, H. C.; Racherla, U. S. *J. Org. Chem,* **1986**, *51*, 427.
156. Eaborn, C.; Hitchcock, P. B.; Lickiss, P. D. *J. Organomet. Chem.*, **1984**, *269*, 235.
157. Bönneman, H.; Bogdanovic, B.; Brinkman, R.; He, D.-W.; Spliethoff, B. *Angew. Chem. Int. Ed. Engl.* **1983**, *22*, 728.
158. Kuchin, A. V.; Nurushev, R. A.; Tolstikov, G. A., *Zh. Obshch. Khim.* **1983**, *53*, 2519; *Chem. Abstr.* **1984**, *100*, 103,426f.
159. Liou, K.-F.; Yang, P.-H.; Lin, Y.-T. *J. Organomet. Chem.* **1985**, *294*, 273.
160. Yang, P.-H.; Liou, K.-F.; Lin, Y.-T. *J. Organomet. Chem.* **1986**, *307*, 273.
161. Boudjouk, P.; Han, B.-H., *J. Catal.* **1983**, *79*, 489.
162. Lukes, R.; Cervinka, O. *Chem. Listy* **1957**, *51*, 2142.
163. Vol'pin, M. E.; Kolomnikov, I. S. *Organomet. React.* **1976**, *5*, 313.
164. Blum, J.; Sasson, Y.; Iflah, S. *Tetrahedron Lett.* **1972**, 1015.
165. Nishiguchi, T.; Tagawa, T.; Fukuzumi, K. *Yukagaku* **1978**, *27*, 174.
166. Cortese, N. A.; Heck, R. F. *J. Org. Chem.* **1985**, *43*, 3985.
167. Boudjouk, P.; Han, B.-H., Jacobsen, J. R. unpublished results. **1987**.
168. Suslick, K. S. "High Energy Processes in Organometallic Chemistry", Suslick, K. S., Ed., American Chemical Society Symposium Series No. 333; American Chemical Society; Washington, D. C., Chap. 12; 1987.
169. Suslick, K. S.; Casadonte, D. J. *J. Am. Chem. Soc.*, **1987**, *109*, 3459.
170. Han, B.-H.; Boudjouk, P., Organometallics, **1983**, *2*, 779.
171. Han, B.-H. *Bull. Korean Chem. Soc.* **1985**, *6*, 320.
172. Suslick, K. S.; Johnson, R. E. *J. Am. Chem. Soc.* **1984**, *106*, 6856.
173. Inouye, Y; Yamashita, J.; Hashimoto, H. *Synthesis* **1984**, 244.
174. Brown, H. C.; Krishnamurthy, S. *J. Org. Chem.* **1969**, *34*, 3918.
175. Han, B.-H.; Boudjouk, P. *Tetrahedron Lett.* **1982**, *23* 1643.
176. Lukevics, E.; Gevorgyan, V. N.; Goldberg, Y. S. *Tetrahedron Lett.* **1984**, *25*, 1415.
177. Sooriyakumaran, R.; Boudjouk, P. *J. Organomet. Chem.* **1984**, *271*, 289.
178. Brown, H. C.; Racherla, U. S. *Tetrahedron Lett.* **1985**, *26*, 2187.
179. Yamawaki, J. Sumi, S., Ando, T., Hanafusa, T. *Chem. Lett.* **1983**, 379.
180. Ando, T.; Kawate, T.; Ichihara, J.; Hanafusa, T. *Chem. Lett.* **1984**, 725.
181. Ando, T.; Kawate, T.; Sumi, S.; Ichihara, J.; Hanafusa, T., *J. Chem. Soc. Jpn., Chem. Ind. Chem.* **1984**, 1731; Abstract No. 029,463, *Curr. Chem. React.* **1985**, *7*.
182. Ando, T.; Sumi, S.; Kawate, T.; Ichihara, J.; Hanafusa, T. *J. Chem. Soc., Chem. Commun.* **1984**, 439.
183. Posner, G. H. *Angew. Chem. Int. Ed. Engl.* **1978**, *17*, 487.
184. Barot, B. C.; Sullins, Sullins, D. W.; Eisenbraun, E. J. *Synth. Commun.* **1984**, *14*, 397.
185. Muzart, J. *Synth. Commun.* **1985**, *15*, 285.
186. Varma, R. S.; Kabalka, G. W. *Heterocycles* **1985**, *23*, 139.
187. Raucher, S.; Klein, P. *J. Org, Chem.* **1981**, *46*, 3558.
188. Julia, S.; Ginebreda, A. *Synthesis* **1977**, 682.
189. Regen, S.; Singh, A. *J. Org. Chem.* **1982**, *47*, 1587.
190. Lash, T. D. *J. Chem. Ed.* **1986**, *64*, 720.
191. Moon, S.; Duchin, L.; Cooney, J. W. *Tetrahedron Lett.* **1979**, 3917.

192. Kristol, D. S.; Klotz, H.; Parker, R. C. *Tetrahedron Lett.* **1981**, *22*, 907.
193 Davidson, R. S.; Patel, A. M.; Safdar, A. *Tetrahedron Lett.* **1983**, *24*, 5907.
194. Ezquerra, J.; Alvarez-Builla, J. *J. Chem. Soc., Chem. Commun.* **1984**, 55.
195. Reeves, W. P.; McClusky, J. V. *Tetrahedron Lett.* **1983**, *24*, 1585.
196. For a general discussion of the effects of ultrasound on surface impurities, see: Agranat, B. A., Bashkirov, V. I., Kitaigorodskii, Yu. I. In "Physical Principles of Ultrasonic Technology"; Rozenberg, L. D., Ed.; Wood, J. S. Transl.; Plenum Press: New York, 1973, Vol. 1, p. 247.
197. For a discussion of cavitation and the factors that affect this process, see Mørch, K. A. In "Treatise on Materials Science and Technology"; Herbert, H., Ed.; Academic Press: New York, 1979, Vol. 16, p. 309.
198. For a discussion effects of ultrasound on metal surfaces, see Preece, C. M. In "Treatise on Materials Science and Technology"; Herbert, H., Ed.; Academic Press, New York, 1979, Vol. 16, p. 249.
199. Kornfeld, M.; Suvorov, L. *J. Appl. Phys.* **1944**, *15*, 495.
200. Naudè, C. F.; Ellis, A. T., *J. Basic Eng.* **1961**, *83*, 648.
201. Benjamin, T. B.; Ellis, A. T. *Phil. Trans. R. Soc. London Ser. A* **1966**, *260*, 221.
202. Plesset, M. S.; Chapman, R. B. *J. Fluid Mech.* **1971**, *47*, 283.
203. Brunton, J. H., "Proc. 3rd, Int. Conf. Rain Erosion"; Fryall, A. A., Ed.; Royal Aircraft Establishment: Farnborough, UK, 1970, p. 821.
204. Lauterborn, W.; Bolle, H. *J. Fluid Mech.* **1975**, *72*, 391.
205. Single cavities far from boundaries remain spherical over the majority of the collapse: Lauterborn, W., In "Finite Amplitude Wave Effects in Fluids"; Bjørnø, L., Ed.; I.P.C. Science and Technology Press: Guildford, Surrey, UK, 1974, 195.
206. Lauterborn, W. *Appl. Phys. Lett.* **1977**, *31*, 663.
207. Vyas, B.; Preece, C. M. *J. Appl. Phys.* **1976**, *47*, 5133.
208. Hansson, I.; Klaestrup-Kristenson, J.; Mørch, K. A. *J. Phys. D* **1978**, *11*, 891.
209. Vaidya, S.; Preece, C. M. *Metall. Trans. Ser. A* **1978**, *9A*, 299.
210. Preece, C. M., Dakshiamoorthy, S.; Prasad, S.; Vyas, B. *Proc. 4th Int. Conf. Strength Met. Alloys*, Nancy, France, **1976**, 1397.
211. Kazantsev, V. F. In "Physical Principles of Ultrasonic Technology", Rozenberg, L. D., Ed.; Wood, J. S. Transl.; Plenum Press: New York, 1973, Vol. 1, 3.
212. Suslick, K. S.; Casadonte, D. J.; Green, M. L. H.; Thompson, M. E. *Ultrasonics* **1987**, *25*, 56.

6

Sonoluminescence

R. E. VERRALL

Department of Chemistry,
University of Saskatchewan

and C. M. SEHGAL

Biodynamics Research Unit,
Department of Physiology and Biophysics,
Mayo Clinic/Foundation

INTRODUCTION

Sonoluminescence (SL) is a weak emission of light observed when ultrasound, sufficiently high in intensity, is passed through a medium containing dissolved gases. Though the phenomenon was reported in 1934 by Zimakov[1] and in 1935 by Frenzel and Schultes,[2] associated effects were first noticed a year earlier by Marinesco and co-workers[3,4] when they observed the fogging of photographic plates submerged in ultrasonically irradiated water. It was not until 1947, when Paounoff[5] and Pinoir and Pouradier[6] showed that the light emission and the fogging of the photographic plates occurred at the pressure nodes of the standing wave, that it became apparent that the darkening of the photographic plates arose from SL and not the direct action of ultrasound.

Following the discovery of the phenomenon of SL many studies were carried out to investigate its nature and origin. However, the progress in understanding sonoluminescence has been relatively slow. One of the reasons for this is that it is a third-order process caused by a second-order effect called cavitation. The latter, in turn, is produced by the displacement and acceleration of fluid particles by ultrasonic waves. Cavitation, in itself, is an exceedingly complex nonlinear process that is not only related to the properties of the ultrasonic field but is also closely tied to the physical condition and properties of the media. Moreover, cavitation as observed in practical situations is not a single-event phenomenon. It is composed of a statistical distribution of bubbles imploding at different points in time and space. The collapsing or oscillating bubbles themselves generates pressure waves that further complicate the acoustic field perceived by neighboring bubbles, thereby significantly altering their motion and behavior. It is under such chaotic conditions that light emission and chemical reactions are produced by ultrasonic waves. Because of the complex and multifaceted nature of SL and, because of a lack of complete understanding and control of the dynamics of bubbles that cavitate to produce light emission, the literature contains a number of conflicting views and observations.

In this chapter the terms "cavity" and "bubbles" are used synonymously. The term "cavitation" is used here in a descriptive sense to include any type of bubble that upon rapid collapse can effectively concentrate sufficient energy into small volumes of the medium to produce luminescence. The term "sonoluminescence" has been extended more recently to include emission of light produced by cavitation voids generated in a liquid by a laser beam.[7] In this discussion sonoluminescence is considered to include only luminescence arising from acoustic cavitation.

In an attempt to understand the phenomenon of SL and the accompanying chemical reactions numerous studies have been conducted and the results explained in terms of several theories. These studies have been lucidly summarized by Flynn,[8] Él' piner,[9] Margulis,[10] Sliwinski[11] and more recently by Walton.[12] To this time there is no general concensus as to the origin of SL. Of the various hypotheses, the "hot-spot" theory of Noltingk and Neppiras,[13,14] in its original or modified form, has been relatively successful in explaining most of the experimental facts. For completeness we first present the various hypotheses and then compare them, collectively, at the end of the next section.

ORIGIN AND MECHANISMS

Initial Theories

Zimakov[1] studied sonoluminescence from aqueous dichloroethanol, dichloroethane, NaI, NH_4CNS, and $Al_2(SO_4)_3$ solutions and concluded that the emission arose from electric discharge between water vapor cavities and the glass walls of the containing vessel. Since no light was emitted from degassed water, Frenzel and Schultes[2] concluded the emission was caused by friction between cavitating bubbles and water. In 1936 Chambers and co-workers[15–17] investigated the visible component of sonoluminescence from 36 liquids at an insonation frequency of 10 kHz and found that the phenomenon of sonoluminescence occurred in both aqueous and nonaqueous solutions. He further determined that there was a qualitative correlation between emission intensity, *I*, and the product of dipole moment and viscosity of the medium. This work also established the existence of an inverse relationship between *I* and ambient liquid temperature. The sonoluminescence was attributed to the sudden destruction of the quasi-crystalline structure of the liquid, a process analogous to triboluminescence observed during the breakdown of a crystal (for example, the emission that occurs when phosphorous or ZnS is ground in a mortar or when sugar crystals are broken).

In 1937 Levshin and Rzhevkin[18] investigated sonoluminescence from water, aliphatic alcohols, benzene, xylene, CCl_4, and $C_6H_5NO_2$ at a frequency of 525 kHz. Using a wedge photometer to measure the luminescence intensity, they concluded that:

1. Potassium iodide and pyrogallol (quenchers of photoluminescence in the liquid phase) do not quench sonoluminescence, whereas CO_2 (a gas quencher) does, indicating the emission to be from the gaseous phase of cavities.
2. The luminescence is caused by high electric potentials that "arise during the very formation of cavities as a result of the drop of pressure up to negative values, in the process of violent ultrasonic vibrations. . . . In the resulting rupture an electric discharge takes place by which the vapor filling the cavities is excited"[18] and produces luminescence.

A more extended study of sonoluminescence was carried out by Harvey.[19] While Levshin and Rzhevkin did not state how the electrical potential developed on the bubbles, Harvey suggested that these electric potentials were balloelectric in nature; i.e., they were produced by an increase in the surface charge of fluids. Such potentials were first used to explain the chemical effects of hydraulic cavitation by Fottinger[20] and Richards.[21] In general, when a bubble compresses, the electrical capacitance decreases and the voltage increases until a discharge occurs, giving rise to a weak emission. The study of sonoluminescence of water saturated with various gases showed it was quenched by hydrogen and this was attributed to the weak balloelectric effect in the presence of hydrogen gas.[19] This result has been taken as evidence in favor of the electrical origin of SL, although such quenching also can be explained by the thermal theory, which is discussed later in this section.

The Microdischarge Theory (1940)

Frenkel,[22] like Levshin and Rzhevkin, and Harvey, also linked the onset of sonoluminescence with the formation of electric charges on the walls of cavities by a mechanism akin to the balloelectric effect. However, Frenkel postulated that the rupture of liquid during the expansive phase of the sound wave formed a lens-shaped cavity, which grew into a sphere by a gradual penetration of vapor from the surrounding liquid or the diffusion of gas from the neighboring region. Simultaneously, ions of opposite charge collected on the opposing sides of the cavity, forming an electric field within the cavity that behaved as a parallel plate capacitor. The field strength, E_c, inside such a cavity at the instant of cavity formation can be shown to be given by the relationship:[22,23]

$$E_c = \frac{1200e}{R_b} [\pi N_i d_c]^{1/2} \tag{6-1}$$

where R_b is the radius of the bubble; d_c, the initial thickness of the cavity; N_i, the number of ions per unit volume; and e the electronic (esu) charge of a monovalent ion. Thus, for example, if it is assumed that $N_i = 10^{18}$ cm^{-3}, $d_c = 5 \times 10^{-8}$ cm, then for a cavity of radius 10^{-4} cm and containing monovalent ions the strength of the electric field is 2280 V/cm. A voltage of this magnitude is sufficient to produce an electrical discharge in a gas at low pressure. However, there are several questionable assumptions made in this model that diminish its validity. First, one can argue that if the cavity is spherical from the very beginning, it may not be possible for an electrical discharge to occur. Second, since pockets of dissolved or entrained gases in a liquid medium are usually spherical with a radius on the order of 10^{-5} to 10^{-4} cm,[23] it follows that Eq. (6-1) is strictly applicable to homogeneous degassed water, in which cavitation is difficult to induce at moderate ultrasonic intensities. Finally, it also follows from this theory that the radiation should be observed from an expanding cavity, which contradicts the experimental observations.

The Mechanochemical Theory (1949)

This theory was first proposed by Weyl and Marboe.[24,25] They suggested that light emission occurred by the photochemical recombination of free ions produced by the "mechanical dissociation" of molecules at the nascent surfaces of growing cavities formed by the rupture of the quasi-crystalline lattice of the liquid. According to this theory the magnitude and nature of chemical reactions produced by cavitation should be independent of the dissolved gases and, second, sonoluminescence should appear during bubble expansion. Neither of these expectations concurs with the experimental facts.

The Anion Discharge Theory (1974)

The first semiquantitative attempt to explain SL on the basis of an electrical theory was made by Degrois and Baldo.[26] According to their hypothesis, the charges on gaseous bubbles arise from the neutralization of anions on the bubble surface by gas molecules adsorbed at the gas–liquid interface. These gas molecules are deformed and polarized by the anisotropic forces they experience because of their asymmetric surroundings. The electrical potential, V_E, produced on the axis of an induced dipole of a gas molecule at a distance r (where $r \gg$ dipole length, L) is given by:

$$V_E = \left(\frac{3\alpha_E R T_0}{\mathcal{N} r^4}\right)^{1/2} \tag{6-2}$$

where α_E is the polarizability coefficient, T_0 is the temperature of the medium, R is the gas constant, $\mathcal{N}$ is the Avogadro number, and r is the interaction distance and is assumed to be on the order of the radius of gas molecules, i.e., $\simeq$ 1–2 Å. Calculated electric potentials for a series of rare gases adsorbed on the inside surfaces of collapsing bubbles are on the order of 0.75 V—sufficient to discharge ions. Thus, the adsorbed gas molecules neutralize the neighboring adsorbed anions to produce electrons, which accumulate inside the bubble. Also, free gas molecules inside the cavity may polarize on striking the interface and transfer electrons to the bubble surface.

Assuming the net charge on a bubble stays constant during the volume change of the bubble, the charge density will increase as R_b^{-3} on compression. When the charge density exceeds a critical value an outflow of electrons occurs from within the bubble into the liquid medium. The microscopic discharge at the bubble surface produces excitation and ionization of molecules in the liquid phase accompanied by luminescence and various other chemical effects. It follows from this hypothesis that if the observed sonoluminescence is a microscopic discharge (i.e., simply an outflow of electrons), the spectral distribution should not vary significantly with the nature of the dissolved gas. On the other hand, if the observed SL is a chemiluminescent process (i.e., caused by radiative deexcitation or the radiative association of radicals formed by the microdischarge), then the emission should occur from the liquid phase or liquid–gas interface.

The quantity of charge accumulated inside the bubble depends upon the extent of gaseous adsorption, which, in turn, depends upon the solubility and the dipole electrical energy, E_d, of the gas molecules. E_d is related to the polarizability coefficient and atomic number $\mathcal{Z}$ according to:[26]

$$E_d = \tilde{k}\mathcal{Z}^3 \alpha_E^{-1/2} \tag{6-3}$$

where $\tilde{k}$ is a constant and is equal to 2.2×10^{-22} at 20°C. For a given compression amplitude the intensity of discharge and hence SL should increase with the gas solubility, α_E, and $\mathcal{Z}$. On the basis of this hypothesis one may expect SL to increase as one descends the periodic table; homonuclear gases should give stronger emission than the corresponding rare gases; and gases with a permanent dipole, NO, NO_2, etc., should give strong emission. Some of these predictions do not match the experimental observations presented later.

The Hot-Spot Theory (1950)

This theory, first proposed by Noltingk and Neppiras,[13,14] is based on the adiabatic heating of a cavity's contents. According to this model a gas nucleus grows slowly and isothermally into a bubble of maximum radius, R_{co}, which subsequently undergoes a rapid compression or collapse. The intracavity temperature at the end of collapse, T_{ad}, is given by the adiabatic equation:

$$T_{ad} = T_0(R_{co}/R_e)^{3(\gamma-1)} \tag{6-4}$$

where T_0 and γ are the initial temperature and the specific heat capacity ratio of the gas inside the collapsing bubble, respectively, and R_e is the minimum radius of the bubble at the end of its collapse. Solving the energy equation for bubble collapse, Noltingk and Neppiras deduced the expression:

$$(R_{co}/R_e)^{3(\gamma-1)} \simeq P(\gamma - 1)/p_{co}, \tag{6-5}$$

where P is the external pressure (equal to the algebraic sum of the hydrostatic and acoustic pressures) and p_{co} is the pressure of noncondensible gas inside the cavity at the radius at which the collapse begins. Combination of Eq. (6-4) and (6-5) yields an expression for the approximate temperature that can be reached during a bubble collapse:

$$T_{ad} = T_0 P(\gamma - 1)/p_{co} \tag{6-6}$$

For example, if T_0 is 300K and $\gamma = 4/3$, T_{ad} is on the order of 10^3K and 10^4K for P/p_{co} values of 10 and 100, respectively. These temperatures are large enough to heat gaseous contents of bubbles to incandescence, which is observed as sonoluminescence.[27,28] According to this theory, SL should increase with an increase in the magnitude of γ.

A variation of the Noltingk and Neppiras model was proposed by Jarman,[29] who suggested that the emission occurred from shockwaves undergoing multiple reflections at the interface of a bubble. Luminescence from such microshocks is primarily thermal in origin with a small contribution from bremsstrahlung and ionic recombinations. According to Trilling[30] a shockwave undergoing multiple reflections leads to the same average behavior as an adiabatic compression. Thus, although the two theories follow different approaches, they are basically in agreement with one another.

An extension of the hot-spot theory was suggested by Griffing and Sette,[31] who studied the visible components of SL and attributed it to chemiluminescence induced by the adiabatic heating of a vapor–gas mixture. According to this mechanism the radiation is a result of emission from excited and reactive chemical species formed by cavitation, a view that has become more popular in a number of studies reported in the last decade and a half.[32–38]

Comparison of Various Theories

Whereas no conclusive evidence supporting any one of the above theories has been obtained to date, in the light of experimental evidence that has accumulated since the discovery of SL it is possible to rule out some of them. This experimental evidence is now briefly reviewed.

Table 6-1. Relation Between SL Flash and Phase of Cavitation Predicted by Various Theories of SL

Sonoluminescence theory	Moment of flash: During cavity growth	Moment of flash: During cavity collapse
Tribomoluminescence	X	
Balloelectric		X
Microdischarge	X	
Mechanochemical	X	
Anion discharge		X
Hot spot		X

Temporal Relation Between SL Flash and Bubble Collapse

Table 6-1 summarizes the relation between the appearance of the SL flash and the phase of cavitation as predicted by various theories. The earliest study of the comparison of the time of light emission with the phase of ultrasonic waves was reported by Gunther and co-workers.[39,40] This was followed by other studies,[41,42] which revealed that the relation between light emission and the phase of ultrasound depends on the properties of the sound field, i.e., its frequency and pressure amplitude. For example, an increase in the acoustic pressure amplitude could cause luminescence to occur throughout the period of the sound wave. On the other hand, at moderate pressures and in the frequency range 30 to 1200 kHz, the luminescence flashes occurred during an extremely short interval of time, on the order of 1/10 to 1/50 of the period of the ultrasonic wave.[43] Consequently, unless an exact relationship between bubble motion and the phase of ultrasonic wave is known under given experimental conditions, it cannot be established unequivocally from the data relating time of light emission to the phase of ultrasound whether SL is emitted during the compressive or the expansive phases of bubble motion.

The problem was overcome by direct comparison of SL pulses, either with the acoustic pressure pulses generated by the collapsing bubbles[44,45] or, with the size of the cavitating bubble measured by recording the light reflected by them.[43] These measurements convincingly established that emission of light occurred during the last stages of bubble implosion. From these results it may be inferred that the electrical theories, except those proposed by Harvey[19] and Degrois and Baldo,[26] are improbable. Balloelectric or frictional charges are highly sensitive to the changes in the electrical conductivity of the liquid,[46] whereas the intensity of SL is not.[29,43] Therefore, even if a balloelectric charge is formed on the bubble surface, it seems unlikely to play a significant role in the onset of luminescence. Furthermore, for a discharge to occur within a cavitating bubble, a nonsymmetrical distribution of charges of opposite sign should be established on the bubble surface. Even with the nonspherical bubbles, this seems to be unlikely and Harvey's mechanism for the onset of SL also seems to be improbable.

Table 6-2. Sonoluminescence in the Presence of Different Gases

Gas	Specific heat ratio, γ	Thermal conductivity $(\text{cal cm}^{-1}\text{s}^{-1}\,°\text{C}^{-1}) \times 10^6$, k_c, (57)[c]	Ionization potential of gas (eV) (9)[c]	SL (Phots $\times 10^{-10}$),[a] 165–650 nm (56)[c]	SL (relative units), 190–280 nm (55)[c]	SL total radiation energy (J) 400–1000 nm (48)[c]	Sonoscintillations[b] (relative units) (95)[c] H_2O	Sonoscintillations[b] (relative units) (95)[c] *n*-Octane
Air	1.33	55.0	15.51	2.40	20		1.0	1.0
Nitrogen	1.40	58.3	12.5	1.22	45		0.7	0.5
Oxygen	1.40	336.5	24.58	2.40	35		0.3	8.0
Helium	1.65	109.2	21.56	1.16	1			
Neon	1.67	38.5	15.75	3.20	18	0.002		
Argon	1.67	21.2	14.0	30	54	0.024	2.7	0.8
Krypton	1.67	12.4	12.3	50	180			
Xenon	1.67			125	540	1.3	7	10
Methane				0.80				
Ethanol				0.85				
Propane				1.00				
n-Butane				1.00				
CO				2.40				
CO_2	1.30			0.86			0	0
N_2O				2.20				
Freon				3.60				
H_2		414		0.86				

[a] 1 phot = 1 ln cm^{-2}.
[b] Sonoscintillation is defined to be equal to the number of photons per unit collapse of all the cavities located in the cavitation zone.[95]
[c] Numbers in parentheses are references.

Evidence from Artificially Collapsing Thin-Walled Spheres

Indirect support for the thermal theory was obtained by Schmid,[47] Muller,[48] and Soloukhin.[49] Schmid observed light from the collapse of a gas-filled glass bulb in a liquid medium and inferred the emission to be from the heating of the gas during the collapse. These studies showed the emission intensity decreased with increasing amounts of H_2, CH_4, CO_2, and He and decreasing amounts of CCl_4 in the sphere. Later Muller measured luminescence by imploding thin-walled glass balls containing Ne, Ar, and Xe, in water. Light emission arising from the heating of gases was most intense from Xe-filled spheres and least intense from Ne-filled spheres. Sonoluminescence intensity also shows the same trend[48] (cf. Table 6-2).

Soloukhin[49] created artificial bubbles using very thin rubber. The bulbs were filled with a self-igniting gas mixture of $C_2H_2 + O_2$. By means of shockwaves the artificial bubbles were induced to collapse and the bubble motion was photographed. The results show that the gas mixture (ignition temperature = 890K) ignites at the minimum bubble radius, indicating that high temperatures are achieved in collapsing bubbles. Additional direct evidence for the thermal quality of cavitation and, hence SL, was demonstrated by Gordeev et al.,[50,51] in experiments in which explosion of solutions of benzene or heptane in tetranitromethane were initiated by the collapse of a large cavitation bubble at atmospheric pressure.

Spectral Investigations

Experimental observations of the ultrasonic luminescence spectra produced by insonation of water containing dissolved gases were first carried out using photographic plates. Srinivasan and Holroyd[27,28] and Gunther and co-workers[52,53] independently found the emission continuum to extend from the infrared to the ultraviolet region and that the spectral distribution was similar to that of blackbody radiation curves at temperatures between 6000 to 11,000K, depending on the nature of the dissolved gas. For example, results[27] indicated that the spectral distribution was close to that of a blackbody at 11,000K when water was saturated with monatomic gases, such as Ar and Ne, and 8800K when saturated with the diatomic gases O_2 and N_2. Gunther et al.[53] obtained blackbody temperatures on the order of 6000K for Xe and Ar. Margulis[10] recalculated Gunther's results of the red/blue ratio and showed that in the case of xenon the equivalent temperatures were not greater than 5000K. As the authors[27,28,52,53] did not observe any lines or bands typical of an inert gas or of free radicals in the gas phase, they concluded that their results supported the hot-spot theory. However, spectral similarities between SL and blackbody radiation cannot be taken as proof of the thermal origin of ultrasonic luminescence[29] because the spectra of certain chemiluminescence and bioluminescence phenomena resemble that of a blackbody. For example, bioluminescence from the chemical reaction of luciferin with the enzyme luciferase has an apparent spectral distribution of a blackbody radiating at 10,000K.[29,54] As well, the resolution of SL spectra often may not be sufficient to reveal line or band spectra.

More recently, evidence has been presented against the blackbody radiation mechanism for SL.[37] Sonoluminescence from aqueous carbon tetrachloride solutions was measured as a function of the CCl_4 concentration and fitted to Wein's Law to obtain color temperatures. The calculated values for pure water, water saturated with CCl_4, and a mixture containing 50% of CCl_4 were estimated to be 5300K, 4700K, and 4200K, respectively. If the SL spectra were blackbody in origin, the total SL intensity should increase as the fourth power of color temperature. That is, the following ratios of SL intensity should result if the blackbody mechanism holds:

$$\frac{I(\mathrm{H_2O\ Saturated\ with\ CCl_4})}{I(\mathrm{H_2O})} = \left(\frac{4700}{5300}\right)^4 = 0.6 \qquad (6\text{-}7a)$$

and

$$\frac{I(\mathrm{CCl_4(50\%) + H_2O\ (50\%)})}{I(\mathrm{H_2O})} = \left(\frac{4200}{5300}\right)^4 = 0.4 \qquad (6\text{-}7b)$$

However, the experimentally measured ratios in the two cases were found to be 2.8 and 6, respectively, in disagreement with the blackbody explanation.

A major step forward in the understanding of SL arose from the studies of Taylor and Jarman[32] and Saksena and Nyborg[33] published in 1970. Using a specially designed spectrophotometer, Taylor and Jarman observed the OH emission band in the SL spectra of various aqueous solutions. They concluded from these studies that light emitted from cavitation arose partly from incandescence and partly from chemiluminescent reactions within a bubble.

Saksena and Nyborg[33] observed SL from the periodic volume oscillation of bubbles of uniform size and concluded that the emission resulted from radiative recombination of H and OH radicals formed by the adiabatic heating of cavity contents during the compression phase of the cycle. Their analysis also showed that emission from bremsstrahlung was negligible. Since the chemiluminescent reactions producing SL are caused by high intracavity temperatures, the dependence of light emission intensity on γ, as observed experimentally and explained by the hot-spot theory, still holds true.

Limitations of the Hot-Spot Theory

According to the thermal theories, gases with the same specific heat ratio γ should have the same final temperatures and the same intensities of emission. The relative intensities of SL from H_2 in the presence of different dissolved gases were obtained by Prudhomme and Guilmart[55] and Young[56] in different spectral regions (see Table 6-2). Although no distinctive radiation bands corresponding to each gas were observed, the emission intensity increased with atomic number, in apparent disagreement with the thermal theory predictions. This disagreement arises because the hot-spot theory[13] does not take into account the heat flow to the surroundings during the short collapse time of the cavity. In a real situation heat is lost from a bubble to the surroundings because of the high temperature gradient at the gas–liquid interfaces. Therefore, the cavity temperatures reached inside a cavity are lower than that predicted by Noltingk and Nepiras' model. Accordingly, for the noble gases the magnitude of the temperature reduction depends on the thermal conductivity of the gas k_c and, since k_c of a gas decreases with an increase in atomic number(cf. Table 6-2), the SL intensity should increase from He to Xe, as observed experimentally.

The first attempt to quantitate these ideas was made by Hickling,[57] who considered a model of a spherical cavity containing a thermally conducting ideal gas with a negligible quantity of liquid vapor in an incompressible liquid. In arriving at numerical solutions for intracavity temperature (T) distributions in bubbles at different stages of collapse, the following assumptions were made:

1. The bubble contents and the liquid are nonviscous.
2. Prior to collapse they have a uniform temperature.
3. During the collapse the increase in temperature in the liquid at the cavity wall is negligible in comparison to the increase in temperature of the gas. This is anticipated because the heat capacity of the liquid is generally about 10 times greater than that of the gas. Assuming the boundary temperature T_B to be equal to that of the surrounding liquid, one may conclude the maximum temperature gradient to occur at the surface.
4. Since the thermal diffusivity of the gas is greater than that of the liquid, the transfer of heat out of the bubble is confined to a comparatively narrow region of the liquid next to the interface.

The numerical results derived by Hickling infer the following conclusions.

1. The temperatures inside a cavity constantly change during its collapse and reach a maximum value at the minimum bubble radius.
2. The true temperatures within a cavity are lower than the adiabatic temperatures (see Table 6-3). For example, a cavity of initial radius 10^{-2} cm and

Table 6-3. Temperature (T) Inside a Bubble at 97% Collapse Estimated by Hickling's Model[a,57]

Bubble radius (cm)	Nature of collapse	T (K) under N_2 gas[b]	T (K) under Ne gas[b]
10^{-1}	Adiabatic	1260	3200
10^{-1}	Nonadiabatic	1110	2600
10^{-2}	Nonadiabatic	1030	1500
10^{-3}	Nonadiabatic	640	600

[a] The bubble is subjected to a constant external pressure of 3 atm and has initial internal pressure of 0.075 atm and an initial temperature of 293K.
[b] Results obtained from Figures 2 and 3 of Ref. 57 at 97% of collapse.

filled with neon at an initial pressure of 0.075 bar has an intracavity temperature of 1500K, as compared to a value of 3200K attained in an identical cavity that has collapsed adiabatically to the same extent.

3. Energy loss through conduction is larger in collapsing cavities of smaller radii (see Table 6-3).

Consequently, because of higher thermal conductivity, a neon gas bubble (cf. Table 6-2) will lose more heat than the corresponding N_2 gas bubble. In small bubbles ($\sim 10^{-3}$ cm or less), the thermal conductivity effect outweighs the dependence on temperature. As a result, neon gas achieves lower temperatures than N_2, even though γ for the former is greater. Such dependence of cavity temperatures on bubble size and thermal conductivity may also be used to explain the difference in relative sonoluminescence intensity values determined by Prudhomme and Guilmart,[55] and Young.[56] The former authors observed that emission from N_2-saturated water is stronger than that from Ne-saturated water, whereas Young's studies show it to be in the reverse order. This difference can be explained if one argues that under the experimental conditions of Prudhomme and Guilmart the average bubble size is on the order of 10^{-3} cm or less, whereas Young's results are obtained from a bubble field with an average bubble radius larger than 10^{-3} cm. In the former case, because of a smaller radius, the thermal conductivity effect would predominate and consequently lower the intracavity temperatures, therefore producing weaker emission from Ne-saturated water as compared to N_2-saturated solutions.

More recently, Young[56] has obtained an analytical expression for the intracavity temperature, T, of a spherical bubble containing a heat-conducting ideal gas. Using assumptions somewhat similar to those made by Hickling, Young showed that T is related to the adiabatic temperature by:

$$T = T_{ad} \exp\left\{ -\frac{5\bar{\alpha}}{4\bar{n}(2p_H/3\rho)^{1/2} R_{co}^{3/2}} \times \left[\frac{T^{1/2} R_e^{1/2}}{[3 \ln (R_e/R_{co}) + 1 + p_A/p_H]^{1/2}} + \frac{T_0^{1/2} R_{co}^{1/2}}{(1 + p_A/p_H)^{1/2}} \right] (R_{co} - R_e) \right\} \quad (6\text{-}8)$$

where $\bar{\alpha}$ is equal to $0.92(3R/M_w)^{1/2}$; R is the gas constant, M_w is the molecular weight of the gas, p_H is the hydrostatic pressure, p_A is the pressure amplitude of the sound wave, ρ is the density of the liquid medium, R_{co} and R_e are the radius of the bubble at which collapse begins and is completed, respectively, and $\bar{n}$ is a constant related to the distance from the bubble surface over which the temperature drops linearly.

The exponential term in Eq.(6-8) can be regarded as a correction for heat loss by conduction. Young estimated intracavity temperatures for bubbles containing different monatomic gases under the condition of $p_H = 1$ atm, $p_A = 6$ atm, $T_0 = 293$K, $\bar{n} = 3$, $R_{co} = 10^{-3}$ cm, and $R_e = 3.33 \times 10^{-4}$ cm. The values of T summarized in Table 6-4 show that, even though monatomic gases have the same value of γ, the bubbles achieve different intracavity temperatures because of different heat losses from the bubble. Therefore, the hydrodynamic models of

Table 6-4. Relationship Between γ and Theoretical Temperatures Estimated by Young's Model[56]

Gas	γ	T (K)
Helium	1.65	815
Neon	1.64	1420
Argon	1.65	1650
Krypton	1.67	1890
Xenon	1.67	2000

Hickling[57] and Young[56] account for the variation of SL intensities amongst gases of the same specific heat ratio, a fact that cannot be readily explained by the original hot spot theory. It should be noted, however, that these mathematical models of cavitational collapse contain many simplifying approximations. Qualitative predictions are reliable, but caution must be applied to their quantitative use.

SONOLUMINESCENCE FROM AQUEOUS SOLUTIONS

In general, an emission continuum may be interpreted by using the following information:

1. The distribution of spectral intensity with wavelength, the most important features being: (a) the wavelength corresponding to maximum intensity, and (b) the low wavelength cutoff of emission.
2. The absolute intensities at particular wavelengths.
3. The effect of the temperature of the emitting system on the nature and intensity of the continuum.
4. The time required for the emitting system to achieve equilibrium as a function of temperature and density of the system.
5. The scavenging of the chemical species that produce luminescence; if the continuum is from more than one source, then the contributions of the various species to the total emission can be determined by selective scavenging.

Because of the random distribution of cavitating bubbles in the insonation field, and because of instantaneous emission from bubbles at different stages of collapse (i.e., at different temperatures), measurement of absolute intensities at different wavelengths is almost impossible. Since the temperature attained within a bubble depends upon the degree and the rate of collapse, and because it is not possible to control the collapse of a bubble precisely under common experimental conditions, it is virtually impossible to regulate intracavity temperatures and to study their effect on the emission continuum. Under these circumstances one can, at best, speak only in terms of average behavior or properties of the bubbles.

The collapse of a bubble occurs over a finite length of time under constantly changing conditions. As a consequence, thermal equilibrium may not be truly attained and the intracavity energy distribution may be non-Maxwell-Boltzmann in nature. Even if one assumes that thermal equilibrium is achieved and uses time-resolved spectroscopy with nano- to picosecond resolution, it is still required that all the bubbles in a cavitation field implode at the same instant for it to be possible to measure the time required to achieve thermal equilibrium. Consequently, this latter property cannot be used to obtain any information on the continuum radiation emitted by a bubble field as a function of intracavity temperature and density.

This leaves only two possible sources of spectral information that are readily obtainable—the average distribution of spectral density with wavelength and the changes that occur because of scavenging of the radiative species.

Spectral Studies

Sonoluminescence spectra of distilled water saturated with various gases have been reported by Taylor and Jarman[32] and Sehgal et al.[36] Representative spectra obtained from argon-saturated water at ultrasonic frequencies of 16, 333, 459, and 500 kHz are shown in Fig. 6-1. Each of the SL spectra shows a peak at 310 nm superimposed upon a strong continuum extending from 240 nm to the near infrared. At lower ultrasonic frequencies the emission at 310 nm is strong and is accompanied by weak bands at ~270 to 290 nm, and at 340 nm. At higher insonation frequencies the 310-nm peak is not as clearly defined because of a stronger continuum. It has been observed that this prominence of the 310-nm peak decreases with dissolved gas in the order argon, krypton, and xenon.[32]

The observed relative intensities of the bands vary slightly from spectrum to spectrum, possibly because of the random and nonequilibrium nature of cavitation.[36] The 310-nm band arises from OH emission caused by the transition $A^2\Sigma^+ \rightarrow X^2\pi$. Similar bands have been observed in flames at 2000–3000K.[58] The broad continuum could arise either from the thermally excited contents of bubbles leading to ionization and recombination and bremsstrahlung or from overemission from chemical reactions.[32] Although it was not explicitly shown, Taylor and Jarman inferred a chemical process of the type:

$$H + H + M \longrightarrow H_2 + M^* \longrightarrow H_2 + M + h\nu \qquad (6\text{-}9)$$

where M represents a rare gas molecule.

In another study, Saksena and Nyborg[33] observed SL from an array of gas bubbles of controlled size. The bubbles were generated in a viscous mixture of glycerol–water and were irradiated with an annular transducer at 30 kHz. Their

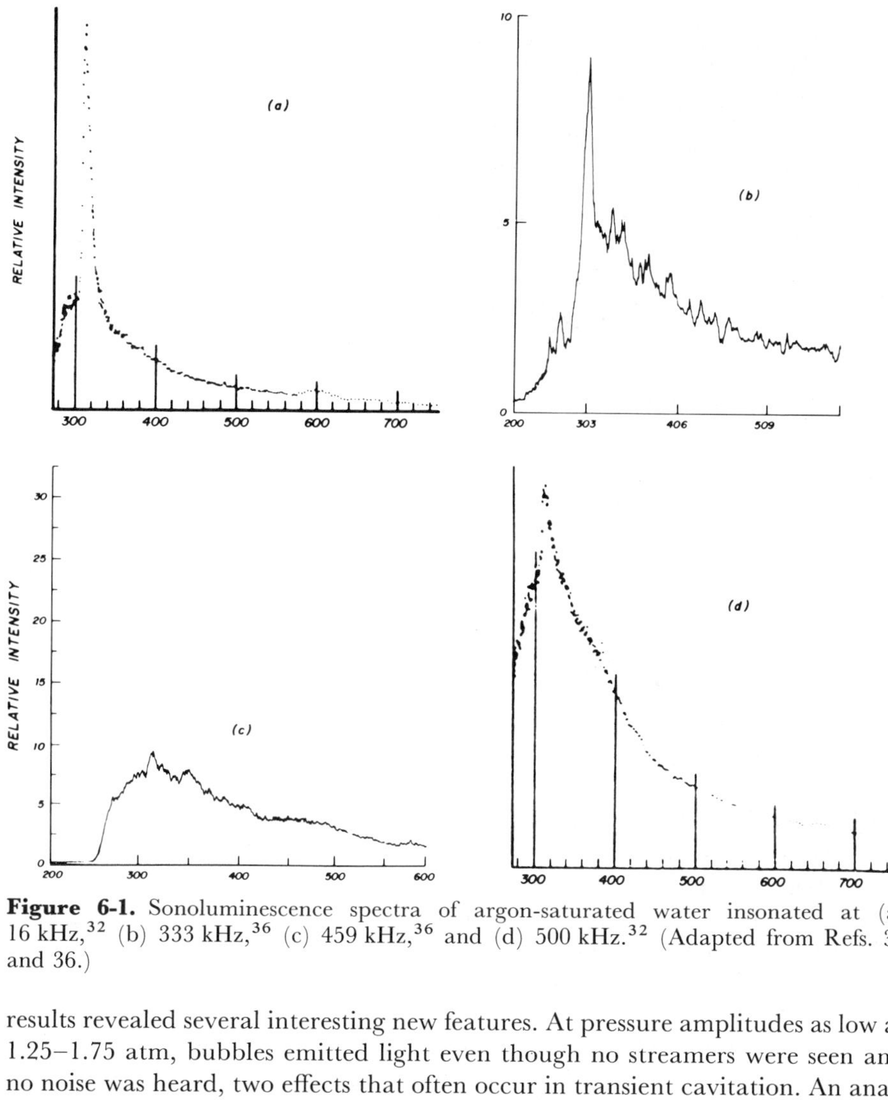

Figure 6-1. Sonoluminescence spectra of argon-saturated water insonated at (a) 16 kHz,[32] (b) 333 kHz,[36] (c) 459 kHz,[36] and (d) 500 kHz.[32] (Adapted from Refs. 32 and 36.)

results revealed several interesting new features. At pressure amplitudes as low as 1.25–1.75 atm, bubbles emitted light even though no streamers were seen and no noise was heard, two effects that often occur in transient cavitation. An analysis of the acoustic spectrum showed no broadband noise at amplitudes below 2.5 atm.

Appreciable luminescence was seen from bubbles of resonant and about twice resonant sizes. The peak intensity of luminescene from resonant size bubbles was about 1.5–3 times greater than that from bubbles of twice the resonant size.

More importantly, high-resolution temporal information of the SL emission revealed light pulses to be 4–5 μs long superimposed on a continuous luminous background (cf. Fig. 6-2a). These light pulses appeared once per time period of the sound wave. Contrary to these observations, in aqueous solutions of luminol the light pulses at sonic frequencies are weak and superimposed on a strong continuous background (cf. Fig. 6-2b).

(a)

(b)

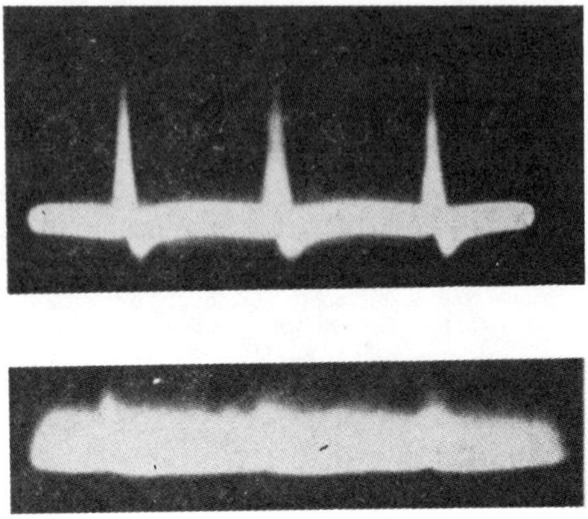

Figure 6-2. (a) High-resolution time spectrum of SL from glycerol–water mixture (95:5 v/v) at a pressure amplitude of 1.5 atm; time between flashes is about 33 μs, the period of sound wave. (b) SL from luminol solution at a pressure amplitude of 2.1 atm. (Reproduced with permission from Ref. 33.)

Because no surface waves were observed, the authors hypothesized that bubble motion was mostly from volume oscillations characteristic of stable cavitation. As the bubble volume oscillates rapidly and adiabatically, or nearly adiabatically, temperature pulses of width Δ and height T_p are generated. This is conceptualized in Fig. 6-3. If the pulse height exceeds a critical value, light emission occurs, appearing as periodic flashes on a high-resolution time spectrum, as shown in Fig. 6-2.

In order to determine whether the light emission arises from chemical reactions involving free radicals $H^{\bullet}$ and $OH^{\bullet}$ or from accelerated electrons formed by thermal ionization of gas in oscillating bubbles, the authors developed equations that estimated the total light output as a function of intracavity temperature for the two types of mechanisms. According to their analysis light emission at 2000K from bremsstrahlung would yield two photons per second as compared to $\sim 4.7 \times 10^9$ photons per second from radiative recombination of the free radicals $H^{\bullet}$ and $OH^{\bullet}$ formed by the decomposition of water inside the oscillating bubbles.

The contention that the emission comes largely from the radiative recombination of radicals was further supported by the results of two experiments. In one case, glycerol–water mixtures were replaced by silicone fluid of similar viscosity. If the emission arose from bremsstrahlung mechanism, only a slight change should have been expected in the SL intensity. However, the experiments showed the SL intensity was reduced by a factor of 20, implying that H_2O was the source of free radicals $H^{\bullet}$ and $OH^{\bullet}$ that led to SL in aqueous media. In the second experiment allyl alcohol was added to water, resulting in a quenching of the light emission. Since allyl alcohol has the same vapor pressure as water, it should not affect the bubble dynamics significantly and the attenuation of SL should result from the scavenging of the free radicals $H^{\bullet}$ and $OH^{\bullet}$. Consequently, the studies by Saksena and Nyborg show SL to occur from stable cavitation caused by radiative recombination of free radicals. This view is also shared by Zhuraler and Tsyler,[59] who have observed SL during stable cavitation. More recently, Margulis and Grundel[60] have also observed light emission from the volume pulsation of bub-

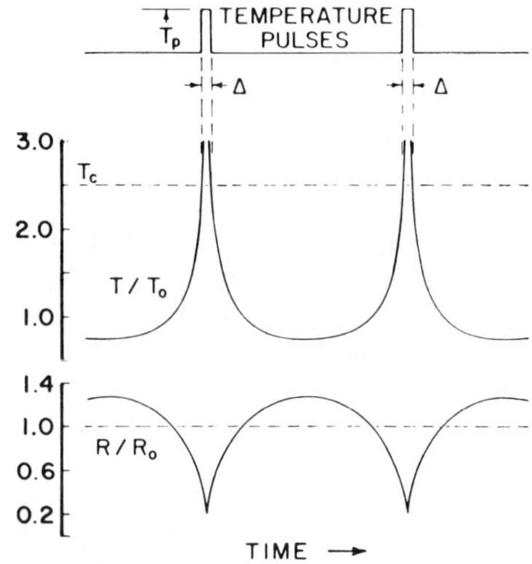

Figure 6-3. Plots of radius R and temperature T versus time for a gas bubble in a sound field; R_0 and T_0 are equilibrium values. Temperature plot obtained from the adiabatic law. Upper plot shows associated fictitious temperature pulses of height T_p, assumed to occur when T exceeds some critical threshold value T_c. (Reproduced with permission from Ref. 33.)

bles at low sound frequencies between 30 and 250 Hz in a 3M KCl solution. Their experiments reveal that at 220 Hz the SL intensity changes sharply with a small change in sound frequency (~ 3–5%). This strongly suggests that SL can be produced from resonance oscillations of bubbles (i.e. stable cavitation). A small change in sound frequency is sufficient to perturb the bubble motion from resonance and thus reduce SL.

Light Emission from a General Bubble Field

When a medium containing dissolved gases is exposed to an ultrasonic field, a wide range of bubble motions occurs of which transient and stable cavities are suitable models of the limiting type of bubble motion. Similarly, gaseous and vaporous cavities define limiting types of bubbles on the basis of intracavity composition. In an experiment in which no deliberate attempt is made to control the size and composition of nuclei that grow into bubbles, one can anticipate a distribution of bubbles between these two limiting types. In fact, a typical bubble may not be described by any one of the limiting cases but may have some properties of each kind. For example, a typical bubble may contain some vapor of the liquid medium along with some gas. It may oscillate linearly or nonlinearly about an equilibrium position for a few cycles before collapsing. It is also possible that during oscillations bubble contents may undergo condensation or evaporation or gas diffusion from neighboring regions, thereby altering the composition. Its walls may execute the dynamics of stable cavitation but heat the cavity contents, a phenomenon often associated with transient cavitation. Therefore, any single definition may only partly define the behavior of bubbles in an experiment involving the measurement of an effect or activity produced by a collection of bubbles at different points of time and space. With this in mind it must be emphasized that the use of these descriptions of bubble character in this chapter is only valid in a rather qualitative sense. Nevertheless, the complexity of the bubble dynamics precludes any better analysis at this time. Keeping this in view, the relationship between sonoluminescence and the type of cavitation is now examined.

Since light emission has been reported to occur from either transient[32] or stable cavitation,[33] it is reasonable to anticipate light emission to take place from both types in a general bubble field.[36] In either case the excitation energy absorbed by chemical species during the adiabatic compression–collapse of a bubble is primarily by Frank-Hertz collisions of the first kind, wherein kinetic energy is converted into vibrational or electronic energy. The nature of the excitation process, i.e., whether it is electronic or vibrational, depends on the intracavity temperature. For a fixed composition, the temperature generated inside a bubble is directly related to the ratio R_{co}/R_e and the rate of compression, dR/dt, which determines the heat loss from the cavity during compression. Since stable cavities have lower values of R_{co}/R_e and dR/dt (for example, see Fig. 6-11), they must have lower temperatures than transient cavities. Thus, the cavitation effect produced by stable cavities is "softer" than that of transient cavities.

Electronic excitation leads to the formation of excited water molecules followed by either radiative relaxation or dissociation into $OH^{\bullet}$ and $H^{\bullet}$ radicals according to the scheme:

$$H_2O + M \xrightarrow{))))} H_2O + M \ \text{(hot molecules)} \longrightarrow M + H_2O^* \ ({}^3B_1)$$

$$H_2O^* \ ({}^3B_1) \xrightarrow{(a)} H_2O \ ({}^1A_1) + h\nu_o \ (270\text{–}290 \text{ nm}) \qquad (6\text{-}10)$$

$$H_2O^* \ ({}^3B_1) \xrightarrow{(b)} H + OH^* \ ({}^2\Sigma^+) \longrightarrow OH \ ({}^2\pi) + h\nu_1 \ (310 \text{ nm}) \qquad (6\text{-}11)$$

where M may be either water or dissolved gas molecules or atoms.

1. *Emission from* H_2O^* (${}^3B_1$): The emission in the near-uv region with a peak maximum in the region 270–290 nm (Fig. 6-1) arises largely from transitions from the triplet ${}^3B_1$ state of water, which was first predicted by Mulliken[61] and was observed experimentally in electron reflection spectroscopy[62] and ionizing radiation studies of water[63–65] producing luminescence. The short-wavelength cutoff at 235–245 nm or 481.2 ± 8.4 kJ mol^{-1} gives the heat of reaction of $H_2O \rightarrow H + OH$ (all in the ground state) at 0K in good agreement with the thermal value of 497.9 kJ mol.$^{-1}$ Selective scavenging agents, such as $[NO_3]^-$, uncover the H_2O emission band, which is overlaid by the stronger OH band system.
2. *Emission from* OH^*: The emission below 400 nm is believed to arise from electronically excited OH (Σ^+) which has been reported to be formed by the fragmentation of water in the excited state.[63–65] The partially resolved bands at

281 and 340 nm, flanking either side of the peak at 310 nm (Fig. 6-1), are from 1 → 0, 0 → 1 transitions, respectively. As noted[32] these peaks are not well defined at higher insonation frequencies.

Because of the predissociation of OH in the Σ^+ state, $O^\bullet(^3P)$ and $H^\bullet(^2S)$ can be formed[66] and $O^\bullet(^3P)$ may carry out further oxidation. The radiative recombination of H and O is a likely process and, should it occur, a low-wavelength cutoff at 282 nm is to be expected. However, it is difficult to assess its contribution in the present cases because it is masked by emission from H_2O^* and OH^*. However, if the $OH^\bullet$ emission occurs mainly from the 1 and 0 vibrational levels of the Σ^+ state, as shown by the presence of peaks at 281, 310, and 340 nm (Fig. 6-1), then a rather low concentration of $O^\bullet(^3P)$ (as compared to OH) would be expected because the $^4\Sigma^-$, $^2\Sigma^-$ and $^4\Pi$ potential energy curves of OH cross that of Σ^+ at higher vibrational levels.[36] Therefore, only a small contribution to the total emission is likely to occur from this path.

Stable cavities produce lower intracavity temperatures, and vibrational excitation leading to the dissociation of water in the ground state should be favored. The radicals formed combine radiatively to give a continuum with a broad maximum around 400 nm.

Griffing[67,68] proposed that this emission arose from the recombination of $H^\bullet$ and $OH^\bullet$ radicals, i.e.:

$$H + OH \longrightarrow (H\cdot\cdot O\cdot\cdot H)^{\#} \longrightarrow H_2O + h\nu_2, \qquad (6\text{-}12)$$

where H and OH approach each other to give a "quasi-molecule" $H_2O^{\#}$ stabilized by radiative processes. However, based on experimental estimates[32,35] of temperatures and pressures, it can be shown that the rather short collision times at these conditions would give rise to a very small probability of emission ($\sim 10^{-12}$) in a single transition.

A more likely possibility, first suggested by Saksena and Nyborg,[33] is the stabilization of a quasi-molecule by weak bond formation with a third body M (gas or water molecule) to give M · · · · OH, which in turn combines with H to give a quasi-molecule as follows:

$$OH + M \xrightarrow{a} M\cdot\cdot OH \xrightarrow{b} [H{-}M{-}OH] \xrightarrow{c} H_2O + M + h\nu_2 \qquad (6\text{-}13)$$

Steps (a) and (b) may take place simultaneously, i.e., in a three-body collision (the number of which increases sharply with pressure), or by consecutive two-body collisions. The last step could be radiative or nonradiative. The quasi-molecules so formed would have a longer lifetime and a greater probability of undergoing emission.

Sonoluminescence in the Presence of Organic Scavengers

Studies of the effect of added radical scavengers on the SL spectra from argon-saturated water have been reported by Sehgal et al.[34] with the view of obtaining evidence for the chemical mechanism of sonoluminescence. The results showed

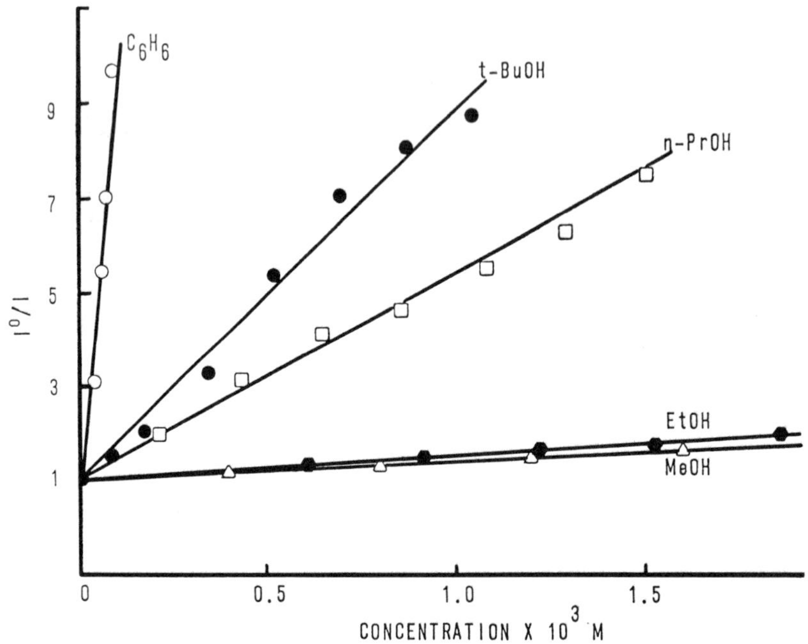

Figure 6-4. Plot of ratio of total sonoluminescence intensity in the absence of scavenger, I_0, to total intensity, I, vs. concentration of scavenger in bulk solution. The insonation frequency is 459 kHz. (Reproduced with permission from Ref. 34.)

that the shape of the SL continuum was invariant, although attenuated in intensity, with the addition of benzene and several aliphatic alcohols. Total SL intensity diminishes with increasing concentration of organic additives in a Stern-Volmer-like fashion over the bulk concentration range 0–2 mM (cf. Fig. 6-4). This decrease is believed to be caused by the scavenging of free radicals. Since both the scavenging of free radicals and the suppression of bubble collapse by volatile alcohols could lead to the diminution of SL, it is difficult to determine their relative contributions. However, at low concentrations the observed effects are not likely to result from suppression of cavitation alone, for the following reasons.

The presence of volatile components, such as CCl_4, is known to increase SL,[43,69] possibly by enhancing nucleation processes. Also, if alcohols suppressed cavitation significantly the intracavity temperatures would be considerably reduced. Under these circumstances one would not only expect an attenuation of SL intensity but also a change in the nature of the SL spectrum. However, experimental results show no measurable change in the spectral distribution. Finally, addition of allyl alcohol, which has the same vapor pressure as water and therefore should not suppress the cavitation phenomenon, has been demonstrated to be an effective quencher of SL.[33]

More recently, the sonochemical oxidation of Fe^{2+} to Fe^{3+} in aqueous media has been observed to be enhanced by the addition of aliphatic alcohols in the concentration range 0 to 10 mM.[70] Such an increase would be difficult to explain if suppression of cavitation by alcohols were a dominant factor. The increase can be conveniently explained by chemical participation of alcohols in the cavitation-induced oxidation–reduction reactions; as in radiation chemistry, alcohols ROH scavenge $OH^{\bullet}$ radicals to form water and free radicals $RO^{\bullet}$. Oxygen readily combines with $RO^{\bullet}$ to yield peroxide radicals $[RO_2]^{\bullet}$. These radicals oxidize Fe^{2+} and provide an additional pathway for alcohols to enhance the oxidation of Fe^{2+}.[70]

Recently, Vaughan et al.[71] have inferred from these results that, if O_2 is replaced by any other gas, no peroxide radicals would be formed; hence no increase in oxidation of Fe^{2+} should be observed. This argument neglects to consider that in the absence of O_2 there are still unreacted alcohol radicals, which are capable of oxidizing ferrous ions either directly ($RO^{\bullet} + Fe^{2+} \rightarrow RO^{=} + Fe^{3+}$) or through intermediate steps. Therefore, the presence of alcohol in the medium will provide additional oxidation–reduction reaction pathways, unless the solution contains a substance that combines with free radicals of alcohol to give end products that are unreactive toward Fe^{2+}.

Also, the results of this study have shown that the average subharmonic emission decreases as the alcohol concentration is increased from 0 to 10 mM. However, it is not clear whether, while carrying out the subharmonic emission measurements, the increased absorption of sound caused by the presence of alcohols was considered. The presence of alcohol in water can easily increase ultrasonic absorption of the medium by an order of magnitude or more,[72,73] especially in the composition range where peak sound absorption caused by concentration fluctuations is observed. It is possible that the decrease in weak subharmonic signal attributed to reduced cavitation results from an increase in sound absorption by the medium. A further fact that has some bearing on the reported results is that subharmonics can be generated in the absence of bubbles owing to the nonlinear behavior of the liquid itself.[74] The manner in which alcohols can alter nonlinearity of a medium is not known but such effects should be considered before any conclusions are drawn from experiments based on subharmonic measurements.[71]

In view of the arguments presented above, the direct effect of alcohols on cavitation are neglected, as in the following analysis. Scavenging by alcohols has been proposed to occur from transient and stable cavities according to the following reaction schemes:[36]

Transient cavities		Stable cavities		
$H_2O \xrightarrow{k_0} H_2O^*$	(A)	$H_2O \xrightarrow{k_6} H_2O^{\#}$	(G)	
$H_2O^* \xrightarrow{k_1} H_2O + h\nu_0$	(B)	$H_2O^{\#} \xrightarrow{k_7} H^{\bullet} + OH^{\bullet}$	(H)	
$H_2O^* \xrightarrow{k_2} H^{\bullet} + [OH^*]^{\bullet}$	(C)	$H^{\bullet} + OH^{\bullet} + M \xrightarrow{k_8} H_2O + M + h\nu_2$	(I)	(6-14)
$[OH^*]^{\bullet} \xrightarrow{k_3} OH^{\bullet} + h\nu_1$	(D)	$OH^{\bullet} + ROH \xrightarrow{k_9}$ products	(J)	
$[OH^*]^{\bullet} + ROH \xrightarrow{k_4}$ products	(E)	$H^{\bullet} + ROH \xrightarrow{k_{10}}$ products	(K)	
$H_2O^* + ROH \xrightarrow{k_5}$ products	(F)			

where the symbol k, with numerical subscripts, represents the various rate constants. Chemical scavenging of free radicals by alcohols can occur by a reaction of the type:

$$R{-}CH_2OH \xrightarrow{OH^\bullet} [R{-}CH{-}OH]^\bullet + H_2O \qquad (6\text{-}15)$$

The possibility of esr spin trapping of the radicals produced by H abstraction from alcohol molecules was suggested as providing evidence in support of the occurrence of chemical quenching of SL.[70] Recently, Makino et al.[75] have reported a detailed study of the trapping of free radicals produced during the sonolysis of aqueous ethanol solutions (using 5, 5-dimethyl-1-pyrroline 1-oxide (DMPO) as a spin trapping agent). According to their results, as the concentration of ethanol is increased, the esr intensity of the OH–DMPO adduct is reduced, whereas that of the $CH_3CH(OH)$–DMPO adduct is enhanced. This result strongly supports the view that scavenging of free radicals plays a significant role in the quenching of SL when alcohol additives are present in aqueous solutions.

There are two possible quenching mechanisms for ROH: it may either react with OH* to give new chemical products, or it may extract excess energy and dissipate it as heat and yield the original chemical species, e.g., ROH and $OH^\bullet$. The latter process represents quenching of excited species by impact of the second kind and is believed by Margulis to be the only mechanism leading to a reduction in SL.[77] In view of the extreme affinity of $OH^\bullet$ for $H^\bullet$, we believe that attenuation[35,36] results largely from the first possibility, i.e., the abstraction of $H^\bullet$ by $OH^\bullet$ from alcohol molecules. Unfortunately the intracavity concentrations of scavengers are not known accurately, because it is not clearly understood how well the equilibrium vapor pressures are established within the bubbles. As well, Arrhenius parameters for $OH^\bullet$ + alcohol reaction and the rate constants for the collisions of the second kind at high temperatures are not known. Therefore, quantitative correlation of relative efficiencies of various alcohols as quenchers is not possible, and it is not yet feasible to choose one possibility of quenching over the other. Nevertheless, the efficiencies of SL attenuation increase in the order MeOH $\simeq$ EtOH $<$ *n*-PrOH $<$ *t*-BuOH $<$ C_6H_6 which, with the exception of *t*-butanol, qualitatively correlates with the relative rates of ROH + $OH^\bullet$ reaction at room temperature.[76]

*Preferential Quenching of Hydroxyl Radicals: Evidence for H_2O**

The light-radiating species involved in SL are extremely reactive and very similar in their chemical behavior; consequently it is difficult to find a scavenger that reacts exclusively with only one of them. The $[NO_3]^-$ ion is a good scavenger of hydrogen and hydroxyl radicals and also of excited water molecules.[78] However, direct interaction with H_2O^* is not expected to occur at $[NO_3]^-$ concentrations $<$1M.[78] Another potential advantage of using nitrate as scavenger is that it has virtually no vapor pressure and does not concentrate in bubbles where it could

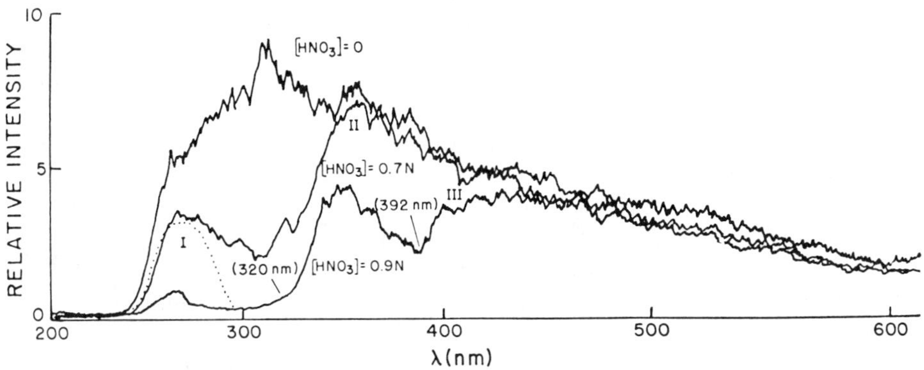

Figure 6-5. Low-resolution sonoluminescence spectra of argon-saturated nitric acid solutions insonated at 459 kHz: I, emission from H_2O^*; II, emission from $H^{\cdot} + [HO_2]^{\cdot}$; and III, emission from NO_2^*. (Reproduced with permission from Ref. 79).

affect the bubble dynamics and intensity of SL, which could be the case with volatile solutes.

Selective scavenging studies with nitric acid have provided chemical evidence for the formation of H_2O^*.[79] Figure 6-5 shows the spectra of rapid scanning of SL from water after known amounts of nitric acid are injected into the insonation cell. The results indicate that $[NO_3]^-$ preferentially scavenges the hydroxyl radicals and uncovers the peak from excited water at ~280 nm, normally embedded under the hydroxyl band system in the absence of an OH scavenger. A continuum extending from 320 nm to the near infrared is observed, which at higher HNO_3 concentrations splits into two broad bands (II and III of Fig. 6-5). These bands are believed to arise from chemical species $[HO_2]^{\cdot}$ and NO_2 resulting from the reaction between $[NO_3]^-$ and primary free radicals.[79] The emission from these two types of chemical species is discussed in more detail in a subsequent section dealing with emission from anions.

SONOLUMINESCENCE IN THE PRESENCE OF ELECTROLYTES

The qualitative effects of electrolytes on SL were reported soon after the discovery of the phenomenon of SL.[18] It was not until some time later that Gunther and co-workers[39] carried out a detailed study of the effect of alkali and alkaline earth salts on SL. Their studies, reviewed earlier by El' piner,[9] show that the addition of electrolytes to water produces significant changes to the SL spectrum of water itself. The following specific effects are observed: the luminescence intensity is dependent on the atomic radius of the cations; an increase in luminescence occurs mainly in the visible region of the spectrum with a smaller effect on the broad band at 425 nm; the visible spectrum contains resonance peaks characteristic of the neutral metal atoms present as ions in the solution; these peaks are accompanied by light emission at lower wavelengths; and the addition of radioactive ^{89}Sr to sodium salt solution leads to the disappearance of sodium D lines without affecting the continuous band. The early studies[39,40,52,53,80] used photographic techniques and long exposure times and did not provide any quantitative estimate of the energy distribution of SL spectra.

In a more recent study Taylor and Jarman[32] measured the change in SL intensity at fixed wavelengths of the emission spectrum from argon-saturated water and with added sodium chloride. Insonation at 500 kHz produced sodium D-line emission that was six times more intense than the original continuum at that wavelength. There was no perceptible change in the remaining part of the spectrum. This observation is consistent with the earlier observations of Gunther and co-workers, who made their measurements at an ultrasonic frequency of 294 kHz. However, at 16 kHz the sodium line intensity was approximately 200 times greater than the original continuum and the remaining spectrum also was altered.[32] The OH band intensity decreased in the presence of alkali salt and this was attributed to quenching of hydroxyl radicals by the metal atoms. Also, it was observed that emission at the extreme red end increased by a factor of 17 to yield a flat plateau in this part of the visible spectrum. Similar results (unreported) have been observed by us in studies of alkali metal halide solutions carried out at an insonation frequency of 459 kHz. However, in addition, it was noticed that the change in emission in the ultraviolet region was much more marked in the presence of nitrates and sulfates than in the presence of halides, suggesting the importance of anions in quenching hydroxyl radicals. On balance, it appears that the presence of soluble salts in water affects the whole sonoluminescence spectrum rather than just causing characteristic emission from excited alkali metal atoms.

Emission from Metal Atoms

An example of emission from an excited state of potassium observed in the SL spectrum from an aqueous potassium iodide salt solution is shown in Fig. 6-6. The spectrum lines result from the transition from the first excited state to the ground state, i.e., $4^2P \rightarrow S$ transition, and appear as a single peak because of poor resolution. The position of this line is independent of the corresponding anion in the salt. As well, diffuse, low-intensity satellite bands appear on each side of the metal atom emission.[35] The lower wavelength band is generally better defined than the one at longer wavelength. Such resonance and satellite peaks are observed for all potassium and sodium salts irrespective of the type of anion. However, the presence of N_2 and air quenches these peaks.[32,35] The results of SL from aqueous solutions of various salts are summarized in Table 6-5.

The mechanism whereby metal ions in solution lead to light emission characteristic of the metal atom has been the subject of some speculation in the last few years. The presence of the satellite peaks, especially in the cases of sodium and

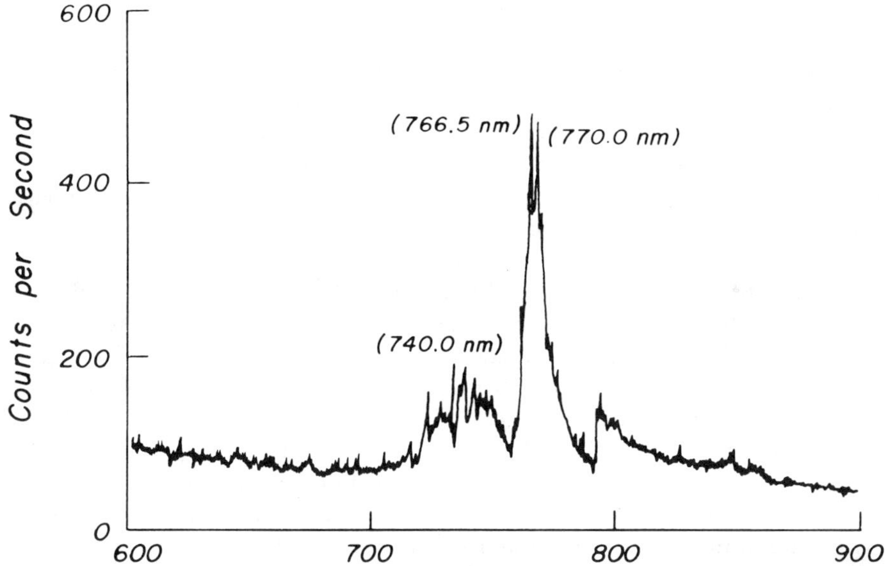

Figure 6-6. Low-resolution sonoluminescence spectrum of argon-saturated aqueous potassium iodide solutions. Time constant = 3 s; resolution = 4 nm; insonation frequency = 459 kHz; pressure = 0.7 atm. (Adapted from Ref. 35.)

Table 6.5. Emission from Metals Atoms in the SL Spectra of Aqueous Electrolytes

	Wavelength of emission (nm)		
Element	Study 1[a]	Study 2[b]	Study 3[c]
		610.4	
Lithium	670.8	670.8	672.0
Sodium	559.0–571.0	554.0	554.0
	589.0–589.6	589.0	589.6–590.1
Potassium	—	—	740.0
			765.5–771.1
Calcium		393.4	
		396.9	
	422.7	422.7	—
		547.0	
	612.5	612.0	
	631.0		
Strontium	460.7		
	602.2–612.0	—	—
	548.5–558.4		
Barium	553.6		
	457.5		
Rubidium	—	No emission	—
Copper	—	—	500–550[d]

[a] From Ref. 9.
[b] From Ref. 32.
[c] From Ref. 35.
[d] From Ref. 83.

potassium resonance lines, has provided some information on cavity properties at the time of SL.[35] Shock tube experiments by Lalos and Hammond[81] indicate that lower wavelength satellite peaks accompany sodium and potassium atom lines when emission occurs from a rapidly compressed mixture of metal vapor in argon. Because rapid compression is realized only during the implosive phase of cavities, it may be deduced that the SL emission occurs at this stage of the bubble's lifetime.

The pathway whereby the metal ions are reduced to atoms and then excited to an upper atomic state from which they are observed to emit light is not clearly understood. Whether the alkali metal ions are vaporized into the cavity and then reduced or are first reduced by chemical processes and subsequently vaporized has not been resolved. According to Taylor and Jarman the ions get into the cavities near the end of the collapse. They based their arguments on the results of SL measurements by using pulsed ultrasound at different repetition rates. A comparison of the intensities of the continuum and sodium D-line peaks, (the intensities being normalized for the actual irradiation time) showed that the continuum emission was not affected by pulsing, whereas the Na line intensity decreased as the repetition pulse rate decreased. This was interpreted as indicating that the sodium ions diffused into the nuclei after cavitation. If the nuclei formed as a consequence of bubble collapse are made to grow again sufficiently soon, i.e., at a higher repetition rate, then these ions inside the growing bubble help to produce light during the subsequent collapse. The mechanism whereby the metal atoms transfer into the cavities are not precisely known, although Margulis and Dmitrieva[82] have attempted to provide some insight into this matter. According to them, metal ions present on the walls of bubbles produce a line spectrum by a mechanism similar to that of light emission from impurities on the walls of shock tubes. In the final stages of bubble collapse, over a time period on the order of $\sim 10^{-7}$ s, the temperature in the bubble–liquid interface of $\sim 10^{-5}$-cm thickness increases sufficiently high to cause substantial vaporization of salts. Small, though sufficient, amounts of salt are trapped in the hot cavity to produce luminescence from metal halides MX by the elementary processes:

$$\mathrm{MX} \longrightarrow \mathrm{M}^* + \mathrm{X} \tag{6-16}$$

$$\mathrm{M}^* \longrightarrow \mathrm{M} + h\nu \tag{6-17}$$

The luminescence from sodium has been suggested[32] to arise from overemission produced by a three-body collission involving two hydrogen atoms:

$$\mathrm{H}^{\bullet} + \mathrm{H}^{\bullet} + \mathrm{M} \longrightarrow \mathrm{H}_2 + \mathrm{M}^* \longrightarrow \mathrm{M} + h\nu \tag{6-18}$$

On the basis of this mechanism it is still not clear how M^+ is transformed into M.

Diffuse bands in the SL spectra from aqueous solutions containing calcium, strontium, and copper cations are ascribed to CaOH, SrOH, and CuOH, respectively,[9,83] although very little is known about the formation of these species by ultrasonic irradiation.

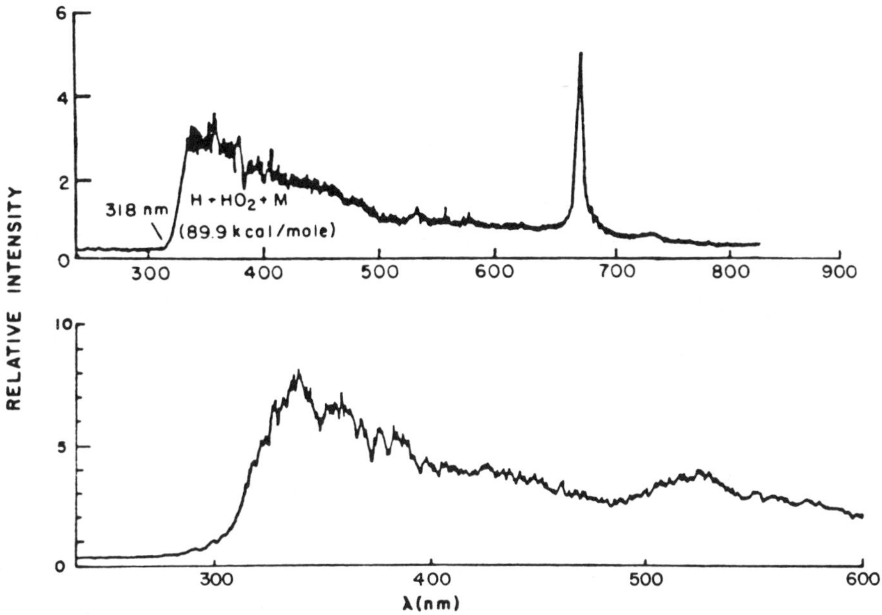

Figure 6-7. Low-resolution sonoluminescence spectra of argon-saturated aqueous (a) 1M lithium nitrate and (b) 0.1M copper sulfate solutions. (Reproduced with permission from Refs. 79, 83.)

Emission from Anions

The effect of different anions on sonoluminescence spectra have not been looked at in great detail. Some studies in this area have been made in our laboratory,[79,83] in the last few years. Low-resolution spectra of 1M lithium, sodium, and potassium nitrate and 0.1M copper sulfate solutions (cf. Fig. 6-7) show that the broad-band continuum observed for pure water is considerably altered when it contains nitrate and sulfate ions. This indicates that the primary emitting species have undergone reaction to produce new emitting species. There is also a close resemblance between the broad continua observed for nitrate and sulfate. A similar broad continuum is observed in the presence of hydrogen peroxide.[79] This suggests that all these systems have a common origin associated with SL. It is believed that nitrate and sulfate react with $OH^{\bullet}$ radicals, in either ground or excited states, to produce hydroperoxyl radicals, which subsequently combine radiatively with $H^{\bullet}$ in the presence of a third body M to give a continuum according to the scheme:

$$[NO_3]^- + OH^{\bullet} \xrightarrow{)))} [NO_2]^- + [HO_2]^{\bullet} \qquad (6\text{-}19)$$

$$[HO_2]^{\bullet} + H^{\bullet} + M \longrightarrow H_2O_2 + M + h\nu \qquad (6\text{-}20)$$

$$[SO_4]^{2-} + OH^{\bullet} \xrightarrow{)))} [SO_3]^{2-} + [HO_2]^{\bullet} \qquad (6\text{-}21)$$

$$[HO_2]^{\bullet} + H^{\bullet} + M \longrightarrow H_2O_2 + M + h\nu \qquad (6\text{-}22)$$

The short-wavelength cutoff of the lithium nitrate and copper sulfate SL spectra are 318 nm and 317.5 nm, respectively. These values yield a dissociation energy for HO_2—H at room temperature of 376 kJ mol^{-1} and 375 kJ mol^{-1} for the two cases in Fig. 6-7, in good agreement with the thermal dissociation energy of 377 $\pm$ 17 kJ mol^{-1} reported for this band.[84]

The details of reactions (6-19) to (6-22) are not known, however, and it is most likely that they involve several intermediate steps. Possible intermediate steps have been suggested[79] for the nitrate–OH reaction but, as yet, no definite conclusions can be made. Prolonged sonication of aqueous KNO_3 solution shows emission from NO_2 above 400 nm.[79] The formation of NO_2 during the sonolysis of aqueous nitrates has been observed by Margulis,[85] which is consistent with the sonoluminescence studies.

THE ROLE OF DISSOLVED GASES

Dissolved gases are the key participants in the phenomenon of cavitation and sonoluminescence. Their presence influences all three stages of cavitation, i.e., nucleation, growth, and collapse. Pockets of gases trapped in a liquid reduce its tensile strength and enable the liquid to cavitate, and hence produce SL at relatively low acoustic pressure amplitudes. The physical properties of gases, such as specific heat capacity, heat conductivity, ionization potential, or compressibility, influence the net SL energy released during rapid bubble compression or collapse. At the same time the gases may affect the SL intensity and its spectral distribution by acting as a catalyst or as a reactant with other constituents of the bubble.

Most of the studies reported in the literature have considered only the physical role the gases play in altering intracavity temperatures. The relative emissions of SL in the presence of various gases are summarized in Table 6-2. There is a significant variation in the magnitudes reported by different authors. In general, with the exception of Ne, monatomic gases give stronger emission than diatomic gases. Also, SL is seen to increase with increasing atomic number of the atom.

Noble Gases

The shapes of the SL spectra do not change significantly with the nature of the dissolved noble gas. Except for the emission from sodium– and potassium–argon complexes formed during the insonation of alkali salt solutions,[35] there has not been, as yet, any evidence for the formation of final products involving noble gases. Also, no emission from the excited states of these gases has been observed within the limits of resolutions. In pure water noble gases appear to be chemically inert. If there is any chemical participation it is purely catalytic, either by providing an alternate route for the production of excited water through an energy transfer reaction of the type:[10,36]

$$H_2O + M^* \longrightarrow H_2O^* + M \qquad (6\text{-}23)$$

or by stabilizing the quasimolecule $(H—M—OH)^{\#}$ formed during the radiative three-body collision.[33,36]

The reverse of reaction (6-23), involving energy transfer from water to the noble gas, cannot be realized because the excitation energy for the gases is greater than that of water.[10] Margulis has also considered the possibility of charge transfer from a noble gas ion to a water molecule according to the reaction:

$$M^+ + H_2O \longrightarrow [H_2O]^+ + M \qquad (6\text{-}24)$$

The $[H_2O]^+$ ion may further decompose to yield OH^* in the $^2\Sigma^+$ state from which it may emit light on deexcitation. On the basis of reaction (6-24) one would anticipate charge transfer to take place from $[H_2O]^+$ to Xe because the ionization potential for Xe is lower (12.13 eV or 1170.8 kJ mol^{-1}) than water (1216.2 kJ mol^{-1}). This would imply weaker SL and less sonochemical reaction in the presence of Xe. It is well known that SL from Xe-saturated solutions is several times greater than that from argon-saturated solution (cf. Table 6-2). Therefore, it may be argued that, at least in the presence of Xe, charge-transfer reactions play an insignificant role. Even in the presence of other noble gases the role of charge-transfer reactions is questionable as it requires ionization energy on the order of 1506.2 kJ mol^{-1} or greater to form noble gas ions of lower atomic number. This may be above the limit of energy density created during cavitation.

Chemically Reactive Gases

A wide variety of gases fall into this category. They include diatomic gases, such as N_2, O_2, NO, Br_2, I_2; triatomic gases, such as CO_2, NO_2, N_2O; and polyatomic gases, such as methane and ethane. These gases can affect SL and sonochemical processes in a number of ways. They can penetrate into the bubbles and behave like the noble gases in energy transfer reactions described above; they may participate chemically by reacting with free radicals to yield end products accompanied by light emission; they may themselves undergo dissociation and subsequent chemical changes; and finally, they may alter the bubble dynamics and the energy released by the cavitating bubbles. With the exception of the first, these possibilities would substantially change the nature of the SL spectra. Keeping these facts in mind we shall now discuss the SL observed from aqueous solutions in the presence of individual gases.

Except for the measurements reported by Prudhomme and Guilmart in the narrow spectral region of 190 to 200 nm, the emission intensities from O_2- and

N_2-saturated aqueous solutions are weaker than that from air-saturated solutions (see Table 6-2). In all three cases emission occurs over a broad spectrum.[36] Air-saturated water shows a stronger emission in the visible region and the continuum consists of three broad bands with maxima around 310, 360, and 430 nm. Insonation in the presence of O_2 leads to the formation of $[HO_2]^\bullet$ by the reaction between H and O_2 and also by the reaction between $OH^\bullet$ and H_2O_2 (formed during sonication). Therefore, the total emission in the presence of O_2 consists not only of light emission from OH^* and H_2O^* but also of the emission from the recombination of $H^\bullet$ and $[HO_2]^\bullet$ which occurs above 320 nm. Since hydroxyl and $[HO_2]^\bullet$ radicals are very reactive toward most reagents, it is difficult to scavenge one of them selectively to resolve the various components better. The luminescence from N_2-saturated solution consists of emission from excited oxides of nitrogen along with that produced by H and OH radicals. The concentration of oxides of nitrogen formed during sonication is low but possibly sufficient to shift the maximum of the band toward the visible region of the spectrum. Because of the direct oxidation of nitrogen by oxygen, a greater concentration of nitrogen oxides would be formed in the presence of air as compared to N_2-saturated solution. This difference probably explains the stronger emission in the visible region in the presence of air.

Sonoluminescence of aqueous bromine and iodine solutions exhibit numerous broadband emissions.[86] The spectrum above 500 nm is weak, probably because of the absorption of radiation by Br_2 and I_2 molecules. The aqueous bromine solution shows a strong continuous band from 300 nm to 400 nm, with a maximum at 367 nm.[86] This band has been observed previously by Parke and Taylor.[87] Kling[88] reported a strong emission band at 668 nm from Br_2 solution. This band was not observed by either Parke and Taylor,[87] or by Sehgal et al.,[86] and the reason for this discrepancy is not known. Bromine solutions also show a weak emission around 249 nm and a broad emission about 490 nm. Sonoluminescence of aqueous iodine solution consists of two broad bands, one of which is centered around 270 nm and the other about 332 nm. Because of the narrow energy gaps between the excited states the precise assignment of the emission peaks to the electronic states of the halogen compound is uncertain. However, comparison of SL spectra with emission from Br_2 and I_2 vapors leads one to suspect that SL occurs largely from the various electronic states of molecular halogens and their oxides. The formation of halogen oxide can occur by the oxidation of anions formed in the reaction between halogen, X_2, and water; for example:

$$X_2 + H_2O \rightleftharpoons 2\,H^+ + X^- + XO^- \qquad (6\text{-}25)$$

$$2\,H^+ + X^- + XO^- \xrightarrow[2\,OH^\bullet]{)))} XO^* + X^* + 2\,H_2O \qquad (6\text{-}26)$$

Sonoluminescence from NO- and NO_2-saturated aqueous solutions is shown in Fig. 6-8.[89] Structureless bands similar to the emission from high-temperature flames containing NO_2 are observed in both cases. The SL from aqueous NO solutions has a continuum from $\sim$ 230– $\sim$ 350 nm, with a maximum at 280 nm, and a second broadband continuum that begins at $\sim$395 nm and rises rapidly

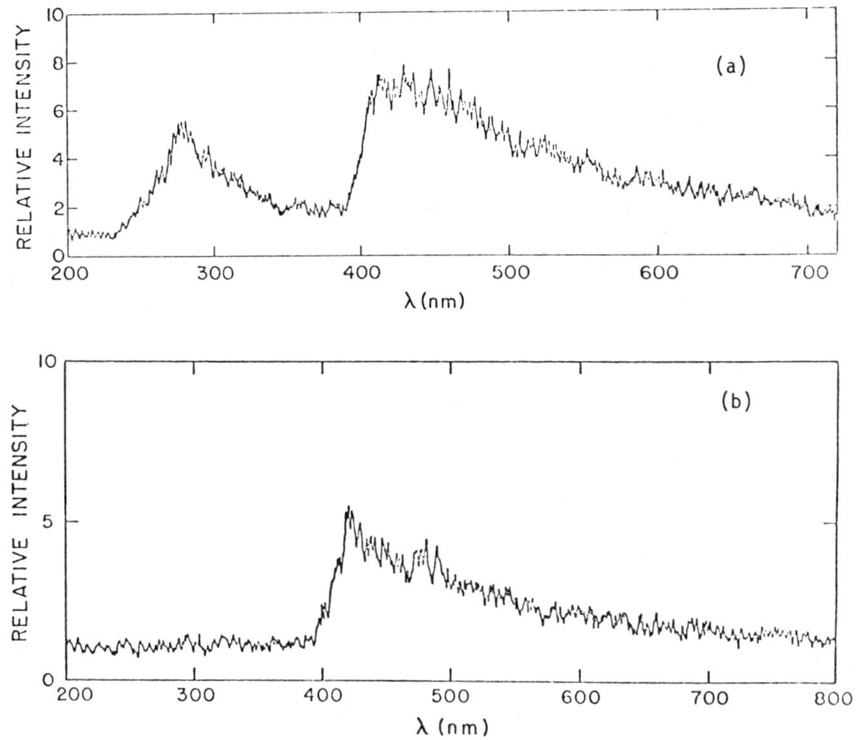

Figure 6-8. Sonoluminescence spectrum of (a) nitric oxide-saturated aqueous solution and (b) nitrogen dioxide-saturated aqueous solution. (Reproduced with permission from Ref. 89.)

to a maximum around 400 nm. The low-wavelength cutoff of this band occurs at 397.5 ± 2.8 nm, which is identical to the value observed for chemiluminescence of NO + O in air afterglow.[90]

Sonoluminescence from NO_2-saturated aqueous solutions on the average is about 3.4 times weaker than the emission from NO solutions under identical conditions of sonication. The characteristics of the long-wavelength continuum are very similar to the emission from NO solution. Because the longer wavelength SL from aqueous NO and NO_2 solutions occurs in the same spectral region as that from NO_2 fluorescence, air afterglow emissions, NO + O_3 chemiluminescence, and thermal emission from NO_2, and since the dissociation energy estimated by the short-wavelength cutoff of SL (299.2 kJ mol^{-1}) is close to the dissociation energy of NO_2 estimated by thermal (301.2 kJ mol^{-1})[91] and spectroscopic methods (299 kJ mol^{-1}),[90] it has been argued[89] that SL arises mainly from emission from electronically excited NO_2. However, there are differences in the spectral distribution of SL from NO_2 and chemiluminescence from NO + O, NO + O_3, and fluorescence from NO_2. For example, in the case of SL there is stronger emission at lower wavelengths. This has been rationalized[89] in terms of the capacity of the various energy sources to populate different excited states of NO_2. In the case of SL the NO_2 molecules preferentially depopulate directly to the ground state instead of crossing over to an intermediate metastable state and then dropping to the ground level, a situation more likely to occur in other forms of luminescence from nitrogen dioxide.[89] These differences between SL and other luminescent processes of NO_2 probably are related to the high-pressure and -temperature conditions arising during cavitation collapse.

The ultraviolet band observed in the SL from NO-saturated aqueous solution is not observed in the SL spectrum of NO_2 solution. This is most likely because of the occurrence of lower intracavity temperatures inside the bubble containing NO_2. The predissociation energy of nitrogen dioxide into $N(^4S) + O_2\ (^3\Sigma_g)$ is 483.7 kJ mol^{-1}.[92] The short-wavelength cutoff of the uv peak of SL from aqueous NO solution is 248 ± 22 nm, i.e., 482.4 ± 41.8 kJ mol^{-1}. Consequently, one is tempted to argue that this part of the emission results from recombination of $N(^4S) + O_2(^3\Sigma^-)$. However, this apparently simple answer is weakened by the fact that the short-wavelength cutoff of the H_2O^* band, which yields a dissociation energy of 493.7 kJ mol^{-1}, is comparable to the 482.4 kJ mol^{-1} value. Therefore, the origin of this band is still unclear. Only if it is assumed that NO has completely quenched H_2O^* can one assign this band to the recombination of $N(^4S) + O_2(^3\Sigma g)$. The quenching efficiency of H_2O^* by NO does not appear to have been reported.

The comparison of SL from NO- and NO_2-saturated aqueous solutions with other forms of emission provides an important reminder. Although there are qualitative similarities between SL and other forms of luminescence in terms of the nature of the molecular processes, there are substantial differences in the details of the process in terms of molecular dynamics (e.g., energy exchange, population distribution) of the radiating chemical species. These differences can be exploited to provide some information about the intracavity conditions at which SL occurs. This is discussed in the section, "SL as a Probe of Acoustic Cavitation." As mentioned earlier, SL occurs in a constantly changing environment inside the bubble, a situation similar to shock tubes. From this point of view shock tube experiments can provide a valuable guide to interpreting SL data. It must be remembered, however, that SL occurs from literally hundreds and thousands of such shock tube events triggered at different points in time and space.

SONOLUMINESCENCE FROM VARIOUS LIQUIDS

In 1937 Chambers[16] observed a visible component of SL from various organic liquids when insonated at 8.9 kHz. He proposed an empirical relation that showed the SL flux to be proportional to the product of dipole moment and viscosity of the liquid and inversely proportional to the temperature. Later, Jarman,[29,93] Gabrielli et al.,[94] and Golubnichii and Goncharov[95] reported measurements of SL intensities, which are summarized in Table 6-6. While there

Table 6-6. Sonoluminescence from Various Liquids

	Relative intensities[29,93] at T:			Relative amplitude of sonoscintillations[95]		Luminous intensity at 20°C phot (10^{10})[94]	
	25°C	40°C	55°C				
Dimethylphthalate	16	6.6	2.4	Glycerine	28	Pentane	0.440
Ethylene glycol	12	3.4	0.5	Dimethylphthalate	12	Hexane	0.074
Tap water	3.6	1.0	—	Ethylene glycol	22	Octane	<0.030
Chlorobenzene	0.84	0.43	0.20	Dibutylphthalate	32	Methanol	<0.030
Isoamyl alcohol	0.54	0.28	0.18	Water	6	Ethanol	0.185
Xylene	0.36	0.24	0.14	Bromobenzene	7	*n*-Butanol	0.680
sec-Butanol	0.30	0.17	0.09	Chlorobenzene	6	Distilled water	200
n-Butanol	0.21	0.10	0.03	Carbon tetrachloride	8		
Isobutanol	0.17	0.09	0.05	Toluene	3		
n-Propanol	0.21	0.08	0.04	Benzene	2		
Toluene	0.15	0.07	0.05	Ethanol	2		
Benzene	0.23	0.06	0.01	Acetic acid	2		
tert-Butanol	—	0.05	0.23	*o*-Xylene	4		
Isopropanol	0.054	0.03	0.01	*n*-Heptane	1		
2*N* NaCl	25			Acetone	1		
2*N* KCl	20			*n*-Octane	2		
2*N* $MgCl_2$	15			*n*-Decane	6		
2*N* $MnCl_2$	5						
1*N* NaCl	10						
Sea water	10						

are some significant differences among the values reported by the various authors, it is apparent from the data that, generally, organic liquids produce weaker SL than aqueous media. Jarman[93] reported an empirical correlation between SL flux and the quantity (surface tension)^2/vapor pressure. According to Golubnichii and Goncharov,[95] the free molecular interaction energy, $\Delta\varepsilon$, defined by:

$$\Delta\varepsilon = RT_0 \ln [RT_0/\tilde{p}_v V_m] - RT_0 \quad (6\text{-}27)$$

is directly related to SL flux. $\Delta\varepsilon$ is a composite parameter that takes into account the physical properties of the medium, e.g., compressibility; saturation vapor pressure, $\tilde{p}_v$; molar volume of the liquid, V_m; surface tension; and viscosity. For reasons that are not understood, values of $\Delta\varepsilon$ determined for various liquids empirically correlate well with measured sonoluminescence.

More recently, Chendke and Fogler[38] have studied SL from $CCl_4 + H_2O$ mixtures as a function of carbon tetrachloride concentration and hydrostatic pressure. The SL intensity was observed to increase linearly with CCl_4 concentration. As well, CCl_4 saturated with water had a luminescence intensity about 6.4 times stronger than pure CCl_4. The addition of CCl_4 to water produces different degrees of enhancement in various spectral regions. For example, with increasing CCl_4 concentration the maxima shifts from 340–370 nm to about 500 nm. Since the pulse radiolysis of pure CCl_4 with 100-MeV electrons produces an absorption spectrum of $[CCl_4]^+$ ion in the spectral region 300–600 nm with a maximum near 500 nm, SL observed near this region has been attributed to $[CCl_4]^+$ and electron recombination,[38] i.e.:

$$M + CCl_4 \rightleftarrows [CCl_4]^+ + e^- + M \quad (6\text{-}28)$$

$$[CCl_4]^+ + e^- \rightleftarrows CCl_4 + h\nu \quad (6\text{-}29)$$

where M is a third body. Application of steady-state kinetics to the above reaction mechanism yields an expression according to which SL intensity is shown to be proportional to CCl_4 concentration, which is consistent with the authors experimental observation. A puzzling feature of this analysis is that although water is present in the mixture there is no apparent emission from it, nor is any quenching process proposed to account for its absence. Second, absorption and emission spectra for complex molecules are generally not superimposable, the emission radiation having a lower frequency than the absorption of radiation. However, this study tacitly assumes absorption and emission to occur at the same wavelengths, as implied by the comparison of the absorption spectrum with the SL spectrum, which is an emission process.

Very little is known about SL from liquefied gases. Attempts to observe SL from liquid He at 1.8–4.2K and static pressures of 0.17–1.5 atm, and from liquid N_2 at 74–77.4K and static pressures of 0.7–1 atm were made by Jarman and Taylor.[96] Although liquid He was found to cavitate, no SL was observed under any circumstance. Liquid nitrogen showed weak luminescence, about 100 to 1000 times weaker than that of water.

DEPENDENCE OF SL ON VARIOUS EXPERIMENTAL PARAMETERS

Time Dependence

According to Alfredsson[97] the intensity of SL at a fixed acoustic power increases with time to a maximum before decreasing and stabilizing at an equilibrium value. The process of achieving equilibrium takes place over a time period of 150 to 200 ms (cf. Fig. 6-9). Eastwood and Watmough[98] observed a time delay of several seconds between the application of ultrasound and SL from water. A similar delay was observed by Jarman[93] and Vaughan et al.[71] Eastwood and Watmough further observed that SL increased quadratically as ultrasonic intensity was increased above a threshold value. On reduction of ultrasonic intensity SL decreased but did not follow the initial curve (cf. Fig. 6-10); i.e., a hysteresis effect was observed. This behavior can be qualitatively explained by the fact that SL, and for that matter cavitation-induced effects, are related not only to ultrasound and the medium's properties but also to the history of sonication. For example, points A and B on the curves of Fig. 6-10 represent the same medium irradiated with equally intense ultrasound but they represent different points in time in the sonication process. At point B the medium has a larger number of and, probably, different sized nuclei generated by the cavitation process during the time interval in going from point A to B along the curve. Because of a change in the number of nuclei that are available for cavitation the SL intensity at point B is not identical to that at point A. It is likely that, because of different nucleating conditions, Alfredsson was able to observe SL

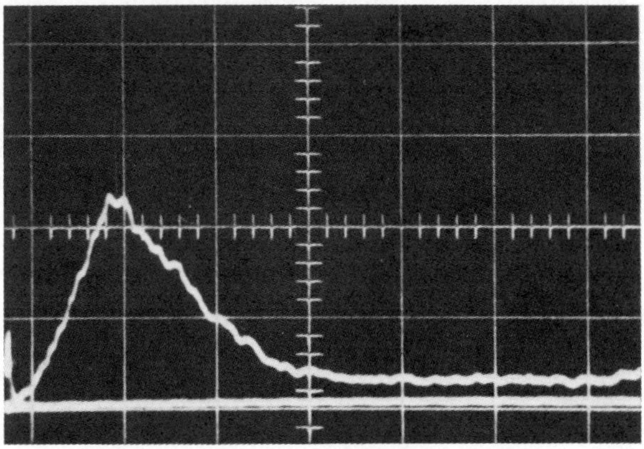

Figure 6-9. Photograph of the oscilloscope screen showing the sonoluminescence intensity versus time. Sensitivity, 0.5 Volts per division. Time scale, 50 ms per div. (Reproduced with permission from Ref. 97.)

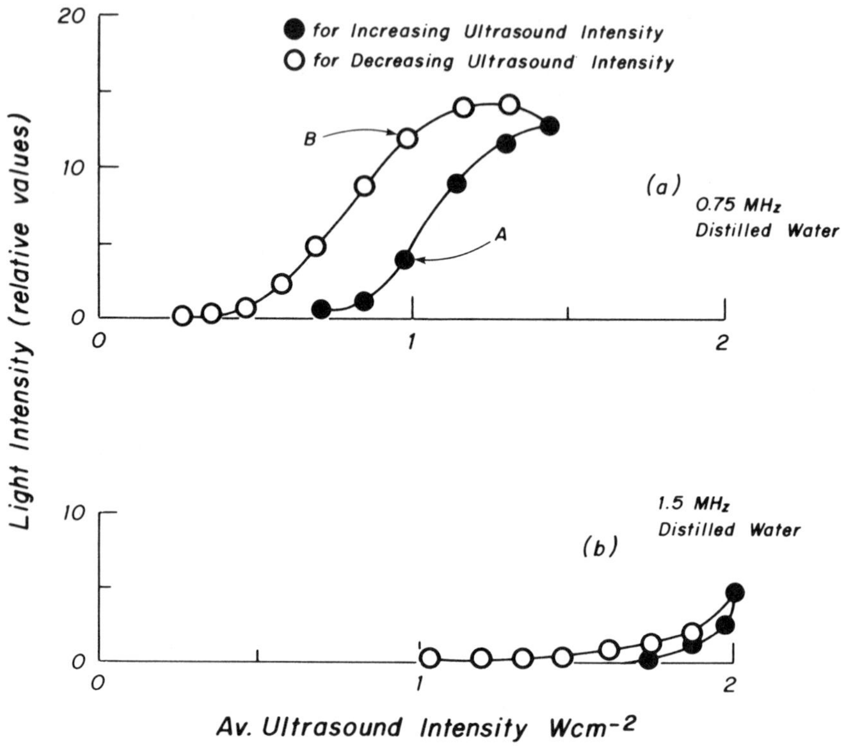

Figure 6-10. Variation of sonoluminescence intensity with ultrasound intensity. (Reproduced with permission from Ref. 98.)

almost immediately after the ultrasonic field was initiated, whereas Eastwood and Watmough required a few seconds before the conditions in the media were conducive for nuclei to grow and collapse. Unfortunately, many reported studies of SL and sonochemical effects have given little attention to control of the nucleation stage of cavitation. This may be one of the primary reasons for the quantitative differences between many reported results of the relative emission intensities of SL from various liquids and in the presence of different gases. Interested readers are urged to consult an excellent review on the inception of cavitation by Apfel.[99]

Frequency Dependence

At a fixed ultrasonic intensity SL decreases with an increase in sound frequency. In water the emission is reduced by as much as 50% as the sound frequency is raised from 0.7 to 1 MHz.[94] Organic liquids show the same behavior but to a greater degree. For example, SL in butyl alcohol and dimethylphthalate is reduced to 1/3 and 1/400, respectively, of the original values as the insonation frequency is increased from 0.175 to 0.7 MHz.[94]

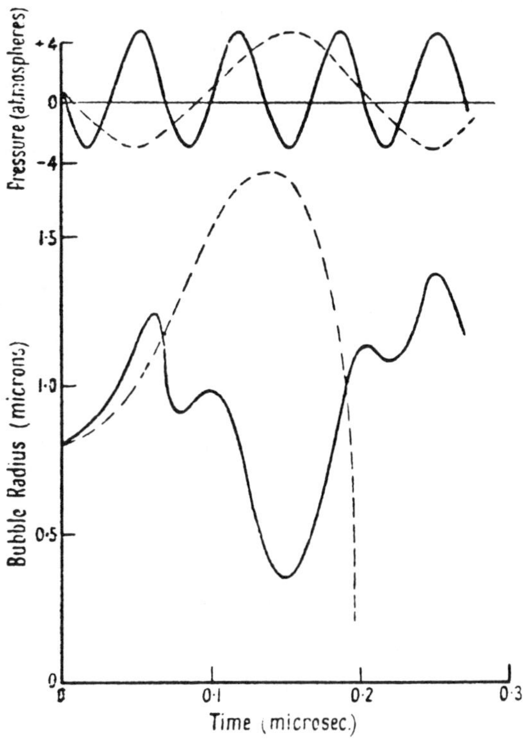

Figure 6-11. Radius–time curves for gas-filled bubble at different frequencies. Pressure amplitude = 4×10^6 cgs units; radius of nucleus = 0.8×10^{-4} cm; ultrasonic frequency = 9×10^7 Hz (full line), 3×10^7 Hz (broken line). (Reproduced with permission from Ref. 13.)

The SL spectral distribution is also altered with sonication frequency.[32,36] At low frequencies the 310 nm band ascribed to OH* ($^2\Sigma^+ \rightarrow {}^2\pi$) transition is pronounced and well defined. At higher frequencies the continuum band from radiative recombination of free radicals predominates and the net spectrum is more diffuse. This variation in SL can be accounted for qualitatively in terms of changes in bubble behavior with sonication frequency.

The variation in bubble motion as a function of insonation frequency has been estimated in a study by Noltingk and Neppiras.[13] The results, shown in Fig. 6-11, illustrate that as the ultrasonic frequency is increased the nuclei tend to grow to a smaller radius and compress, slowly, to a minimum radius. This indicates that before the collapse is complete the inward motion is arrested and the bubble oscillates nonlinearly about the mean position. Thus, at higher insonation frequencies a bubble in an ultrasonic field has greater stable cavity character. Assuming that such bubbles also compress adiabatically, the temperatures achieved in them will be lower than in bubbles undergoing violent collapse. Since broadband emission from radiative recombination is associated with relatively lower energy stable cavitation, the emission spectrum becomes more diffuse with an increase in frequency as is observed experimentally.

It must be emphasized that other factors related to variations in sonic field patterns arising from changes in ultrasonic frequency may also contribute to changes in SL. Also, at lower frequencies the compression phase of a bubble may last longer and the diffusion of free radicals from the bubble to the medium may be more important. How and to what extent these factors contribute to variations in SL is still not well understood.

Temperature Dependence

Sonoluminescence intensity from liquids is also dependent upon the bulk temperature of the insonated medium. The relative emission intensities from various liquids measured at temperatures of 25°C, 40°C, and 55°C are summarized in Table 6-6. These results of Jarman[93] indicate that over this temperature range SL decreases monotonically with increasing temperature.

Temperature studies of the SL from water in the presence of air and noble gases have been reported by Gabrielli et al.[94] and Sehgal et al.,[100] respectively. The gas-saturated solutions were allowed to stand overnight before any measurements were carried out. In both cases SL was found to decrease with increasing temperature in the range 15–70°C. In contrast, Alfredsson[99] observed SL from air-saturated water to increase from 3°C to about 25°C, after which it decreased with an increase in temperature. The reason for the difference between the work of Alfredsson and Gabrielli et al. is not clear, although it may be speculated to result from different experimental protocols followed in the studies, which has led to the presence of different amounts of gas and nucleating conditions.

Sonoluminescence intensities, I, normalized with respect to sonoluminescence at 30°C, I_r, for various temperatures in the presence of various gases and frequencies of sonication are plotted in Fig. 6-12. The slope of ln (I/I_r) vs. $1/T_0 - 1/(T_0)_r$ is virtually the same for solutions containing noble gases: $\sim 8 \times 10^3$K. $(T_0)_r$ represents reference temperature, which in this case was taken to be 30°C. The slopes of solutions containing air at insonation frequencies of 0.7 and 2.17 MHz are 5×10^3K, and 3.9×10^3K, respectively. It appears that there may be a dependence of the magnitude of the slope on the insonation frequency. It must be emphasized that it is not uncommon to have such quantitative differences between the results of various laboratories in this area of research. Therefore, this apparent frequency dependence should be confirmed before any final conclusions are made.

In order to explain the decrease in SL with temperature, it was proposed that SL intensity (and possibly other sonochemical reaction rates) was related to the Helmholtz free energy, $\Delta\tilde{A}$, that bubbles gain during formation and growth, and which they release, subsequently, during the collapse stage.[100] Under the assumptions that gas–vapor intracavity contents can be described by van der Waals relation and that the growth of the bubbles from their initial to their final stage is slow and is in isothermal equilibrium with the surroundings, the molar energy gained during the bubble growth by nuclei containing a mole of gas is given by:[100]

$$\Delta\tilde{A} = RT_0 \ln\left[\left(\frac{p_g + p_v}{p_{go} + p_v}\right)\cdot\frac{Z_0 RT_0 - b(p_{go} + p_v)}{ZRT_0 - b(p_g + p_v)}\right] + \frac{a}{RT}\left[\frac{(p_{go} + p_v)}{Z_0} - \frac{p_g + p_v}{Z}\right] \tag{6-30}$$

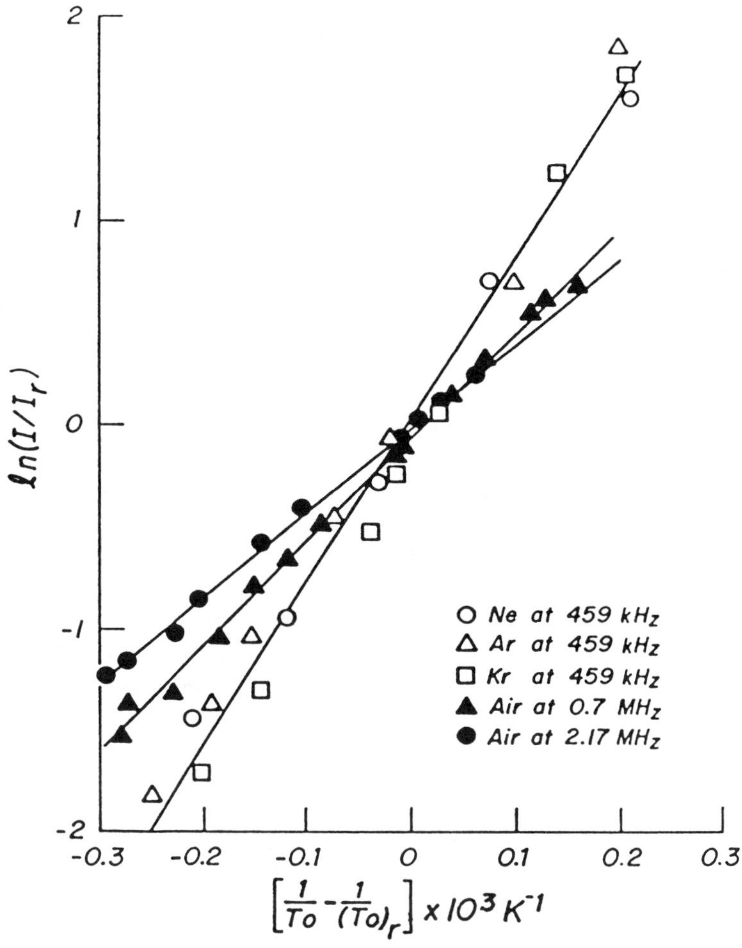

Figure 6-12. Plot of ln I/I_r vs. $[(1/T_0 - 1/(T_{0_r})]$ for neon-, argon-, and krypton-saturated water. I_r is the emission intensity at the ambient temperature of 30°C. Data for SL from rare gas-saturated solutions was obtained from Ref. 101 and the data for air-saturated solutions was obtained from Fig. 5 of Ref. 94.

where T_0 is the ambient temperature; $\mathcal{Z}$ is the compressibility factor of the gas; p_v and p_g are the vapor and the gas pressures, respectively, inside the cavity; a and b are effective van der Waals constants for the gas–vapor mixture; and R is the gas constant. The subscript o on p_g and $\mathcal{Z}$ represents the values of the gas pressure and compressibility factor of the contents of the nuclei at the beginning of bubble growth. Equation (6-30) can be approximated to varying degrees, depending upon the magnitude of the intracavity pressure of the growing nuclei. The latter is a function of hydrostatic pressure, pressure amplitude of the sound wave, surface tension of the liquid, and average radius of the nuclei.[100] When the intranucleus pressure is comparable to or less than atmospheric pressure the cavity contents can be assumed to approach ideal gas behavior and Eq. (6-30) approximates to the form:

$$\Delta\tilde{A} = RT_0 \ln [(p_g + p_v)/(p_{go} + p_v)] \tag{6-31}$$

In a nucleus that expands isothermally to approximately 10 times its initial size and contains comparable amounts of gas and vapor, one can write $p_g = p_{go}(R_0/R)^3 \ll p_{go}$ and $p_g + p_v \sim p_v$ so that Eq. (6-31) simplifies to:[100]

$$\Delta\tilde{A} = RT_0 \ln [p_v/(p_{go} + p_v)] \tag{6-32}$$

According to the conditions under which Eqs. (6-31) and (6-32) apply, $\Delta\tilde{A}$ is independent of constants a and b and therefore independent of the nature of the dissolved gas. Relative sonoluminescence intensities of aqueous solutions containing different noble gases at atmospheric pressure, moreover, do not show a marked dependence on the nature of the dissolved gas.[100] Since the SL was measured after allowing the solution to equilibrate overnight, the larger bubbles containing excess amounts of undissolved gases were effectively removed. For all practical purposes it may be assumed that the nuclei that grew into bubbles on exposure to ultrasound were in equilibrium with the surrounding fluid and contained comparable amounts of gas and vapor. Under these circumstances it may be inferred that experimental conditions reported by Sehgal et al.[100] and earlier by Gabrielli et al.,[94] closely correspond to the conditions for which Eq. (6-32) is suited. Consequently, for $p_{go} + p_v$ equal to atmospheric pressure, $\Delta\tilde{A}$ can be expressed:

$$\Delta\tilde{A} \simeq RT_0 \ln p_v \tag{6-33}$$

There are two ways to determine $\Delta\tilde{A}$: by using the experimentally measured values of p_v,[100] or by using the Clausius-Clapeyron equation.[101] Both methods are illustrated here.

The Helmholtz energy was determined by using the values of p_v, obtained from the literature,[102] in Eq. (6-33). Sonoluminescence intensity normalized with respect to the emission intensity at a reference temperature of 30°C is plotted in Fig. 6-13 as a function of $\Delta\tilde{A}(T_0) - \Delta\tilde{A}(T_0)_r$, where the subscript r designates the reference state. Figure 6-13 shows a linear correlation between the two parameters, which can be expressed mathematically as:

$$\ln I/I_r = k[\Delta\tilde{A}(T_0) - \Delta\tilde{A}(T_0)_r] \tag{6-34}$$

where k is equal to 2.9×10^{-3} mol cal^{-1} and refers to the slope of the line in Fig. 6-13.

The second approach is to use the Clausius Clapeyron equation, which can be written in the form:

$$\ln p_v = -(\Delta H_v/R)(1/T_0 - 1/T_b) \tag{6-35}$$

where ΔH_v is the molar heat of vaporization and T_b is the boiling point of water. When Eqs. (6-33) and (6-35) are combined, one obtains a direct relationship between free energy ΔA and the heat of vaporization, i.e.:

$$\Delta\tilde{A} = -\Delta H_v[1 + T_0/T_b] \tag{6-36}$$

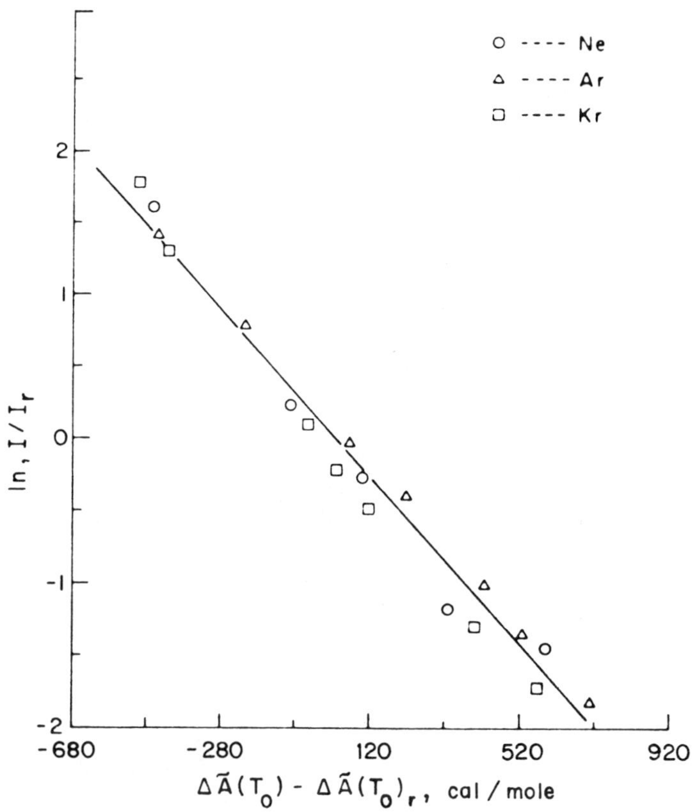

Figure 6-13. Plot of ln I/I_r vs. $\Delta\tilde{A}(T_0) - \Delta\tilde{A}(T_0)_r$ for neon-, argon-, and krypton-saturated aqueous solutions. (Adapted from Ref. 100.)

Equation (6-36), when substituted into Eq. (6-34), yields:

$$\ln (I/I_r) = -(k\Delta H_v/T_b)[T_0 - (T_0)_r] \tag{6-37}$$

Therefore, the slope of a plot of ln (I/I_r) vs. T_0 should yield the heat of vaporization of the liquid provided the constant k is known. The value of the slope determined from Fig. 6-14 is $8.6 \times 10^{-2}\text{K}^{-1}$. Using the value of k as 2.9×10^{-3} mol cal^{-1} and T_b as 373K, one obtains $H_v = 46$ kJ mol^{-1}. In view of the approximations made to arrive at this result this value is in good agreement with the known value of 41 kJ mol^{-1} for water.

It must also be emphasized that Eqs. (6-32)–(6-37) are valid only for nuclei that do not contain significantly larger amounts of gas than vapor. In experiments where dissolved gases are in excess, for example, tap water, or solutions freshly saturated with gases, or when there is external bubbling of gases during insonation, Eq. (6-32) may not be applicable. Under these circumstances $p_v \leq p_{go}$ and Eq. (6-31) can be written:

$$\Delta\tilde{A} = RT_0 \ln [p_g/p_{go}] + RT_0 \ln [(1 + p_v/p_g)/(1 + p_v/p_{go})] \tag{6-38}$$

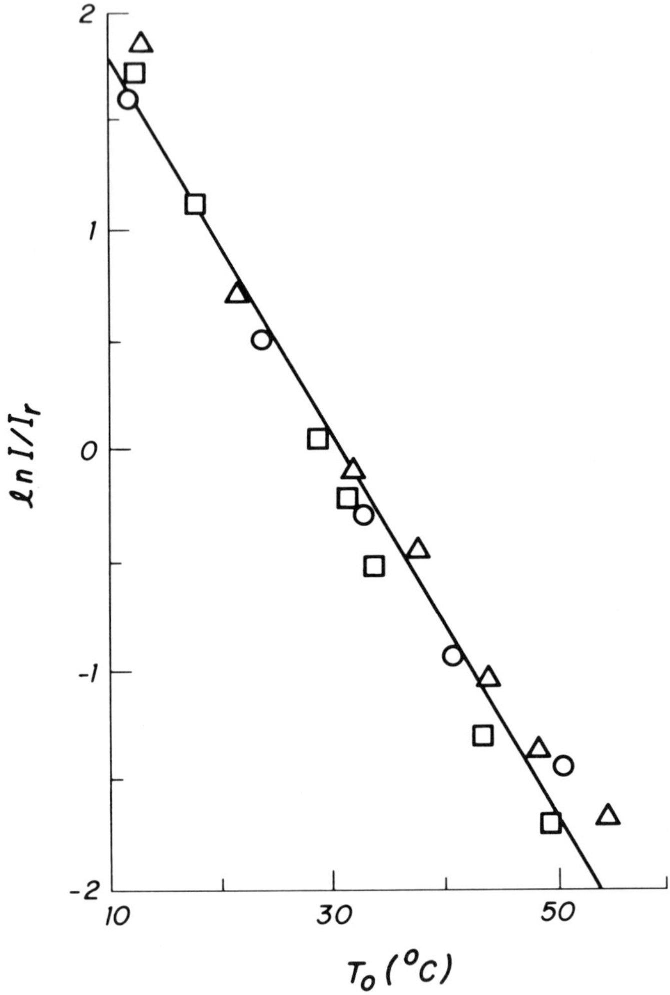

Figure 6-14. Plot of ln I/I_r vs. T_0 for neon-, argon-, and krypton-saturated aqueous solutions. ○ = Neon; △ = argon; □ = krypton.

Expanding the second term of Eq. (6-38) by a Taylor series and neglecting the higher order terms, one obtains:

$$\Delta\tilde{A} = RT_0 \ln (p_g/p_{go}) + RT_0 p_v[(1/p_g) - (1/p_{go})] \tag{6-39}$$

If the bubble grows to a size significantly larger than its initial size (cf. 10 times its initial radius) so that $p_g \ll p_{go}$, Eq. (6-39) can be expressed:

$$\frac{\Delta\tilde{A}}{RT_0} = 3 \ln (R_{co}/R_0) + (1/RT_0\tilde{x})p_v \tag{6-40}$$

where $\tilde{x}$ represents the concentration (moles per unit volume) of gas inside the bubbles when they are fully expanded. Since ln$I$ is proportional to $-\Delta\tilde{A}/RT$, ln$I$ will vary directly as $-p_v/T_0$ according to Eq. (6-40). Although Eqs. (6-39) and (6-40) have been derived for SL, it is reasonable to expect them also to hold true for other rate constants of cavitation-induced reactions. Such a dependence of ln (rate constant) on p_v at a fixed temperature has been observed for various nonaqueous liquids.[103]

Effect of Hydrostatic Pressure

Changes in hydrostatic pressure have been shown to alter the sonoluminescence by an order of magnitude. Early studies by Harvey,[19] Prudhomme,[104] Bunsel and Degrois,[105] Polotskii,[106] and Finch[107] have been reviewed by El' piner,[9] and Chendke and Fogler.[37] The results from these studies show that as the hydrostatic pressure is raised from a few millimeters of mercury to several atmospheres the SL increases to a maximum before decreasing to zero. The results of Weissler's early investigation[108] of the pressure dependence of SL are summarized in Fig. 6-15. They clearly indicate that the static pressures at which

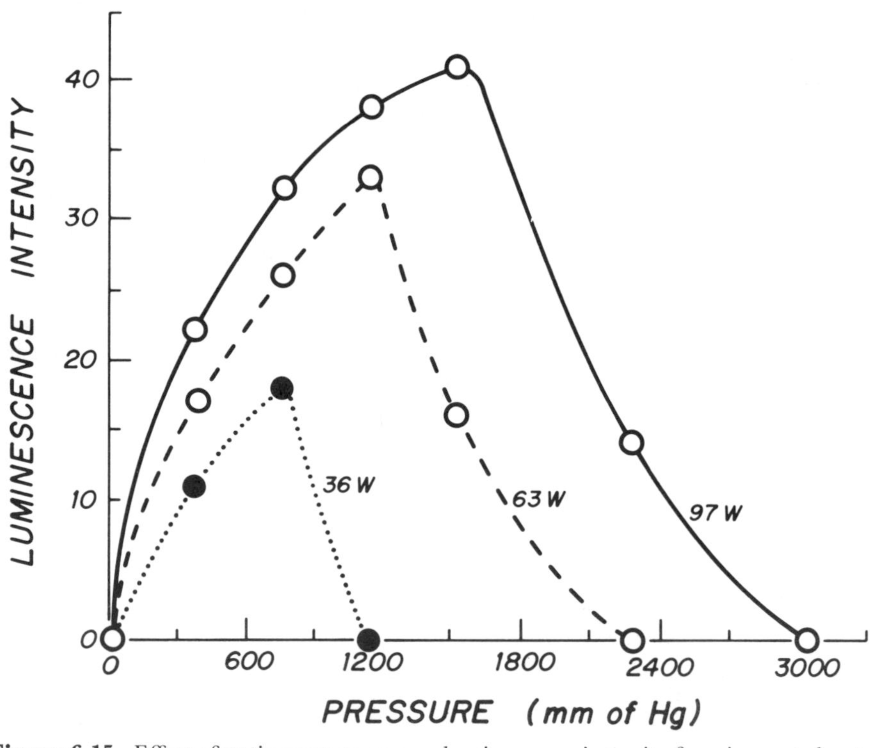

Figure 6-15. Effect of static pressure on sonoluminescence intensity from insonated water, at ultrasonic powers of 36, 63, and 97 W. (Reproduced with permission from Ref. 108.)

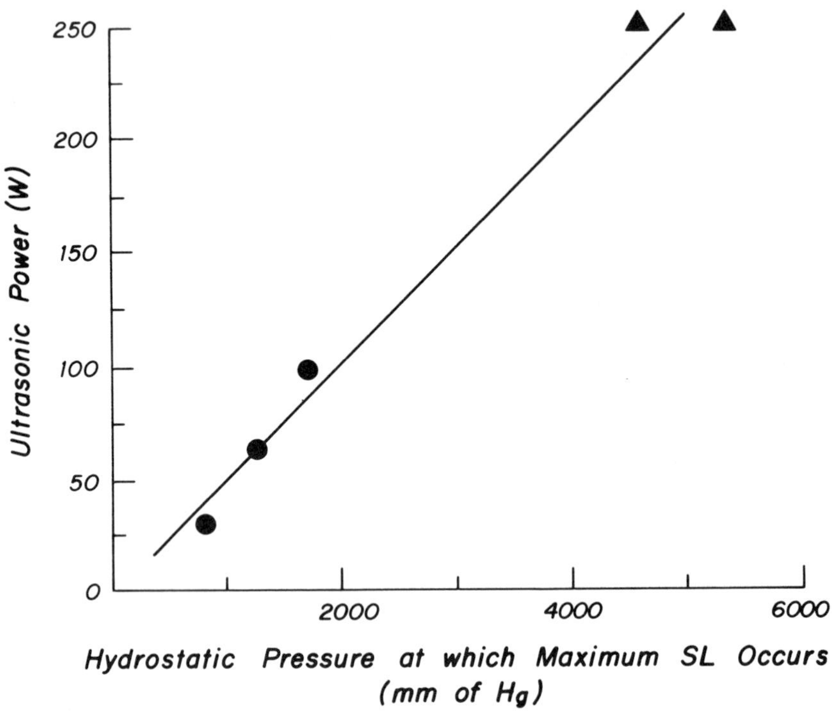

Figure 6-16. Hydrostatic pressure at which maximum SL occurs vs. ultrasonic power. ● represents the data obtained from Fig. 6 of Ref. 108 and ▲ represents the data obtained from Ref. 37.

maximum SL occurs depend upon the ultrasonic power used in the experiments. A recent study by Chendke and Fogler[37] carried out at 250 W shows that a maximum in SL occurs around 6–7 atm. Using these two sets of data, a plot of ultrasonic power vs. hydrostatic pressure, p_H, at which maximum SL occurs shows a near linear dependence (Fig. 6-16). The pressure dependence of SL is also sensitive to whether the pressure is applied hydraulically or by increasing the gas pressure above the insonated liquid.[107] In either case the SL intensity goes through a maximum with an increase in pressure; however, the absolute intensity in the latter case is 10–20 times greater.

Weissler's results[108] are explicable on the basis of numerical studies by Neppiras,[109] who showed that for a fixed sound pressure amplitude, p_A, and fixed frequency f there is an optimal static pressure p_H(max) at which a maximum pressure (p_L) is developed in the liquid surrounding an imploding cavity (Fig. 6-17). If one takes pressure developed on collapse as an indicator of the magnitude of energy released during bubble collapse and the concomitant chemical change, then it is argued that SL should show similar behavior. This expectation is qualitatively supported by experimental results (compare Figs. 6-15 and 6-17).

Recently, Chendke and Fogler[37] have reexamined the pressure dependence of SL and have specifically considered whether the increase in SL is caused by

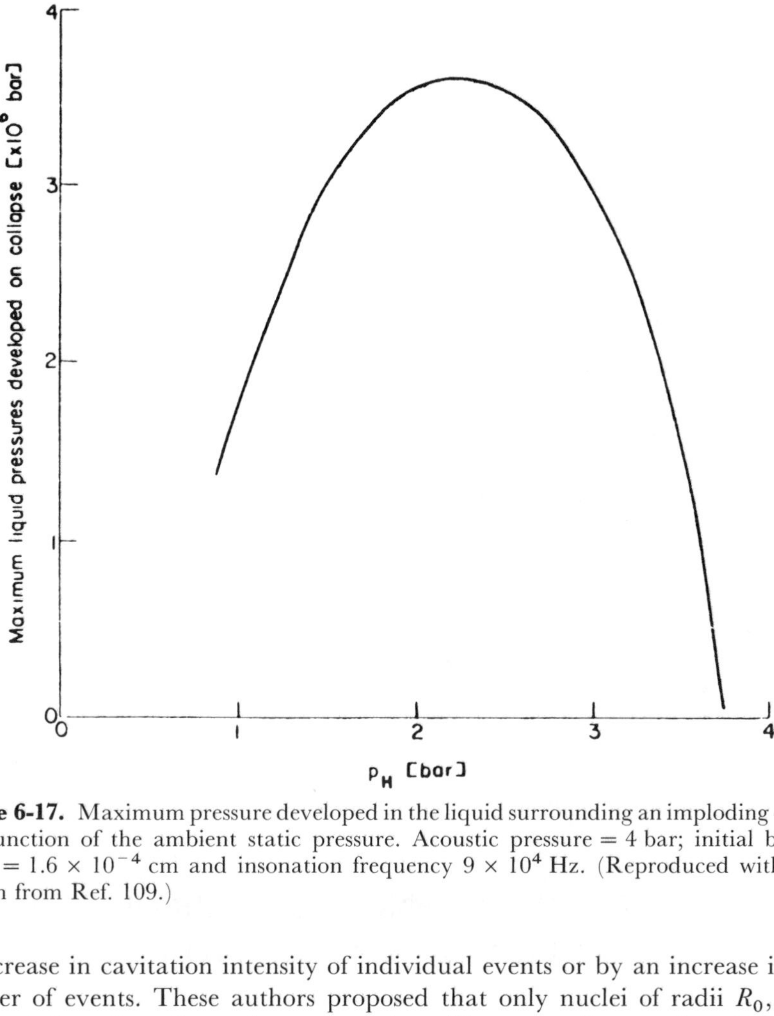

Figure 6-17. Maximum pressure developed in the liquid surrounding an imploding cavity as a function of the ambient static pressure. Acoustic pressure = 4 bar; initial bubble radius = 1.6×10^{-4} cm and insonation frequency 9×10^4 Hz. (Reproduced with permission from Ref. 109.)

an increase in cavitation intensity of individual events or by an increase in the number of events. These authors proposed that only nuclei of radii R_0, with values of $R_c \le R_0 \le R_r$, undergo cavitation. The upper limit radius, R_r, is the radius of the bubble that is resonant at the applied angular frequency and is given by:[37]

$$R_r^2 = (3b)(p_H + 2\sigma/R_r) - (2\sigma/R_r)/(\omega_0^2\rho) \qquad (6\text{-}41)$$

where b is the polytropic index and is equal to C_p/C_v and unity for adiabatic and isothermal conditions respectively, σ is the surface tension of the medium, p_H is the hydrostatic pressure, ω_0 is the resonance frequency of a bubble of radius R_0, and ρ is the density of the liquid.

The lower limit radius, R_c, of the cavitating nuclei is given by the relation:[37,110]

$$R_c^3 + [2\sigma R_c^2/p_H] - [(32\sigma^3)/27p_H(p_H - p_A)^2] = 0 \qquad (6\text{-}42)$$

One of the problems in solving Eq. (6-42) is that the magnitude of p_A necessary for cavitation to occur is a function of p_H. This problem is overcome by using

Strasberg's equation,[111] which interrelates p_A and p_H over the hydrostatic pressure range of 1–4 atm, i.e.:

$$p_A = \tilde{\alpha} p_H + \beta \tag{6-43}$$

where the constants $\tilde{\alpha}$ and β are 2 and 1.5, respectively.[23,111] Chendke and Fogler[37] determined the values of R_r and R_c by using Eqs. (6-41)–(6-43) for 1 and 6.4 atm, respectively. They used these values in conjunction with Gavrilov's semiempirical relationship,[112] which interrelates nuclei radius and their number distribution, to determine the total number of nuclei that grow into bubbles. From their study they have concluded that the number of nuclei that undergo cavitation at 6.4 atm is about five times greater than that at 1 atm. According to their measurements the SL intensity increases by a factor of 6 times when the pressure is raised from 1 to 6.4 atm. Therefore, the authors concluded that the increase in SL with pressure results from an increase in the number of cavitation events rather than an increase in the intensity of the individual events, in contradiction to the earlier explanation offered by Flynn.[8] Although the incremental part of the curve can be readily and quantitatively explained by Chendke and Fogler's proposal, the decrease in SL after the maximum has been reached can be accounted for only qualitatively. That is, at high hydrostatic pressures, when $p_H > p_A$, the acoustic pressure is insufficient to initiate cavitation and so inhibits SL.

The effect of static pressure on SL intensity as discussed above appears to be strictly applicable when water is the medium. In the presence of CCl_4, SL intensity vs. pressure curves for the aqueous solution show maxima at about 6 and 12 atm. The cause of this complex variation in SL intensity with pressure is not well understood.

SONOLUMINESCENCE AS A PROBE OF ACOUSTIC CAVITATION

A majority of studies of SL in the past have focused on the elucidation of its mechanism. Very little attention has been devoted to the use of this phenomenon to probe characteristic parameters of acoustic cavitation. Since SL is caused by cavitation in principle, it can be used to extract some information about the properties of cavitating bubbles. Sonoluminescence is a measurement of an observable effect (light emission) and is determined by a set of laboratory operations. This, in terms of Flynn's terminology,[8] is an "activity measure" in that it measures the average effect of a bubble field rather than what happens in the individual bubbles.

It is well known that the physical properties, such as pressure and temperature, of a medium containing an emitting species are reflected in the parameters of the spectral emission of the species, i.e., the wavelength of maximum emission and bandwidth at half the maximum intensity (FWHM). Thus, radiating atoms or molecules inside cavities could serve as probes to determine physical properties such as pressure and temperature within the cavities. An initial attempt in this type of investigation was made by Taylor and Jarman,[32] who measured the FWHM of the sodium D line at an insonation frequency of 16 kHz. From their measurements they estimated intracavity pressures and temperatures to be 2×10^3 atm and 10^4K, respectively. A similar peak broadening was observed earlier by Heim.[80]

More extensive studies to estimate cavitational parameters by use of SL were carried out by Sehgal et al.[35,89] In one case they measured, with reference to zero acoustic field conditions, the change in the FWHM and the position of the peaks of the sodium and potassium resonance lines from insonated aqueous solutions of alkali metal halide salts saturated with argon; and in the second case they measured relative SL intensities of NO and NO_2. These measurements provided an estimate of the apparent temperatures generated by acoustic cavitation in the presence of mono-, di-, and triatomic gases. A caveat needs to be made: since SL arises from many bubbles at different stages of collapse, FWHM and peak displacement can arise from the inhomogeneity of the bubble field. In addition, emission need not arise in, or even near, the original collapsing cavity. but may occur after diffusion into or secondary reactions in, the bulk solution.

Figures 6-18 and 6-19 show typical high-resolution SL spectra of argon-saturated solutions of NaI and KI. The splitting is more marked for potassium because of a larger energy difference between the ^2P states. The lines show

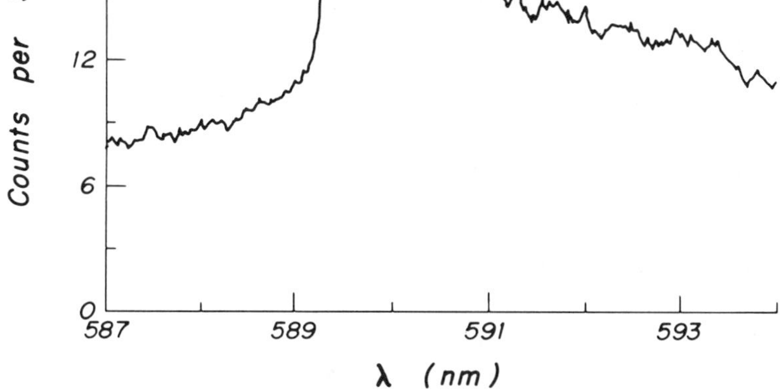

Figure 6-18. High-resolution sonoluminescence spectrum of argon-saturated aqueous sodium iodide solution showing the sodium atom resonance line split into a doublet. Time constant = 30 s; resolution = 0.2 nm; insonation frequency = 459 kHz, and pressure = 0.7 atm. (Adapted from Ref. 35.)

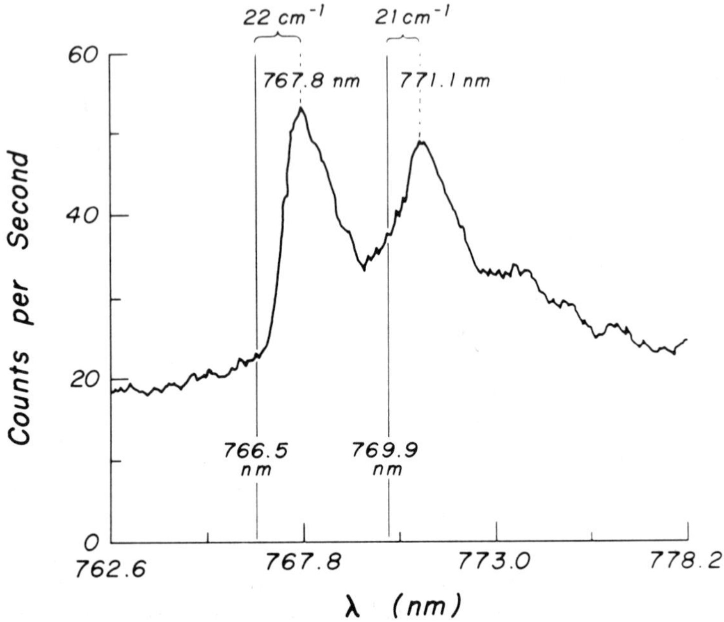

Figure 6-19. High-resolution sonoluminescence spectrum of argon-saturated aqueous potassium iodide solution showing potassium atom resonance line split into a doublet. Time constant = 30 s; resolution = 0.8 nm; insonation frequency = 459 kHz pressure = 0.7 atm. (Adapted from Ref. 35.)

broadening, asymmetry and a shift to longer wavelengths. The sodium peaks are not fully resolved and the measurement of their peak position is less accurate. Peak displacements and bandwidth broadening were determined from spectra of the type shown in Figs. 6-18 and 6-19. At an insonation frequency of 459 kHz the pressure broadening of the bandwidths and shifts in the peak position range between 28–47 cm^{-1} and 22–34 cm^{-1}, respectively, for the various alkali salt solutions.[35]

The results of studies by Chen and Takeo[113] provide for a relationship between either peak shift or bandwidth broadening of Na and K lines as a function of the density of the medium relative to the density at STP. Using the results of this study and the correlation of relative density (RD) with the band parameters according to Chen and Takeo, RD values of the luminescent medium were determined to be between 31 and 50 for the various sodium and potassium halides. Relative density values can also be shown to be related to the collapse ratio (R_{co}/R_e), by the expression:[35]

$$RD \simeq (\rho_e/\rho_{co}) = (R_{co}/R_e)^3 \tag{6-44}$$

where ρ and R represent the density of the intracavity contents and bubble radius, respectively. The subscripts *co* and *e* represent the conditions at which collapse begins and is completed, respectively. Equation (6-44) is obtained under the following assumptions:

1. No condensation of cavity contents occurs during the collapse. Condensation processes tend to reduce the amount of vapor and thus reduce the intracavity pressure. Also, since condensation is accompanied by heating, the temperature of the medium rises. Consequently, condensation produces dual effects. On the one hand, it will tend to reduce the number of vapor molecules and thus reduce the intracavity pressure and temperature. On the other hand, because condensation is accompanied by heating, the temperature of the medium rises. Therefore, to a first approximation, these two opposing effects may be assume to cancel one another.
2. There is no significant increase in pressure resulting from an increase in the number of species caused by the dissociation of the vapor molecules in the cavities.
3. The bubbles are sufficiently large (so that surface tension can be ignored) and in thermal and mechanical equilibrium with the surroundings before they become unstable and collapse. This means that before collapse begins the gas–vapor contents of bubbles, on average, are at ambient temperature and hydrostatic pressure, and their motion is modulated by the external pressure, which is of the order of magnitude of the ultrasonic pressure amplitude. Such a situation would especially occur in bubbles that grow to a large radius and oscillate over a few cycles before eventually collapsing.

Because of the rather complicated statistical nature of cavitation events in space and time and, because very little information is available on cavity composition, the validity of these assumptions is not fully demonstrated. Recently,[101] it has been suggested that the pressure inside the bubbles at a maximum radius should be considerably less than atmospheric pressure, in contradiction to the last assumption listed above. The authors are of the opinion this argument may be true only for gaseous bubbles that grow, without any quantitative change in their contents, from nuclei whose internal pressure is on the order of atmospheric pressure. While such types of bubbles were often modeled in early theoretical studies, it seems less likely that they occur under most practical conditions. For the latter, it is conceivable that during the slow bubble growth there is a constant input of water vapor into the bubble from the surroundings. As well, there can be a significant increase in the gas content of the bubble because of rectified diffusion. On balance, it seems reasonable to assume that the bubble is in equilibrium with the ambient temperature and pressure before the onset of its unstable dynamics, caused by the modulation of the sonic field, leads to its eventual collapse.

The intracavity temperatures, T_{ad}, and pressures, P_{ad}, for an adiabatic compression are related to the radius ratio R_{co}/R_e by the relations:

$$T_{ad} = T\left(\frac{R_{co}}{R_e}\right)^{3(\gamma-1)} \tag{6-45}$$

and

$$P_{ad} = P\left(\frac{R_{co}}{R_e}\right)^{3\gamma} \tag{6-46}$$

where T and P represent, respectively, the intracavity temperature and pressure of the cavity contents just before the collapse begins. These values are taken to be equal to the ambient temperature T_0 and pressure p_H. In argon-saturated alkali halide solutions the adiabatic temperatures and pressure, as estimated from peak displacements and FWHM measurements, are $\simeq$3400K and $\simeq$310 atm, respectively.[35] However, the cavity collapse is not strictly adiabatic because some of the energy is lost by thermal conduction to the surrounding medium during the process. Young[56] has shown that the temperatures achieved inside the bubble can be related to the adiabatic temperatures according to Eq. (6-8). Using a value of T_{ad} equal to 3400K, the true temperature inside the cavity was estimated from Eq. (6-8) to be 2450K.

The magnitudes of the adiabatic pressures and temperatures are less than what one would expect on the basis of the Noltingk and Neppiras model for gaseous cavitation in incompressible liquids. One of the main reasons for this difference is that the model considers cavities to be exclusively composed of gas, whereas the experimental bubbles contain vapor plus gas. The presence of vapor inside the bubble can impede the collapse either by postponing the initiation of collapse or by providing a cushioning action against the inward motion of the bubble wall. In either case, the severity of collapse is reduced, thereby reducing intracavity temperatures. Also, as a result of the assumption that the liquid is incompressible, the maximum bubble wall velocity assumed by theory may exceed the velocity of sound in the liquid. Under such circumstances the momentum of the converging bubble walls, as predicted by the theory, may exceed the true value and the collapse may appear to be more severe than it actually is. There are other complex factors that also may contribute to the reduction in severity of the bubble collapse. For example, when the inward acceleration of a bubble wall exceeds a critical value of surface tension, $4\sigma/R^2$, then the liquid surface breaks up to yield a number of small bubbles.[114] This means that the collapse of a bubble to a minimum radius is interrupted by bubble fragmentation. Once again in these circumstances the intracavity temperatures and pressures achieved will be lower. Alternatively, the site of sonoluminescence may not be in the cavity, but may be the surrounding heated shell, whose temperature is observed to be significantly lower.[115,116] In the light of these arguments it is not surprising that the experimental estimates of cavity parameters are smaller than the theoretical expectations.

A second independent attempt to estimate cavity properties was based on measurements of the relative distribution of SL intensities from NO- and NO_2-saturated aqueous solutions. As described previously, in the spectral region above 400 nm the emission from these solutions is believed to be associated with the excited NO_2^* species formed during bubble collapse. The formation, quenching, and dissociation of NO_2^* are described by the following reactions:[89]

$$NO_2 + M \xrightarrow{k_f} NO_2^* + M \tag{6-47}$$

$$NO_2^* + M \xrightarrow{k_q} NO_2 + M \tag{6-48}$$

$$NO_2^* + M \xrightarrow{k_d} NO + O + M \tag{6-49}$$

where M is an energized chemical species, such as H_2O^*, OH^*, OH, that collisionally activates NO_2, and k_f, k_q, and k_d are the rate constants of the excitation, quenching, and dissociation processes, respectively. Using steady-state treatment, the NO_2^* concentration can be expressed as:

$$[NO_2^*] = k_f[NO_2]/(k_q + k_d) \tag{6-50}$$

It has been shown[89] that the rate constants can be expressed in terms of the following parameters: collision numbers; proportion of energetically favorable encounters for which electronic–vibrational excitation occurs, P_E; energy level, E_λ, to which NO_2^* is excited; and the effective temperature T. These relations when substituted in Eq. (6-50), yield the expression:

$$[NO_2^*] = \frac{P_E \exp\,(E_\lambda/RT)[NO_2]}{\delta P_E + \frac{1}{2}\left[\frac{E_D - E_\lambda}{RT}\right]^2 \exp -\left[\frac{E_D - E_\lambda}{RT}\right]}, \tag{6-51}$$

where δ is a correction factor whose magnitude is a measure of the degree of deviation from the equilibrium distribution, and E_D is the dissociation energy of NO_2 molecules. Levitt[117] has derived a equation that interrelates emission intensity at wavelength λ to NO_2^* concentration. Combining Eq. (6-51) with Levitt's equation, one obtains, after algebraic manipulation:

$$Y = \frac{I_{\lambda D}}{I_\lambda}\frac{\lambda_D^2}{\lambda^2} = 1 - \frac{2}{RT}[E_D - E_\lambda] + \left(2 + \frac{1}{\delta P_E}\right)\frac{(E_D - E_\lambda)^2}{(RT)^2} - \cdots \tag{6-52}$$

where $I_{\lambda D}$ and I_λ are the emission intensities at wavelengths λD and λ, arising from the transitions from energy levels E_D and E_λ to the ground state, respectively. According to Eq. (6-52) the parameter Y, which is a weighted ratio of emission intensities at two wavelengths, shows a complex dependence on $E_D - E_\lambda$. In the region where E_λ approaches E_D such that $E_D - E_\lambda/RT < 1$, the higher order terms can be neglected and the Y vs. $E_D - E_\lambda$ relationship should show linear behavior.

Figures 6-20 and 6-21 show a plot of Y vs. $E_D - E_\lambda$ obtained from the SL spectra of NO- and NO_2-saturated aqueous solutions. E_D was taken to be 299.2 kJ mol^{-1} and λ_D was taken as the wavelength at which maximum emission occurred in the band above 400 nm. This wavelength corresponds approximately

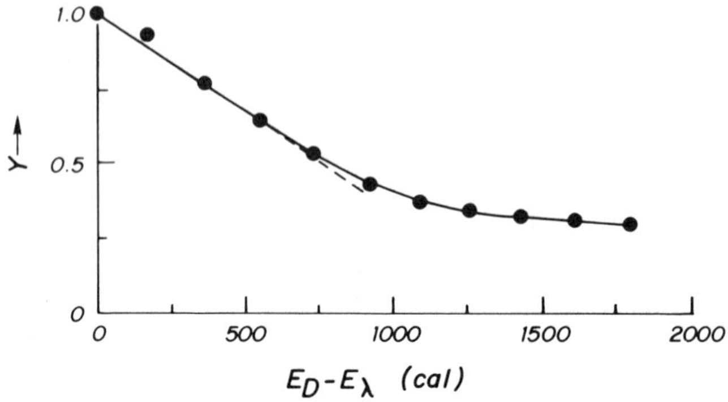

Figure 6-20. Plot of Y vs. $(E_D - E_\lambda)$ for a nitric oxide-saturated aqueous solution. (Reproduced with permission from Ref. 89.)

to the highest energy level that a molecule acquires before dissociating. As E_λ approaches E_D the Y vs. $E_D - E_\lambda$ curve becomes linear, as expected from Eq. (6-52). The linear portion of the curve extends over a wider range of $E_D - E_\lambda$ in the case of NO- then for NO_2-saturated solutions. This derives from the fact that the terminal cavity temperature in bubbles containing the triatomic gas NO_2 is smaller than that in bubbles containing diatomic NO. As a consequence, the neglect of higher order terms in Eq. (6-52) for the case of NO_2 is less valid.

Initial slopes of the curves in Fig. 6-21 and 6-22 were determined to be -7.5×10^{-4} cal^{-1} mol and -1.18×10^{-3} cal^{-1} mol, respectively. From these slopes intracavity temperatures were estimated to be 1350 ± 50K and 860 ± 100K in the presence of NO and NO_2, respectively. These temperatures are the actual temperatures within a cavity and not the adiabatic temperatures. Using these values in Young's equation, R_{co}/R_e ratios were solved for by inspection for NO- and NO_2-saturated solutions. The values for the two cases were 5.9 and 9.7,

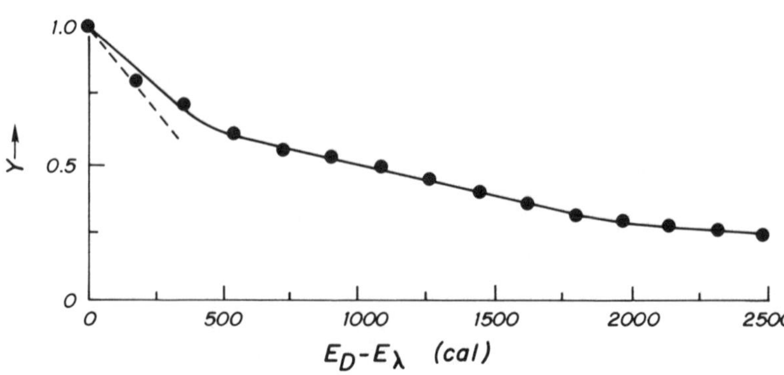

Figure 6-21. Plot of Y vs. $(E_D - E_\lambda)$ for a nitrogen dioxide-saturated solution. (Reproduced with permission from Ref. 89.)

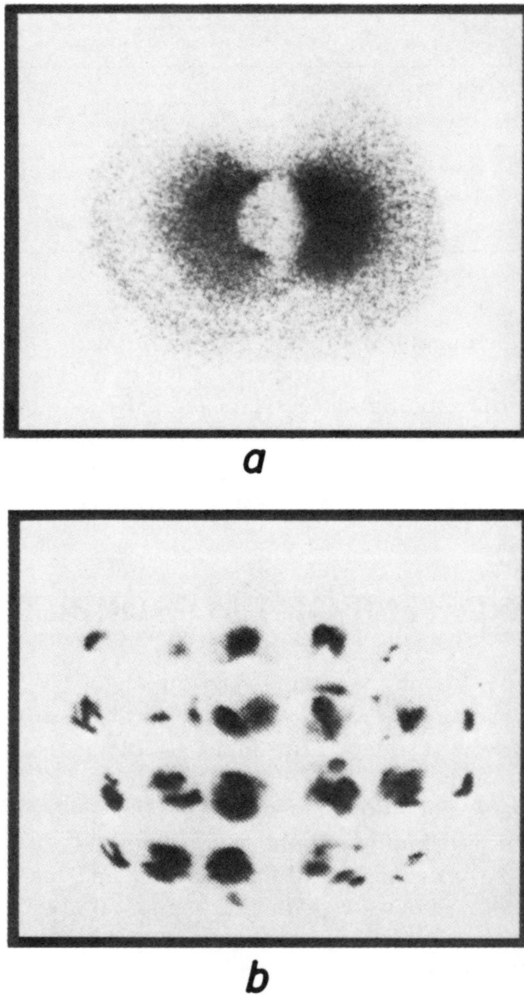

Figure 6-22. Spatial distribution of SL. (a) View from below the transducer and (b) side view. (Reproduced with permission from Ref. 119.)

respectively. For adiabatic compression these values would yield temperatures of ~ 2000K.

The R_{co}/R_e values calculated (Eq. 6-5) on the basis of the Noltingk and Neppiras model for $P/p_{co} = 100$ are 8.1, 21, and 50 for γ equal to 5/3, 7/5, and 9/7, respectively. If R_{co} is the same for the three cases, it follows that the absorption of energy by a molecule with additional degrees of freedom, that is, as the magnitude of γ is decreased, causes the bubbles to collapse to a smaller radius. A similar conclusion can be arrived at by using the radius vs. time relationship for bubbles derived by Hickling[57] based on Gilmore's theory.[118] A comparison of experimental and theoretical values of R_{co}/R_e (normalized to the R_{co}/R_e value of argon) are given in Table 6-7.

Table 6.7. Relative Values of R_{co}/R_e of Nitric Oxide- and Nitrogen Dioxide-Saturated Aqueous Solutions Calculated with Respect to Argon-Saturated Solution

Medium	γ	NN model (R_{co}/R_e)	Hickling-Gilmore (R_{co}/R_e)[a]	Experimental (R_{co}/R_e)	T_{ad} (K)	T (K)
Ar/H_2O	5/3	1	1	1	3400	2450
NO/H_2O	7/5	2.6	2.1	1.7	2400	1350
NO_2/H_2O	9/7	6.2	4.0	2.8	2000	860

[a] These values were read from Figure 1 of Ref. 57 for monoatomic, diatomic, and triatomic gases.

According to experimental results R_{co}/R_e increases with a decrease in γ. This is in agreement with the theoretical predictions. The differences between the magnitudes of the experimental and theoretical values are not unexpected. Noltingk and Neppiras' model,[13] strictly speaking, is true only for an incompressible liquid. Calculations by Hickling,[57] based on Gilmore's work, take into account liquid. compressibility and fluid velocity and give R_{co}/R_e values closer to experimental estimates.

SPATIAL AND TEMPORAL DISTRIBUTION OF SL

Very little is known about the spatial distribution of bubbles that produce sonoluminescence. Some of the earlier studies performed using photographic plates[5,6] show SL to occur at pressure nodes of standing waves set up in the irradiated aqueous medium.

Flynn[8], using image intensification techniques, was able to obtain a photograph of the spatial distribution of bubbles producing SL. More recently, Reynolds et al[119,120] have extended this earlier study and have used the visible component of SL to map out the spatial and temporal distribution of bubbles. Views of planes transverse and longitudinal to the direction of propagation are shown in Fig. 6-22. The transverse view shows that most of the light comes from the sides of the probe and only a slight amount from the end. The side view obtained at a gain of $\sim 2 \times 10^6$ shows a standing wave pattern that is in accordance with the driving frequency of the probe. This is analogous to the information obtained by photographic plates. However, because of better sensitivity, image intensifier techniques have a discrete advantage over the use of photographic plates. The SL spatial distribution can be measured in a few seconds of exposure as opposed to minutes, or longer, for photographic methods. As a result, image intensifiers can be used to trace the temporal variation of sonoluminescence in the localized region of a bubble field, information that cannot be obtained by the use of either photographic plates or a photomultiplier tube. Recently Crum and Reynolds[121] have observed sonoluminescence by using an image intensifier tube. There results indicate that large amounts of light are emitted from regions near pressure antinodes in a standing wave field, and secondly the light emission was most likely from "stable" cavitation rather than "transient" cavitation.

SOME APPLICATIONS OF SL

According to a recent review by Greguss,[122] one of the earliest applications of SL was suggested by Spengler,[123] who used this phenomena to record sonic images employing an alkaline luminol solution as sonoluminescent material and photographic plates as a detector. Although the sonic images obtained showed good resolution, the exposure times and sound intensities needed were so large that no direct application of the sonoluminescent image detector could be considered. Later, using an image intensifier in place of photographic plates, Greguss[122] was able to record sound images at lower ultrasonic intensities and shorter exposure times. A photograph of a sound-absorbing cross, employing this technique, is shown in Fig.6-23.

More recently, an application of SL in the area of analytical chemistry has been suggested by Yamada et al.[124] Preliminary investigations suggest the possibility of using SL from water to estimate the water content in organic solvents. The technique consists of allowing a methanol solution containing water to flow continuously into an insonation cell, where it is irradiated by continuous-wave (CW) ultrasound at 28 kHz. Ultrasound is alternately turned on and off to obtain optimum reproducible signals. The on/off times were chosen to be 5 and 30 s, respectively. The SL intensity was found to vary linearly in the range 0–0.2 g L^{-1} of water. The detection limit, as defined by a signal to noise ratio of 3, was 0.005 g L^{-1}. However, this lower limit depends on how thoroughly the methanol is dried before use and to what degree it absorbs moisture from the surroundings

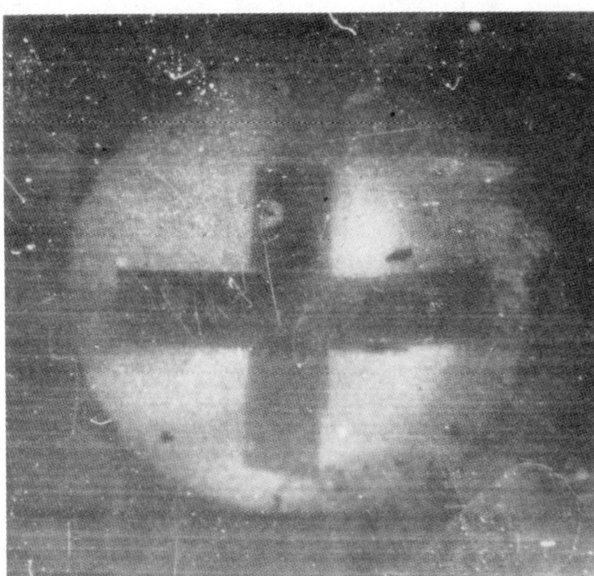

Figure 6-23. Image of sound-absorbing cross recorded with a sonoluminescent detector. (Reproduced with permission from Ref. 122.)

during the measurements. The sensitivity of this method is low as compared to the traditional Karl Fisher technique for estimating water content of nonaqueous solvents; however, the sonoluminescent method has an advantage in that it is simple, is rapid, and does not require the use of any chemical reagents.

SONOLUMINESCENCE AND EMISSION FROM IONIZING RADIATION

Sonolysis and radiolysis, although two different phenomena—the former involving material waves and the latter electromagnetic waves—produce similar chemical changes. The strong analogy between sonochemistry and radiation chemistry was suggested in early reports of Lindstrom,[125] Weissler,[126] and Prudhomme et al.[127] Since then Todd[128] has shown more explicitly that cavitation bubbles deposit chemical energy in a manner very similar to slow α particles. Margulis[129] proposed a cavitation-diffusion model and showed ultrasonic cavitation to be similar to fission ionizing radiation in terms of the concentration of energy and density distribution of radicals. Sehgal and Wang[130] more recently used the method developed by Todd[128] to estimate transient cavitation energy and the fraction that is converted into chemical energy. The close similarity between sono- and radiation chemistry is probably a result of the formation of similar free radicals and excited molecules. In this regard, SL may in fact be regarded as the acoustic counterpart of the luminescence[63] produced in water by ambient ionizing radiation.

Ionizing radiation excites water molecules by transferring momentum in local hot spots, called spurs. Similarly, the rapidly converging walls of a cavitating bubble transfer momentum to the cavity contents, giving rise to high nonequilibrium temperatures, which in turn lead to excitation of water molecules. The high-energy particles cause molecular excitation of H_2O, which either radiates around 260 nm or decomposes to give OH^* (Σ^+) and e^- with emissions in the regions of 350 nm and 460 nm, respectively. Sonoluminescence in water occurs from similar chemical species produced during cavitation.

Although the two types of emissions are analogous, there are discrete differences that must be emphasized. In the case of ionizing radiation there is a direct electrical interaction between the high-energy particles and the molecules of water, whereas in the case of sonoluminescence the interaction occurs via a mechanical process of wall motion. Whether the wall motion leads to electrical and/or thermal excitation of molecules is still not unequivocally determined, although the current view, as far as SL from water is concerned, favors the thermal mechanism. A further distinction is that dissolved gases play a much more important role in SL than they do in the emission induced by ionizing radiation. Finally, to date there is no convincing evidence for the formation of hydrated electrons during sonolysis. This is in contrast to radiolysis, which frequently involves the formation of such particles.

CONCLUDING REMARKS

Although an increasing number of studies of SL have been reported, the area is still in its early stages of development. Unfortunately, in many of these studies, there has been little or no attempt to control the acoustic conditions of insonation. Ultrasonic intensities are quoted without qualification and this makes it difficult to make proper comparisons among various results. As well, there has been a greater emphasis on explaining the many variations in the observed SL in terms of bubble collapse, but very little attention has been devoted to correlating SL with the inception stage of cavitation. This requires a proper control of the nucleation process, which not only depends on the nature and purity of the medium but also on the history and the sonicating conditions. Specifically, it depends on such factors as whether the medium has been insonated before; the length of time it has been left standing to the atmosphere; and degassing procedures, bubble size, and bubbling rate during sonication, if the gases are bubbled through a medium to enhance cavitation.[131] Incomplete specification of acoustic conditions and the state of the liquid at the time of the experiment has resulted in considerable variability and uncertainty in the reported results.

Another factor that would greatly aid the understanding of SL is the knowledge of gas–vapor composition of bubbles and its effect on bubble dynamics. According to Margulis[132] the composition of bubbles undergoing pulsation and collapse is not constant. Also, most of the analysis of SL is based on a simple model of independent bubbles, whose motion is controlled by a sinusoidal pressure field. The analysis does not take into account any interaction between them either by coalescence and fragmentation, or through the sound fields that they radiate. Proper consideration of these factors could resolve some of the differences that appear to exist among the results reported by various authors.

Finally, with respect to the various theories of SL, there appears to be considerable evidence that supports the theory that the emission arises from chemical processes induced largely by thermal mechanisms. Although our present understanding[133] of the electrical contributions to SL in aqueous media indicates that it is not the primary mechanism, there may be conditions in other media where electrical contributions may play a dominant role.

We have restricted our discussion in this chapter to more recent developments in the area of sonoluminescence. Early studies have been thoroughly covered in the excellent reviews by Flynn[8] and El'piner.[9] We have selected a cross section of material that illustrates the broadest aspects of the phenomenon of sonoluminescence. One area not included, specifically, relates to studies of pulsed ultrasound. Interested readers are referred to the work of Iernetti[134] and Ceschia and Iernetti.[135] It is hoped that the material in this chapter serves to further stimulate research efforts to extend both our understanding and our application of sonoluminescence.

Since the initial writing of this chapter a few more studies in SL have been reported that may be of interest to readers. Margulis[136] has proposed a new electrical theory for cavitation as an alternative to the thermal theory. Crum and Fawlkes[137] have observed sonoluminescence from microsecond long pulses and have used this emission to determine threshold for transient cavitation. Suslick and Flint[138] have observed SL from dodecene and have concluded it to be from various chemical species formed during the high temperature and pressure of acoustic cavitation.

ACKNOWLEDGMENTS

The authors would like to thank Ms. Elaine Quarve for typing this chapter and Messrs. Steve Richardson and Jim Hanson for their assistance in preparing the figures.

REFERENCES

1. Zimakov, P. V. *C. R. Acad. Sci. USSR* **1934**, *3*, 450.
2. Frenzel, H.; Schultes, H. *Z. Phys. Chem.* **1935**, *27B*(5/6), 421.
3. Marinesco, N.; Trillat, J. J. *C. R. Hebd. Seances Acad. Sci. Paris* **1933**, *196*, 858.
4. Marinesco, N.; Reggani, M. *C. R. Hebd. Seances Acad. Sci. Paris* **1935**, *200*, 548.
5. Paounoff, P. *C. R. Hebd. Seances Acad. Sci. Paris* **1947**, *44*, 261.
6. Pinoir, R.; Pouradier, I. *J. Chim. Phys.* **1947**, *44*, 261.
7. Akmanov, A. G.; Benkovskii, V. G.; Golubnichii, P. I.; Masslennikov, S. I.; Shemanin, V. G. *Sov. Phys. Acoust.* **1979**, *19*(5), 417.
Golubnichii, P. I.; Dyadyushkin, P. I.; Kalyuzhnyi, G. S.; Korchikov, S. D. *Sov. Phys. Tech. Phys.* **1979**, *14*(8), 100.
8. Flynn, H. G. In "Physical Acoustics"; Mason, W. P., Ed.; Academic Press: New York, 1964, p. 57.
9. El' piner, I. E. "Ultrasound: Physical Chemical and Biological Effects"; Consultants Bureau: New York, 1964.
10. Margulis, M. A. *Sov. Phys. Acoust.* **1969**, *15*(2), 135.
11. Sliwinski, A. In "Acoustics and Vibration Progress"; Stephens, R. W. B.; Leventhall, H. G., Eds.; John Wiley and Sons: New York, Chapter 3, pp. 93.
12. Walton, A.; Reynolds, G. T. "Advances in Physics" 1984, 33(6), 595.
13. Noltingk, B. E.; Neppiras, E. A. *Proc. Phys. Soc.* (*London*) **1950**, *B63*, 679.
14. Neppiras, E. A.; Noltingk, B. E. *Proc. Phys. Soc.* (*London*) **1950**, *B63*, 1032.
15. Chambers, L. A. *Phys. Rev.* **1936**, *49*, 881.
16. Chambers, L. A. *J. Chem. Phys.* **1937**, *5*, 290.
17. Flosdorf, E. W.; Chambers, L. A.; Malisoff, W. *J. Am. Chem. Soc.* **1936**, *58*, 1069.
18. Levshin, V. L., Rzhevkin, S. N. *Doklady Akad. Nauk SSR* **1937**, *16*, 407.
19. Harvey, E. N. *J. Am. Chem. Soc.* **1939**, *61*, 2392.
20. Fottinger, H. In "Hydraulische Problems"; Verein Deutscher Ingenieure, Verlag: Berlin 1936, p. 14.
21. Richards, W. T. *Rev. Mod. Phys.* **1939**, *11*, 36.
22. Frenkel, Ya. I. *Zh. Fiz. Khim.* **1940**, *14*, 305.
23. Sirotyuk, M. G. *Sov. Phys. Acoust.* **1966**, *11*, 318.
24. Weyl, W. A., Marboe, E. C. *Research* (*London*) **1949**, *2*, 19.
25. Weyl, W. A. *J. Colloid Sci.* **1951**, *6*, 389.
26. Degrois, M.; Baldo, P. *Ultrasonics* **1974**, *12*, 25.
27. Srinivasan, D.; Holroyd, L. V. *Phys. Rev.* **1955**, *94*, 633.
28. Srinivasan, D.; Holroyd, L. V. *J. Appl. Phys.* **1961**, *32*, 446.

29. Jarman, P. *J. Acoust. Soc. Am.* **1960**, *32*, 1459.
30. Trilling, L. *J. Appl. Phys.* **1952**, *23*, 14.
31. Griffing, V.; Sette, D. *J. Chem. Phys.* **1955**, *23*, 503.
32. Taylor, K. J.; Jarman, P. D. *Aust. J. Phys.* **1970**, *23*, 319.
33. Saksena, T. K.; Nyborg, W. L. *J. Chem. Phys.* **1970**, *53*, 1722.
34. Sehgal, C. M.; Steer, R. P.; Sutherland, R. G.; Verrall, R. E. *J. Phys. Chem.* **1977**, *81*, 2618.
35. Sehgal, C. M.; Steer, R. P.; Sutherland, R. G.; Verrall, R. E. *J. Chem. Phys.* **1979**, *70*, 2242.
36. Sehgal, C.; Sutherland, R. G.; Verrall, R. E. *J. Phys. Chem.* **1980**, *84*, 388.
37. Chendke, P. K.; Fogler, H. S. *J. Phys. Chem.* **1983**, *87*, 1644.
38. Chendke, P. K.; Fogler, H. S. *J. Phys. Chem.* **1983**, *87*, 1362.
39. Gunther, P.; Zeil, W.; Grisar, U.; Langmann, W.; Heim, E. *Z. Naturforsch.* **1956**, *11a*, 882.
40. Gunther, P.; Heim, E.; Schmidt, A.; Zeil, W. *Z. Naturforsch.* **1957**, *12a*, 521.
41. Meyer, E.; Kuttruff, H. *Z. Agnew. Phys.* **1959**, *11*, 325.
42. Kuttruff, H.; Plass, K. *Acustica* **1961**, *11*, 224.
43. Negishi, K. *J. Phys. Soc. Jpn.* **1961**, *16*, 1450.
44. Gunther, P.; Heim, E.; Eichkorn, G. *Z. Phys.* **1959**, *11*, 279.
45. Kuttruff, H. *Acustica* **1962**, *12*, 230.
46. Loeb, L. B. In "Static Electrification"; Springer-Verlag: Berlin, 1958.
47. Schmid, J. *Acustica* **1962**, *12*, 70.
48. Muller, H. H. *Acustica* **1965/66**, *16*, 22.
49. Soloukhin, R. I. *Sov. Phys. Dokl.* **1961**, *6*, 16.
50. Gordeev, V. E.; Serbinov, A. I.; Troshin, Ya. K. *Dokl. Akad. Nauk SSR* **1967**, *172*, 383.
51. Gordeev, V. E.; Serbinov, A. I.; Troshin, Ya. K. *Prikl. Mekh. i Tekh. Fiz.* **1967**, *1*, 45.
52. Gunther, P.; Zeil, W.; Gisar, U.; Heim, E. *Z. Electrochem.* **1957**, *61*, 188.
53. Gunther, P.; Heim, E.; Borgsted, H. O. *Z. Electrochem.* **1959**, *63*, 43.
54. Harvey, E. N. "Bioluminescence"; Academic Press: New York, 1952.
55. Prudhomme, R. O.; Guilmart, T. *J. Chem. Phys.* **1957**, *54*, 336.
56. Young, F. R. *J. Acoust. Soc. Am.* **1976**, *60*, 100.
57. Hickling, R. *J. Acoust. Soc. Am.* **1963**, *35*, 967.
58. Pearse, R. W. B.; Gaydon, A. G. "The Identification of Molecular Spectra"; John Wiley and Sons: New York, 1976.
59. Zhuraler, A. I.; Tsyler, O. P. *Biophysics* **1968**, *13*, 1187; translated from *Biofizika* **1968**, *13*, 1018.
60. Margulis, M. A.; Grundel, L. M. *Dokl. Phys. Chem. Proc. Acad. Sci. USSR* **1983**, *269*, 160.
61. Mulliken, R. S. *J. Chem. Phys.* **1935**, *3*, 506.
62. Lewis, D.; Hamill, W. H. *J. Chem. Phys.* **1969**, *51*, 456.
63. Quickenden, T. I.; Que Hee, S. S. *Radiat. Res.* **1971**, *46*, 28.
64. Sitharmamarao, D. N.; Duncan, J. F. *J. Phys. Chem.* **1963**, *67*, 2126.
65. Maria, H. J.; McGlynn, S. P. *J. Chem. Phys.* **1970**, *52*, 3402.
66. Suchard, S. N. *Spectroscopic Data 1* (Pt B), IFI/Plenum Data Company, 1975.
67. Griffing, V. *J. Chem. Phys.* **1952**, *20*(6), 939.
68. Griffing, V. *Phys. Rev.* **1952**, *87*, 234.
69. Griffing, V. *J. Chem. Phys.* **1950**, *18*, 997.
70. Sehgal, C.; Sutherland, R. G.; Verrall, R. E. *J. Phys. Chem.* **1980**, *84*, 2920.
71. Vaughan, P. W.; Graham, E.; Leeman, S. *Appl. Sci. Res.* **1982**, *38*, 45.
72. Verrall, R. E.; Nomura, H. *J. Solution Chem.* **1977**, *6*, 1.
73. Burton, C. J. *J. Acoust. Soc. Am.* **1948**, *20*, 186.
74. Yen, N. *J. Acoust. Soc. Am.* **1975**, *57*, 1357.
75. Makino, K.; Mossoba, M. M.; Riesz, P. *J. Phys. Chem.* **1983**, *87*, 1369.
76. Spinks, J. W. T.; Woods, R. J. "An Introduction to Radiation Chemistry"; John Wiley and Sons: New York, 1976.
77. Margulis, M. A. *Russ. J. Phys. Chem.* **1980**, *54*, 859.
78. Stein, G. "The Chemistry of Ionization and Excitation"; Johnson, G. R. A.; Scholes, G., Eds.; Taylor and Francis: London, 1967.
79. Sehgal, C.; Sutherland, R. G.; Verrall, R. E. *J. Phys. Chem.* **1980**, *84*, 529.
80. Heim, E. *Z. Angew Phys.* **1960**, *12*, 423.
81. Lalos, G. T.; Hammond, G. L. *Astrophys. J.* **1962**, *135*, 616.
82. Margulis, M. A.; Dmitrieva, A. F. Russ. *J. Phys. Chem.* **1982**, *56*, 530.
83. Sehgal, C.; Sutherland, R. G.; Verrall, R. E. *J. Phys. Chem.* **1980**, *84*, 227.
84. Gray, P. *Trans. Faraday Soc.* **1959**, *55*, 408.
85. Margulis, M. A. *Russ. J. Phys. Chem.* **1979**, *48*, 1653.

86. Sehgal, C.; Sutherland, R. G.; Verrall, R. E. *J. Phys. Chem.* **1981**, *85*, 315.
87. Parke, A. V. M.; Taylor, D. *J. Chem. Soc.* **1956**, *4*, 4442.
88. Kling, R. *Rev. Sci.* **1947**, *85*, 363.
89. Sehgal, C.; Sutherland, R. G.; Verrall, R. E. *J. Phys. Chem.* **1980**, *84*, 396.
90. Broida, H. P.; Schiff, H. I.; Sugden, T. M. *Trans. Faraday Soc.* **1961**, *57*, 259.
91. Rossini, F. D.; Wagman, D. D. "Selected Values of Chemical Thermodynamics Properties"; U.S. Government Printing Office: Washington, D.C., 1952.
92. Herzberg, G. "Electronic Spectra of Polyatomic Molecules"; Van Nostrand: Princeton, NJ, 1967.
93. Jarman, P. *Proc. Phys. Soc.* (*London*), **1959**, *723*, 628.
94. Gabrielli, I.; Iernetti, G.; Lavenia, A. *Acustica* **1967**, *18*, 173.
95. Golubnichii, P. I.; Goncharov, V. D. *Sov. Phys.* (*Acoust.*) **1971**, *16*, 323; **1971**, *15*, 464.
96. Jarman, P.; Taylor, K. J. *J. Acoust. Soc. Am.* **1966**, *39*, 584.
97. Alfredsson, B. *Acustica*, **1965/66**, *16*, 127.
98. Eastwood, L. M.; Watmough, D. J. *Ultrasound Med. Biol.* **1976**, *2*, 319.
99. Apfel, R. E. In "Methods of Experimental Physics: Ultrasonics"; Edmonds, P., Ed.; Academic Press: New York, 1981, Vol. 19. 356.
100. Sehgal, C.; Sutherland, R. G.; Verrall, R. E. *J. Phys. Chem.* **1980**, *84*, 525.
101. Walton, A. Personal communication, 1984.
102. "Handbook of Physics and Chemistry"; 53rd Edn., Chemical Rubber Publishing Co.: Cleveland, 1972/73.
103. Suslick, K. S.; Gawienowski, J. J.; Schubert, P. F.; Wang, H. H. *Ultrasonics* **1984**, *1*, 33.
104. Prudhomme, R. O. *J. Chim. Phys.* **1949**, *46*, 318.
105. Bunsel, R. G.; Degrois, M. *J. Chim. Phys.* **1955**, *53*, 279.
106. Polotskii, I. G. *Zh. Fiz. Chim., USSR* **1969**, *22*, 787.
107. Finch, R. D. *Br. J. Appl. Phys.* **1955**, *16*, 1543.
108. Weissler, A. *J. Acoust. Soc. Am.* **1953**, *225*, 651.
109. Neppiras, E. A. *Ultrasonics*, **1980**, *18*, 201.
110. Macleay, R. Q.; Holroyd, L. V. *J. Appl. Phys.* **1961**, *32*, 449.
111. Strasberg, M. Ph.D. Dissertation, Catholic University of America, Washington, DC, 1956.
112. Gavrilov, L. R. In "Physical Principles of Ultrasonic Technology"; Rosenberg, L. D., Ed.; Plenum Press: New York, 1973, Vol. 2, Part IV.
113. Chen, S. Y.; Takeo, M. *Rev. Mod. Phys.* **1957**, *29*, 20.
114. Binnie, A. M. *Proc. Camb. Phil. Soc.* **1953**, *49*, 151.
115. Suslick, K. S.; Hammerton, D. A. *IEEE Transactions on Ultrasonics, Ferroelectrics, and Frequency Control* **1986**, *33*, 143.
116. Suslick, K. S.; Cline, R. E., Jr.; Hammerton, D. A. *J. Amer. Chem. Soc.* **1986**, *108*, 5641.
117. Levitt, B. P. *Trans. Faraday Soc.* **1962**, *58*, 1789.
118. Gilmore, F. R. California Inst. Technol. Eng. Div.; Pasadena, Ca.: 1952, Report number 26-4.
119. Reynolds, G. T.; Walton, A. J.; Gruner, S. M. *Rev. Sci. Instrum.* **1982**, *53*, 1673.
120. Reynolds, G. T.; Walton, A. J.; Gruner, S. M. *IEEE Trans. Nucl. Sci.* **1983**, *NS-30*, 440.
121. Crum, L. A.; Reynolds, G. T. *J. Acoust. Soc. Am.* **1985**, *78*, 137.
122. Greguss, P. "Ultrasonic Imaging"; Focal Press: New York, 1980. pp 48
123. Spengler, G. *Nachrichtentechnik*, **1953**, *3*, 399.
124. Yamada, M.; Toshyuki, H.; Suzuki, S. *Chem. Lett. Chem. Soc. Jpn.* **1983**, *3*, 283.
125. Lindstrom, O. *J. Acoust. Soc. Am.* **1955**, 27, 654.
126. Weissler, A. *J. Acoust. Soc. Am.* **1960**, *32*, 283.
127. Prudhomme, R. O.; Ricard, D.; Busnel, R. G. *J. Chim. Phys.* **1953**, *50*, 1078.
128. Todd, J. H. *Ultrasonics* **1970**, *8*, 234.
129. Margulis, M. A. *Russ. J. Phys. Chem.* **1976**, *50*, 534.
130. Sehgal, C. M.; Wang, S. Y. *IEEE Trans. Sonics Ultrason.* **1983**, *SU-30*, 374.
131. Sehgal, C. M.; Wang, S. Y. *J. Am. Chem. Soc.* **1981**, *103*, 6606.
132. Margulis, M. A. *Russ. J. Phys. Chem.* **1978**, *52*, 1711.
133. Sehgal, C. M.; Verrall, R. E. *Ultrasonics* **1982**, *1*, 37.
134. Iernetti, G. *Acustica*, **1970**, *23*, 189.
135. Ceschia, M.; Iernetti, G. *Acustica*, **1973**, *29*, 127.
136. Margulis, M. A. Ultrasonics **1985** (July) 157.
137. Crum, L. A.; Fawlkes, J. B. Nature **1986**, *319*(12), 6048, pp 52.
138. Suslick, K. S.; Flint, E. B. *Nature*, **1987**, In press.

7

Biological Effects of Acoustic Cavitation

LEON A. FRIZZELL

*Bioacoustics Research Laboratory,
Electrical and Computer Engineering Department,
University of Illinois*

INTRODUCTION

Acoustically induced cavitation has numerous beneficial uses in industry and laboratories. Ultrasonic cleaning, which depends directly upon acoustic cavitation, has been one of the primary industrial uses of ultrasound for many years. Cell disruptors generate acoustic outputs sufficient to induce cavitation, which breaks up biological cells so that their contents may be used or analyzed. The use of acoustic cavitation to induce chemical reactions is a more recent and growing application that is discussed at length in Chapter 6. This chapter deals with the biological effects of ultrasonically induced cavitation and its potential for harm in the clinical applications of ultrasound. See Chapter 8 for a discussion of other mechanisms involved in the production of biological effects.

The same effects that make acoustic cavitation such a useful phenomenon in various types of processing make it a potential mechanism of damage to patients. Yet relatively little is known regarding the details of ultrasonic cavitation in vivo. The best understood mechanism of damage is probably ultrasonic heating. Temperature rises can be calculated, at least approximately, from a knowledge of the ultrasonic absorption coefficient, the ultrasonic intensity, and the thermal characteristics of the body, including blood perfusion. On the other hand, the threshold for, and the type of, cavitation that can occur in vivo and the resulting biological effects are not well understood.

The first extensive investigation of the effects of ultrasound on living systems was reported in 1927.[1] Since that time ultrasound has developed as a useful tool in medical practice and receives widespread clinical application.[2] Ultrasound is employed clinically, for example, in medical diagnosis as a pulse-echo technique (time-averaged intensities typically less than 100 mW cm^{-2} and 0.1% duty cycle, i.e., percentage of time that sound is on) for obtaining information regarding the static and dynamic state of gross tissue structures. As a therapeutic tool, continuous-wave ultrasound is employed as a deep-heating agent; low intensities (usually 1 W cm^{-2} or less) are used. As a means of generating localized hyperthermia for oncological treatment, ultrasound intensities of several to tens of watts per square centimeter are used. As a selective tissue-modifying agent, relatively high intensities (on the order of 10^3 W cm^{-2}) are used in the form of short-duration pulses.

In this chapter a critical review is made of the effects of ultrasonic cavitation on biological materials. This discussion follows a useful progression from consideration of the effects of ultrasound on solutions and suspensions, to those on plants, on insects, and finally on mammals. The body of evidence from studies of suspensions, plants, and insects suggests the potential for harmful effects in mammals at clinically relevant intensities. Additionally, Flynn[3,4] and Apfel[5] have calculated thresholds for transient cavitation events in water from microsecond pulses, which are well within the levels emitted by clinical instruments. Heretofore the short pulses used in pulse-echo diagnostic systems were assumed too short for the generation of cavitation. Their results have generated renewed interest in the potential role of cavitation in mammals. However, there is also strong evidence that the behavior of bubbles within mammals may be substantially different (and less effective at producing damage) than in other systems.

The review of the literature in each category is not meant to be exhaustive but only to include representative examples. The reader interested in a more complete listing of relevant literature is referred to other reviews of biological effects of ultrasound that have been published recently.[6–9a] Flynn,[10] Coakley and Nyborg[11], Neppiras[12] and Apfel[13] have each presented a development of the fundamentals of acoustic cavitation, which are covered thoroughly in Chapter 1.

Although the acoustic pressure may be more directly applicable to cavitation phenomena, the acoustic intensity is used throughout this chapter as the parameter defining the acoustic level of exposure. This choice has been made because the bioeffects literature deals almost exclusively with intensity, and an effort has been made to aquaint the clinical user with its various expressions. When dealing with traveling plane waves (an approximation applicable to the typical situation with clinical ultrasound) the intensity I is given as $I = p^2/2\rho c$, where p is the acoustic pressure amplitude, ρ is the material density, and c is the speed of sound. A very useful approximation for water and soft tissues is that an acoustic pressure of 1 bar (100,000 Pa or approximately 1 atm) corresponds to an intensity of 1/3 W cm^{-2}. In situations where standing waves are present, it is more straightforward to use pressure as opposed to intensity, although it can still be useful to calculate an intensity at a position of interest equivalent to that in a plane wave with the pressure amplitude existing at that position.

Finally, the standard abbreviations are used here for the various intensities encountered.[14,15] The subscripts SP and SA refer to the peak intensity in the acoustic field within the volume of interest (spatial peak) and the intensity averaged over the acoustic beam or source (spatial average), respectively. The subscripts PA, TP, and TA refer to the intensity averaged over the duration of the pulse (pulse average), the temporal peak intensity (applicable to pulsed waves), and the temporal average intensity, respectively. For example, using I as the symbol for intensity, I_{SATP} corresponds to the spatial average, temporal peak intensity.

CAVITATION

Definition

Acoustically induced cavitation is considered here to involve the behavior of gas or vapor bubbles associated with acoustic exposure, which may include their formation, growth, and collapse. "Cavitation phenomena are produced in liquid media subjected to acoustic disturbances when the acoustic pressure during the rarefaction phase of the cycle reduces the total pressure to a specific threshold value." As discussed in Chapter 1, "this threshold value of the acoustic pressure amplitude is a function of a number of physical parameters that describe the state of the medium. These include temperature, pressure, frequency, kind and amount of dissolved gas, and previous history of the medium. Two types of cavitation phenomena can be distinguished, stable and transient cavitation, each of which exhibits different kinds of behavior of a bubble in response to an acoustic field. Stable cavitation of a medium containing dissolved or entrained gas occurs when a bubble oscillates for a number of cycles without collapsing or otherwise leaving the field."[9a] The bubble may grow by rectified diffusion to resonance size or somewhat larger over a period of many cycles. The term "gas body activation" has been used to describe the behavior of bubbles trapped in plant tissues and the pores of hydrophobic membranes. In this chapter, these phenomena are included in the general term, stable cavitation.

Transient cavitation occurs during the compression phase, in media that experience a tension stress during a portion of the rarefaction phase of the acoustic disturbance. Here the bubble collapse is very rapid, causing the bubble to disintegrate from the inertia associated with the collapse or by surface instability, such that it is transient in nature.[9a] Also, it is possible for a bubble to oscillate and grow in a stable manner for a period of time before undergoing a rapid, transient collapse at a specific size.

Mechanisms of Damage

Cavitation phenomena may cause damage to biological materials in several important ways. Transient cavitation generates very high pressures and temperatures, which theoretically can reach thousands of bars and degrees kelvin, respectively, in water during the final stages of the collapse. The high-pressure shockwave that emanates from the location of the bubble is capable of causing mechanical damage to surrounding material. As discussed in earlier chapters, in cases where the bubble is adjacent to a solid surface a high-velocity liquid jet may shoot through the bubble, impacting on, and damaging, the solid. The high temperatures can cause bond dissociations in molecules, producing free radicals that can react with biomolecular species in much the same way as those produced by ionizing radiations. Apfel has calculated that a submicron bubble that grows to 5 μm in one acoustic cycle deposits approximately 300×10^6 eV of energy upon collapse.[16] Sonically generated emission of light, sonoluminescence, is also believed to be associated with free-radical formation, as discussed in Chapter 6.

Violent stable cavitation activity has also been shown to generate sonoluminescence. The inhomogeneous cyclic field established around stably oscillating bubbles can cause steady flow of the fluid medium surrounding the bubble in a process known as microstreaming. The effect of microstreaming near bubbles can be one of mixing of the surrounding fluid. If the streaming velocities are great enough, shear stresses (resulting from the decreasing velocity with distance from the bubble) can be sufficient to damage cells.

Nuclei

The negative pressures associated with the rarefaction phase of the acoustic pressure and required to generate a bubble in a pure liquid would be several hundreds of bars. Yet cavitation occurs in typical media at much lower pressures, often <1 bar. This is because most liquids contain dirt or other inhomogeneities that trap small bubbles or serve as sites where bubbles may form at far lower pressures. These cavitation nuclei and theories relating to the stabilization of small microbubbles within a liquid are discussed in detail in Chapter 1.

Of interest to the question of biological effects of cavitation is the fact that these nuclei are expected to occur in systems in vitro that have not been specially processed to remove them. Also, certain parts of plants and insects contain trapped gas. The occurrence and type of nuclei present in mammalian systems is less well understood, but there is evidence for their existence, as discussed in the relevant sections of this chapter.

Predictions of Cavitation Thresholds

The problems associated with predicting cavitation thresholds or the effects of cavitation in mammals are many. The type, location, and size of cavitation nuclei in vivo are all largely unknown. Additionally the role of the surrounding tissues is not easily modeled, except perhaps in cases of simple low-viscosity body fluids. Various investigators, however, have predicted thresholds for the occurrence of rectified diffusion[17] and transient cavitation in water or aqueous media.[3–5,16,18]

In general, the threshold for cavitation increases with increasing frequency, sample pressure, and sample viscosity, and decreases with increasing gas content and temperature. The reader is referred to Chapter 1 for details of cavitation prediction. Of particular interest here are the results of several investigations that suggest the potential for transient cavitation from microsecond diagnostic pulses.[3–5,16] Since the threshold for transient cavitation increases with decreasing pulse duration, until recently most investigators have dismissed the possibility of transient cavitation associated with diagnostic ultrasound. However, Flynn[3,4] and Apfel[5] calculated that transient cavitation could occur in water at clinical intensity levels. Apfel has also considered theoretically the potential for acoustic cavitation in animals and developed some simplified equations for predicting thresholds, which consider the effects of viscosity.[16] His results suggest the possibility of transient cavitation with diagnostic ultrasound in low-viscosity body environments, such as blood or amniotic fluid. On the other hand, above 2 MHz, cavitation in the higher viscosity tissues (0.1 poise or greater) appears unlikely. Crum and Hansen have shown that the thresholds for rectified diffusion and the predicted bubble growth are reasonably consistent with experimentally observed bubble formation.[19]

IN VITRO SYSTEMS

The biological models and exposure systems used to study biological effects induced in vitro are both many and varied. Various solutions of proteins and other biological materials, cells in suspension, cells in spheroids, and cultured cells attached to surfaces have all been used. Exposure vessels include test tubes, specially constructed chambers for plane wave exposure, Petri dishes, and acoustically transparent bags. Many effects have been observed, and in most cases they appear to be related to a cavitational mechanism, whether induced intentionally or not.

The literature is abundant in reports of in vitro studies showing an effect related to either demonstrated or likely occurrence of cavitation. Only a small selection of reports is presented here to help justify the conclusion that, in these systems, nonthermal effects are most often related to cavitational activity. In addition, there are a few examples of effects from nonthermal mechanisms other than cavitation (see Chapter 8). By and large, however, the overwhelming conclusion must be that cavitation is the primary nonthermal mechanism for production of in vitro biological effects reported so far.

Additionally, other exposure systems have been employed that have been specifically designed to introduce air bubbles into the exposed medium. Biological effects associated with these bubbles have occurred at very low intensities and serve as good examples of the potential for induced damage.

Cells in Suspension

Insight into the importance of stable cavitation has been provided by the results of experiments involving exposures of cell suspensions in test tubes. When the test tube is exposed from the side, the walls of the tube cause a complex standing wave pattern to be established within the medium contained in the tube. It has been observed that the threshold for effects on cells is considerably lower when the tube is rotated than when it is held stationary. Several investigators report a threshold on the order of 1 W cm^{-2} (I_{SPTA} measured with the exposure vessel absent) for the alteration of plating efficiency (survival) of V-79 and HeLa cells exposed to 1 MHz continuous-wave (CW) ultrasound.[20–22] About the same threshold was reported for survival of mouse L5178Y cells using similar exposure conditions.[23] However, without tube rotation no effect was observed. Others have also shown that it is necessary to rotate the tube in order to obtain effects.[24,25] The explanation appears related to the trapping of the small biologically effective bubbles at the pressure antinodes and the cells at the pressure nodes.[26] Thus, for the stationary tube the cells and bubbles are separated, but during rotation the weakly held cells rotate with the medium, while the strongly held bubbles remain fixed in the field. During rotation they come together and a biological effect is produced that is associated with stable cavitation.

Morton et al.[27] found a strong correlation between cell damage and a measure of the total emitted subharmonic energy. Suspensions of Chinese hamster lung fibroblasts (V79-379A cells) were irradiated with 1 MHz CW ultrasound and emissions at the half harmonic were monitored using a 0.5-MHz receiver focused on the irradiation volume. The received half harmonic signal was integrated over the 30-min duration of exposure to yield the amplitude–time integral, which was subsequently squared to yield a "subharmonic index" related to the half harmonic energy emitted. They assayed for intact cells, the ability of the cells to take up the vital dye trypan blue, and reproductive integrity. Not only was there a correlation between each of the three measures of cell damage and the subharmonic index, but the intensity threshold for that damage corresponded with the threshold for half harmonic emission. The threshold ranged from 75 to 125 mW cm^{-2} spatial average intensity, a variation that is consistent with the random nature of cavitation activity.

It is also well understood that cavitation, usually transient cavitation, is capable of producing free radicals that can react with the medium. Armour showed that lethality of intact cells was reduced in the presence of a free-radical scavenger which penetrated the cells.[28] This occurred despite the fact that there is no concomitant reduction in cell lysis.[29] These results suggest that cavitation occurs and generates free radicals that have a significant biological effect.

Solutions

Macleod and Dunn associated the degradation of proteins in aqueous solution with the presence of cavitation in the frequency range 1–27 MHz.[30] Larger molecules such as DNA may be degraded in the absence of transient cavitation presumably because of shear stresses that may arise from stable cavitation.[31]

Ultrasonic irradiation of base and nucleoside solutions has shown that glycols, similar to those observed in *x*- and γ-radiolysis, are formed at 5 W cm^{-2} and 0.8 MHz.[32] A sonochemical mechanism was proposed for changes in uracil produced by sonically generated hydroxyl radicals.[33] Wang showed that the reactivity of nucleic acid components is Thy > Ura > Cyt > Gua > Ade.[34] It was also shown that the gaseous environment had an effect and that the order of effectiveness was Ar > O_2 > air > N_2 > He > H_2. No sonolytic products were observed when nitrous oxide, a free-radical scavenger, was present.

Recently other investigators used spin trapping and electron spin resonance to detect formation of $^{\bullet}$OH radicals and $^{\bullet}$H atoms in aqueous solutions sonicated with 1 MHz ultrasound.[35,36] Edmonds and Sancier demonstrated free-radical formation in aqueous solutions of biological molecules and whole human blood exposed to therapeutic levels of CW ultrasound for 10 and 30 min.[35] Carmichael et al. found a threshold for radical production between 0.7 and 1.4 W cm^{-2} (I_{SPTA}) for CW exposure.[36] They also found that for the case of a 6.5-μs pulse duration, 4-kHz repetition frequency pulsed exposure, the threshold was 2.4 W cm^{-2} (I_{SPTA}) and 140 W cm^{-2} (I_{SPTP}). These results indicate that free radicals are produced in aqueous solutions for pulse durations on the order of those used clinically but at slightly higher intensity levels.

Systems Designed to Induce Cavitation

Though the biological effects observed in most suspensions are attributable to acoustic cavitation, the random nature of cavitation and the uncertainties regarding the type and size of nuclei make it nearly impossible to characterize clearly the processes that occur. To obviate these problems investigators have employed several methods for introducing stable gas bubbles into suspensions. Hughes and Nyborg[37] trapped bubbles in small holes drilled into the end of a cell disruptor probe at 20 kHz; Rooney[38] used a single bubble stabilized at the end of 2 stainless steel tube; and more recently Miller, Nyborg, and other colleaques[39–41] used bubbles trapped in the pores of hydrophobic thin plastic membranes.

Rooney formed a 250-μm diameter hemispherical bubble at the end of a tube immersed in a suspension of canine red blood cells sonicated at 20 kHz.[38] He monitored the hemolysis by using an optical absorbance technique to determine the amount of hemoglobin released. He observed a rather sharp threshold at a radial oscillation amplitude of 18 μm. This corresponds to a viscous stress of approximately 4300 dyn cm^{-2} in the microstreaming field adjacent to the oscillating bubble. This is a higher stress than obtained for hemolysis of red cells subjected to steady hydrodynamic shearing for 100 s, a difference that likely results from the shorter duration of exposure for individual cells streaming by the bubble.

Miller and co-workers exposed human platelet-rich plasma to 2.1-MHz ultrasound in the presence of 4-μm diameter gas bubbles stabilized in the hydrophobic pores of a membrane.[40] Platelets form aggregates as part of the clotting process. Two different sources (both a laboratory system, which was used to generate spatial peak intensities of 8–250 mW cm^{-2}, and a commercial Doppler device, producing a spatial peak intensity of 80 mW cm^{-2}) were used to expose the plasma and immersed membrane for 10 min. After sonication at or above 32 mW cm^{-2} platelet aggregates were seen near the pores, and they increased in size with increasing intensity. At 8 mW cm^{-2} the platelets were observed to gather around the pores but not to form permanent aggregates. The authors determined that the threshold for aggregate formation was about 16–32 mW cm^{-2} spatial peak intensity. Aggregate formation was also observed after sonication with the Doppler device.

Williams and Miller used the nucleopore membranes to introduce bubbles trapped in 4-μm diameter pores into suspensions of human erythrocytes during exposure to ultrasound.[41] Cellular damage was determined by monitoring ATP release from the cells by using the firefly luminescence assay. The luminescence assay is very sensitive and allowed the detection of ATP release at spatial peak intensities as low as 20–30 mW cm^{-2} during CW sonication at 1.6 MHz. They later conducted studies, with an improved luminescence technique, that showed that light emission was proportional to the square of the intensity for CW exposure.[42] A similar dependence on temporal average intensity was observed for pulse mode exposures with 1-μs pulses and a 10% duty cycle when the intensity was changed by varying the intensity of the pulses. They also observed a proportionality to the exposure duration which suggested the product I^2T as a dose parameter, where I is the temporal average intensity and T is the exposure duration. This is more significant when it is noted that the threshold for lesion production in vivo exhibits a similar dependence on intensity and exposure duration, as discussed later in this chapter under Mammalian Systems.

PLANTS AND INSECTS

Many studies have been conducted on plants and insects. They are considered separately from the in vitro studies because these systems are complex organized living tissues. As such they are an intermediate step between in vitro systems and mammalian systems, which are considered in the next section. The study of higher plants and insect eggs and larvae has proved particularly relevant to the determination of the effects of cavitation on living tissue because they often contain naturally occurring gas-filled intercellular channels and air sacs, respectively.[43,44] In several instances it has been possible to make quantitative comparisons between the results of theory and experiment.

Plants

The leaves of aquatic plants and the root tips of peas and beans have provided particularly useful systems for studying the effects of bubbles trapped within their structures.

Elodea Leaves

Elodea was used by Harvey and Loomis[45] and Harvey et al.[46] to observe microscopically the effects of irradiation at 4 kHz. They were able to observe stirring and tearing of cellular structures that were believed to result from nonthermal mechanisms since the leaf temperature never exceeded 30°C during sonication.

Nyborg et al.[47] noted that motions induced by pulsation of gas-filled channels of *Elodea* when irradiated with 1 MHz ultrasound were similar to those induced near the tip of a needle vibrating at 85 kHz. The vibrating needles are known to induce microstreaming similar to that adjacent to an oscillating bubble. Still other investigators observed effects associated with the spacing of pressure maxima when leaves were irradiated with a standing wave field.[48] The effects observed included rotation of single chloroplasts or pieces of protoplasm and swelling of the nucleus at lower amplitudes and disintegration at higher amplitudes.

The threshold for cell death for 100-s CW exposures of *Elodea* leaves has been determined as a function of frequency from 0.45 to 10 MHz.[49] Cell death was identified by staining such that lysed cells were easily seen with a light microscope. A minimum in the threshold of approximately 0.2 W cm^{-2} was found around 5 MHz. The range for the resonance frequency, based on the range of bubble dimensions and a simple linear theory of radial oscillation of a cylindrical bubble, was calculated to be 2.9–7.5 MHz.[50] This is very good agreement considering that the linear theory is of very limited application at high amplitudes and for bubbles whose motion is likely severely constrained by the surrounding tissues. Miller also examined the threshold for cell death in different parts of the leaves containing bubbles of different size.[49] He again found rough agreement between the predicted and experimentally determined thresholds. Thus there appears to be a good correlation between the predicted resonance frequency and the minimum intensity for cell death.

Miller also observed effects on the intracellular organization of *Elodea* leaves at levels below those necessary for cell lysis.[51] Sonication freed the chloroplasts, allowing them to move to the side of the cell.

Further evidence for the association of damage in these leaves with bubble activity comes from Pfirsch, who showed that far greater damage occurred in

Elodea leaves than in leaves of *Mnium*, which have no gas-filled channels.[52] Martin and Gemmel showed that the streaming velocity in cells of *Elodea* varied with frequency and that the maximum again showed rough agreement with the resonance frequency of the bubbles based on linear pulsation theory.[53]

Root Tips

The tip of a plant root is another system that contains intercellular gas.[43] The meristem in the tip is the region of root growth, i.e., where cells divide, elongate, and differentiate. Portions of this region may contain 3–4% of its volume as gas.

In 1954 Lehmann et al. exposed *Allium cepa* root tips to 0.8- and 1-MHz ultrasound and observed the development of small elliptical zones of destruction.[54] This effect was produced by spatial peak temporal average intensities of 2–4 W cm^{-2}, but disappeared when the root was pressurized to 6.7 bar during sonication. This is very strong evidence for cavitation damage from the trapped intercellular gas.

Several investigators have examined the effects of ultrasound on growth rate of roots.[55–57] Bleaney and Oliver showed that 1.5-MHz ultrasound at 3 W cm^{-2} spatial average for 15 min reduced the growth rate of *Vicia faba* (bean) roots to 40% of controls, but that growth rate later returned to normal.[55] Gregory et al. showed reduction in growth rate and mitotic index of these roots from exposure to 2-MHz ultrasound for 1 min at intensities as low as 2 W cm^{-2} spatial peak temporal average.[56] Histological studies showed more damage to cells on the side of the roots toward the transducer.[58] This apparent lack of penetration of the ultrasound is consistent with the high-attenuation coefficient measured for the root tissue, which is in turn associated with the high gas content.[59] The preferential damage on the side facing the source also explains why some roots tend to curl toward the source as they grow after insonation.

Miller et al. reported that DNA, RNA, and protein synthesis and mitotic index were suppressed in *Pisum sativum* (pea) roots after exposure to 2-MHz, 30-W cm^{-2} (I_{SPTA}) ultrasound for 1 min.[60] Miller also showed that the transition through the G_2 phase of the cell cycle for the pea root cells was delayed.[61] The effects of pulsed (10% duty cycle) and continuous wave ultrasound on growth rate were compared at 10 W cm^{-2} (I_{SPPA} and I_{SPTA}, respectively) for 2-MHz, 1-min exposures.[62] For the same total on time the pulsed ultrasound was as effective as CW until the pulse duration was decreased below 1 ms. Arcuri et al. showed a similar dependence on pulse length for reduction in mitotic index.[63]

Morris and Coakley detected subharmonic and other acoustic emissions when bean roots were sonicated at 20 W cm^{-2} (I_{SPTA}) at 1 MHz.[64] The emissions that are indicative of cavitation activity correlated better with cell death than the growth rate reduction, which occurs at lower intensities.[65] Carstensen et al. showed that 1- and 2-MHz exposures produced greater effects in pea roots than 3 and 5 MHz.[66] They also found that increased ambient pressure up to 20 bar reduced the effects on growth rate and reduced subharmonic emissions but did not eliminate them. Additionally, Graham et al. did not find a correlation between subharmonic emission, sonoluminescence, or iodine release, as tests for the presence of cavitation, and the observed growth rate reductions.[67] Thus some of the results clearly indicate cavitational involvement in the observed effects, whereas other results indicate a lack of correlation with some of the standard indicators for cavitation activity. However, Carstensen and co-workers conclude that reduced growth rate and reduced mitotic index in plant roots resulting from irradiation with ultrasound most likely arise from cavitation associated with the small pockets of entrapped gas.[62,68]

Insects

Drosophila (fruit flies) at various stages of development have provided a convenient system, by virtue of their size and submersibility, for studying biological effects for many years. In 1953 Selman and Counce exposed fruit fly eggs to 1-MHz, 0.5-W cm^{-2} ultrasound and observed several developmental abnormalities.[69] Later studies by Thacker and Baker showed no chromosomal effects associated with adult flies surviving exposures up to 2 W cm^{-2} for 5–10 min.[70]

In recent studies cavitation, or some other nonthermal mechanism, has been suggested for the killing of fruit fly eggs by exposure to continuous[71] and pulsed[72] ultrasound and for the killing of fruit fly larvae by exposure to pulsed ultrasound, similar to the diagnostic regime.[73] These investigators showed that the effects on the larvae, which are known to contain trapped gas, were more directly related to the temporal peak than to the temporal average intensity. These results indicated a threshold for killing by 2-MHz ultrasound between 10 and 20 W cm^{-2} temporal peak intensity at very low temporal average intensities (below 5 mW cm^{-2}). They showed, by changing the pulse repetition frequency, that the rate of killing varied little with changing temporal average intensity at constant temporal peak intensity. Additionally, subthreshold peak intensities at temporal average intensities as high as 80 mW cm^{-2} showed no effect. This is very conclusive evidence for a nonthermal mechanism of damage, most likely cavitation associated with the gas in the trachea of the larvae.

The role of gas in the effects on insects was amply demonstrated by Child and Carstensen using fruit fly eggs.[72] In early stages, the eggs were quite insensitive to the low-level ultrasound; just prior to hatching, however, the respiratory apparatus is filled with air, at which time the sensitivity increased dramatically.

MAMMALIAN SYSTEMS

Although studies on suspensions, plants, and insects have provided information on the cavitation mechanism, it is difficult or impossible to extrapolate to mammalian systems the levels at which effects of cavitation occur. Much less work has been done to determine the role of cavitation as a mechanism of damage to mammalian systems in the megahertz frequency range. Nevertheless, several studies of cavitation in mammals have provided valuable insights regarding the levels for gross effects of cavitation, and some information concerning the existence and sites of cavitation nuclei is available.

Biological Effects Studies

Esche reported the occurrence of transient cavitation within thick pieces of beef muscle tissue in vivo at peak intensities of 1.7 kW cm^{-2} (about 400 W cm^{-2} by his calculations when absorption and scattering are considered).[74] Fry et al. concluded that paralysis of the hind limbs of frogs was not caused by cavitation since an increased hydrostatic pressure did not eliminate the effect, although the process observed did appear to have a pressure dependence.[75] The effect was the same whether pressurization was accomplished with oxygen or with nitrogen.

Other studies have reported the strongly destructive effects of transient cavitation caused by single-pulse exposure to high-intensity ultrasound, on mammalian systems in vivo.[76–78] Throughout the exposure conditions the threshold relationship for microscopically visible lesions is given by $IT^{1/2} = 200$, where I is the intensity (W cm^{-2}) and T is the exposure duration in seconds. Exposures greater than this produce lesions, whereas lesser exposures do not. In the brain at intensities above about 1500 W cm^{-2} cavitation is known to occur.[76,77] The resulting lesions differ greatly from those appearing in response to lesser dosages. The lesions generally exhibit a greater degree of destruction; they do not always occur at the location of the focal region; and the damage is apparent at the earliest time the tissues can be examined.

Chan and Frizzell have reported that the threshold for focal lesions in the cat liver at 3 MHz is 100% greater than that for brain for intensities below approximately 1000 W cm^{-2}.[78] Above this intensity the threshold decreases to the level found for the brain.[79] At these levels many of the lesions produced exhibit an irregular boundary and a total homogenization of the tissues within the lesion, whereas at lower intensities the lesions tend to have a uniform elliptical boundary and to contain intact though necrotic cells.[78] The change in the threshold and the damage observed both occur in the region where cavitation appears to be involved in brain lesioning.

Lele attempted to correlate the detection of cavitation with observed damage in the brain in vivo and the liver in vitro.[80,81] Both the half harmonic and anharmonic signals were monitored as a means of detecting steady and transient cavitation, respectively. It was reported that the time-averaged energy from subharmonic signals was directly related to intensity to 1500 W cm^{-2} and was observed at the lowest intensities used (100 mW cm^{-2}). On the other hand, the time-averaged output from the anharmonic detector was erratic, exhibiting no clear relationships among "cavitation power," intensity, and duration; an anharmonic signal was never observed below 1000 W cm^{-2}. The subharmonic emission increased with the onset of anharmonic emission, and the threshold for anharmonic emission was reduced if acoustic emissions had previously been generated. Histological studies revealed good correlation between anharmonic emission and gross tissue disintegration in the damaged zone. This is consistent with studies by other groups as discussed above. However, Lele reported no correlation between subharmonic emission, which was presumed to indicate steady cavitation, and histological evidence of damage. Subharmonic emission was observed at intensities that were orders of magnitude below the threshold for histological damage, and the damage, when present, was indistinguishable from that caused by heat. He concluded that there was no threshold for the steady cavitation and that it was inconsequential in organized tissue.

Taylor and Pond studied the production of paraplegia and hemorrhagic injury in the rat spinal cord by pulsed ultrasound of peak intensities 25 and 50 W cm^{-2}.[82] The results reported are consistent with cavitation as the mechanism of damage. They also showed that the endpoint occurred at lesser dosages when the rat was placed under hypoxic conditions. Lehmann and Herrick have shown that petechial hemorrhages develop in the abdomen of the mouse when exposed using a therapy unit.[83] Dependence upon frequency, pressure, and temperature was consistent with a cavitational mechanism.

Frizzell et al. determined the exposure conditions that produce hind limb paralysis in 10, 50, and 90% of mouse neonate specimens exposed to 1-MHz ultrasound at 10°C and at 1- and 16-bar hydrostatic pressure for the intensity range 86–289 W cm^{-2}.[84] The change in the exposure duration, for 50% of the specimens paralyzed, resulting from pressurization to 16 bar was negligible at 86 and 144 W cm^{-2} but was approximately a 100% increase at 289 W cm^{-2}. These results suggest that cavitation may influence the production of the hind limb paralysis produced in the neonate at 1 bar and at 289 W cm^{-2}. Sommer and Pounds reported that transient cavitation in pork muscle in vivo could be detected by half harmonic emission above 75 W cm^{-2} exposure with continuous-wave ultrasound.[85]

Recently Gross and co-workers used resonant bubble detectors to search for both preexisting bubbles and bubbles created by cavitation within the cardiovascular system of dogs.[86] They found no preexisting bubbles of 3.8 μm diameter or larger and found that no bubbles were created by exposing the left ventricle to 0.51–1.61 MHz ultrasound of up to 16 W cm^{-2} spatial peak intensity. The authors conclude that their essentially negative results "may limit the expected or potential risk of this mechanism in regard to medical applications of ultrasound."

Williams used a 20-kHz vibrating wire, which generates microstreaming similar to that induced by an oscillating bubble, to demonstrate in vivo thrombus formation by acoustic microstreaming.[87] Frizzell et al. have recently demonstrated intravascular microstreaming of blood and thrombus formation adjacent to the tip of a micropipette during 1-MHz sonication at therapeutic levels or higher.[88] The shear stresses calculated for the two cases were approximately the same, indicating that the same biophysical phenomenon may be involved, but at two widely different frequencies.[88] One case involved a wire oscillating in a fixed fluid and the other case involved a fixed micropipette in an oscillating fluid. It is not known whether similar intravascular microstreaming could result at these intensity levels from small microbubbles.

Bubble Growth at Low Intensities

Reports by ter Haar and associates have provided evidence of bubble growth in vivo during exposure to CW ultrasound at 0.75 MHz and intensities above 80 mW cm^{-2}.[89,90] They exposed the hind legs of guinea pigs to ultrasound while imaging with an 8-MHz scanned pulse-echo ultrasonic system capable of detecting bubbles 10 μm or larger in diameter. Bubbles were detected as new echoes generated after the beginning of irradiation. The number of detected bubbles increased as the intensity was raised from 80 to 680 mW cm^{-2}. Additionally, the number of detected bubbles was suppressed by increased ambient pressure.

Crum and Hansen have shown that it is possible to account for the presence of the bubbles by using a theory for growth of microbubbles (nuclei below detectable size) by rectified diffusion.[19] Recently ter Haar et al. determined the threshold intensity for the formation of bubbles 10 μm or larger in diameter during 10-min exposures.[91] They reported thresholds of approximately 110 mW cm^2 (I_{SA}) and 240 mW cm^2 (I_{SATP}) for CW and pulsed (2-ms pulse duration, 50% duty cycle), respectively.

Although these studies did not test for biological effects related to the bubble phenomena, they do suggest the potential for effects. In particular, they demonstrate the presence of cavitation nuclei in mammalian tissues and that low-level continuous-wave ultrasound can cause these to grow.

DISCUSSION

It is quite clear that acoustic cavitation is the primary mechanism for the production of biological effects in most solutions, suspensions, plants, and insects. Some of these effects occur at levels lower than used clinically. It has also been demonstrated that cavitation nuclei exist in mammals and that 10-μm and larger bubbles are developed during sonication at low therapeutic levels. Some theorists predict that transient cavitation can occur in low-viscosity fluids from exposure to diagnostic pulsed ultrasound.

Yet attempts to detect biological effects from acoustic cavitation in mammals have suggested a remarkable lack of effects of bubbles. In almost all cases where cavitation does appear to be involved, the levels are high, several hundred W cm^{-2} or more. In cases where the effects occur at lower intensities (a few W cm^{-2}) exposures of several minutes are required. This seems to suggest that, for similar exposure durations, the activity of bubbles may be inhibited within the mammalian system such that greater intensities are required to cause damage (at least gross damage involving several cells). Thus, it would seem that one area of research needing further investigation is the physical factors affecting bubble motion and its effects in mammalian tissues. Also, more subtle intracellular effects that may involve only a single cell should be examined. However, based on current knowledge it appears that clinical systems as they are presently used pose little or no risk of harmful biological effects associated with acoustic cavitation.

ACKNOWLEDGMENTS

The author wishes to thank A. Atchley, R. Apfel, E. Carstensen, S. Goss, and F. Kremrau for reading the manuscript and making helpful suggestions.

REFERENCES

1. Wood, R. W.; Loomis, A. L. *Phil. Mag.* **1927**, *4*, 417.
2. O'Brien, W. D., Jr. In "Ultrasound: Its Applications in Medicine and Biology"; Fry, F. J., Ed.; Elsevier: New York, 1978, Part I, Ch. V, pp. 343–391.
3. Flynn, H. G. *J. Acoust. Soc. Am.* **1982**, *72*, 1926.
4. Carstensen, E. L.; Flynn, H. G. *Ultrasound Med. Biol.* **1982**, *8*, L720.
5. Apfel, R. E. *Br. J. Cancer* **1982**, *45*, Suppl, V, 140.
6. World Health Organization "Ultrasound, Environmental Health Criteria 22"; WHO: Geneva, 1982.
7. National Council on Radiation Protection and Measurements "Biological Effects of Ultrasound: Mechanisms and Clinical Implications"; NCRP: Bethesda, MD, 1983, Report No. 74.
8. Williams, A. R. "Ultrasound: Biological Effects and Potential Hazards"; Academic Press: New York, 1983.
9. Nyborg, W. L.; Ziskin, M. C. "Biological Effects of Ultrasound"; Churchill Livingstone: New York, 1985.

9a. Dunn, F.; Frizzell, L. A. In "Therapeutic Heat and Cold" 3rd Edition, Lehman, J. F. Ed. Williams and Wilkins: Baltimore, 1982, Ch. 9, p. 386.

10. Flynn, H. G. In "Physical Acoustics"; Mason, W. P., Ed.; Academic Press: New York, 1964, Vol. 1A, Ch. 2, p. 57.
11. Coakley, W. T.; Nyborg, W. L. In "Ultrasound: Its Applications in Medicine and Biology"; Fry, F. J. Ed.; Elsevier: New York, 1978, Part 1, Ch. II, pp. 77–159.
12. Neppiras, E. A. *Phys. Rep.* **1980**, *61*, 159.
13. Apfel, R. E. In "Ultrasonics (Methods of Experimental Physics Series)"; Edmonds, P. D., Ed.; Academic Press: New York, 1981, Vol. 19, Ch. 7, pp. 355–411.
14. American Institute of Ultrasound in Medicine/National Electrical Manufacturers Association *J. Ultrasound Med.* **1983**, 2, Suppl. (Safety Standard for Diagnostic Ultrasound Equipment. AIUM/NEMA Standards Publication/No. UL 1-1981).
15. American Institute of Ultrasound in Medicine Bioeffects Committee. "Safety Considerations for Diagnostic Ultrasound"; AIUM: Bethesda, MD 1984, Publication 316.
16. Apfel, R. E. *IEEE Trans. Ultrason. Ferroelectrics Freq. Control* **1986**, UFFC-33, 139.
17. Crum, L. A. *J. Acoust. Soc. Am.* **1980**, *68*, 203.
18. Apfel, R. E. *J. Acoust. Soc. Am.* **1981**, *69*, 1624.
19. Crum, L. A.; Hansen, G. M. *Phys. Med. Biol.* **1982**, *27*, 413.
20. Kaufman, G. E.; Miller, M. W.; Griffiths, T. D.; Ciaravino, V.; Carstensen, E. L. *Ultrasound Med. Biol.* **1977**, *3*, 21.
21. Kaufman, G. E.; Miller, M. W. *Ultrasound Med. Biol.* **1978**, *4*, 139.

22. Fu, Y. K.; Miller, M. W.; Lange, C. S.; Griffiths, T. D.; Kaufman, G. E. *Ultrasound Med. Biol.* **1980**, *6*, 39.
23. Clarke, P. R.; Hill, C. R. *J. Acoust. Soc. Am.* **1970**, *47*, 649.
24. Bleaney, B. I.; Blackburn, P.; Kirkley, J. *Br. J. Radiol.* **1972**, *45*, 354.
25. Maeda, K.; Murao, F. In "Ultrasound in Medicine"; White, D. N.; Brown, R. E., Eds.; Plenum Press: New York, 1977, Vol. 3B, p. 2045.
26. Church, C. C.; Flynn, H. G.; Miller, M. W.; Sacks, P. G. *Ultrasound Med. Biol.* **1982**, *8*, 299.
27. Morton, K. I.; ter Haar, G. R.; Stratford, I. J.; Hill, C. R. *Br. J. Cancer* **1982**, *45*, Suppl. V, 147.
28. Armour, E. P. Nonthermal Effects of Ultrasound on Melanized and Nonmelanized Cells *in vitro*; Ph.D. Dissertation, University of Texas, Houston. 1980.
29. Fu, Y. K.; Kaufman, G. E.; Miller, M. W.; Griffiths, T. D.; Lange, C. S. *Radiat. Res.* **1979**, *80*, 575.
30. Macleod, R. M.; Dunn, F. *J. Acoust. Soc. Am.* **1968**, *44*, 932.
31. Peacocke, A. R.; Pritchard, N. J. *Biopolymers* **1968**, *6*, 605.
32. Gupta, A.; Wang, S. In "Ultrasonics Symposium Proceedings", IEEE Cat. No. 76, CH1120-5SU, **1976**, p. 92.
33. McKee, J. R.; Christman, C. L.; O'Brien, W. D., Jr.; Wang, S. Y. *Biochemistry* **1977**, *16*, 4651.
34. Wang, S. Y. In "Symposium on Biological Effects and Characterization of Ultrasound Sources", Hazzard, D. G.; Litz, M., Eds.; U.S. Dept. Health, Education and Welfare (FDA) 78-8948; U.S. Government Printing Office: Washington, D.C., **1977**, p. 196.
35. Edmonds, P. D.; Sancier, K. M. *Ultrasound Med. Biol.* **1983**, *9*, 635
36. Carmichael, A. J.; Mossoba, M. M.; Riesz, P.; Christman, C. L. *IEEE Trans. Ultrason. Ferroelectrics Freq. Control* **1986**, *UFFC-33*, 148.
37. Hughes, D. E.; Nyborg, W. L. *Science* **1962**, *138*, 108.
38. Rooney, J. A. *Science* **1970**, *169*, 869.
39. Miller, D. L.; Nyborg, W. L.; Whitcomb, C. C. In "Ultrasound in Medicine"; White, D.; Lyons, E. A., Eds.; Plenum Press: New York, **1978**, Vol. 4, pp. 545–553.
40. Miller, D. L.; Nyborg, W. L.; Whitcomb, C. C. *Science* **1979**, *205*, 505.
41. Williams, A. R.; Miller, D. L. *Ultrasound Med. Biol.* **1980**, *6*, 251.
42. Miller, D. L.; Williams, A. R. *Ultrasound Med. Biol.* **1983**, *9*, 297.
43. Gershoy, A.; Miller, D. L.; Nyborg, W. L. In "Ultrasound in Medicine"; White, D.; Barnes, R., Eds.; Plenum Press: New York, **1976**, Vol. 2, pp. 501–511.
44. Carstensen, E. L.; Child, S. Z.; Lam, S.; Miller, D. L.; Nyborg, W. L. *Ultrasound Med. Biol.* **1983**, *9*, 473.
45. Harvey, E. N.; Loomis, A. L. *Nature (London)* **1928**, *121*, 622.
46. Harvey, E. N.; Harvey, E. B.; Loomis, A. L. *Biol. Bull.* **1928**, *55*, 459.
47. Nyborg, W. L.; Miller, D. L.; Gershoy, A. In "Fundamental and Applied Aspects of Nonionizing Radiation"; Michaelson, S. M.; Miller, M. W.; Carstensen, E. L.; Magin, R. L., Eds.; Plenum Press: New York, **1975**, p. 277.
48. Goldman, D. E.; Lepeschkin, W. W. *J. Cell. Comp. Physiol.* **1952**, *40*, 225.
49. Miller, D. L. *Ultrasound Med. Biol.* **1979**, *5*, 351.
50. Miller, D. L. *J. Acoust. Soc. Am.* **1979**, *65*, 1313.
51. Miller, D. L. *Environ. Exp. Botany* **1983**, *23*, 1.
52. Pfirsch, R. *C. R. Acad. Sci. Paris* **1957**, *245*, 2361.
53. Martin, C. J.; Gemmell, H. G. *Phys. Med. Biol.* **1979**, *24*, 600.
54. Lehmann, J. F.; Herrick, J. F.; Krusen, F. H. *Arch. Phys. Med. Rehab.* **1954**, *35*, 141.
55. Bleaney, B. I.; Oliver, R. *Br. J. Radiol.* **1972**, *45*, 358.
56. Gregory, W. D.; Miller, M. W.; Carstensen, E. L.; Cataldo, F. L.; Reddy, M. M. *Br. J. Radiol.* **1974**, *47*, 122.
57. Law, W. K.; Carstensen, E. L.; Miller, M. W. *Environ. Exp. Botany* **1978**, *18*, 207.
58. Miller, M. W.; Voorhees, S. M.; Carstensen, E. L.; Eames, F. A. *Radiat. Botany* **1974**, *14*, 201.
59. Eames, F. A.; Carstensen, E. L.; Miller, M. W.; Li, M. *J. Acoust. Soc. Am.* **1975**, *57*, 1192.
60. Miller, M. W.; Ciaravino, V.; Allen, D.; Jensen, S. *Int. J. Radiat. Biol.* **1976**, *30*, 217.
61. Miller, M. W. *Environ. Exp. Botany* **1979**, *19*, 357.
62. Child, S. Z.; Carstensen, E. L.; Miller, M. W. *J. Acoust. Soc. Am.* **1975**, *58*, 1109.
63. Arcuri, K. B.; Pizzutiello, R. J.; Miller, M. W.; Kaufman, G. E.; Carstensen, E. L. *Radiat. Res.* **1976**, *65*, 458.
64. Morris, J. V.; Coakley, W. T. In "Ultrasonics International"; I.P.C. Science and Technology Press: Guildford, Surrey, UK, **1977**, p. 206.
65. Morris, J. V.; Coakley, W. T. *Ultrasound Med. Biol.* **1980**, *6*, 113.

66. Carstensen, E. L.; Child, S. Z.; Law, W. K.; Horowitz, D. R.; Miller, M. W. *J. Acoust. Soc. Am.* **1979**, *66*, 1285.
67. Graham, E.; Hedges, M.; Leeman, S; Vaughn, P. *Ultrasonics* **1980**, *18*, 224.
68. Carstensen, E. L.; Miller, M. W.; Linke, C. A. *J. Biol. Phys.* **1974**, *2*, 173.
69. Selman, G. G.; Counce, S. J. *Nature (London)* **1953**, *172*, 503.
70. Thacker, J.; Baker, N. *Br. J. Radiol.* **1976**, *49*, 367.
71. Child, S. Z.; Carstensen, E. L.; Smachlo, K. *Ultrasound Med. Biol.* **1980**, 6, 127.
72. Child, S. Z.; Carstensen, E. L. *Ultrasound Med. Biol.* **1982**, *8*, 311.
73. Child, S. Z.; Carstensen, E. L.; Lam, S. *Ultrasound Med. Biol.* **1981**, *7*, 167.
74. Esche, R. *Akust. Beih.* **1952**, *4*, 208.
75. Fry, W. J.; Tucker, D.; Fry, F. J.; Wulff, V. J. *J. Acoust. Soc. Am.* **1951**, *23*, 364.
76. Fry, F. J.; Kossoff, G.; Eggleton, R. C.; Dunn, F. *J. Acoust. Soc. Am.* **1970**, *48*, 1413.
77. Dunn, F.; Fry, F. J. *IEEE Trans. Biomed. Eng.* **1971**, *BME-18*, 253.
78. Chan, S.; Frizzell, L. A. In "Proceedings of the IEEE Ultrasonics Symposium"; Cat. No. 77CH1264-1SU **1977**, pp. 153–156.
79. Frizzell, L. A. *IEEE Trans. Ultrason. Ferroelectrics Freq. Control* **1988**, In press.
80. Lele, P. P. In "Proceedings of the Symposium on Biological Effects and Characterizations of Ultrasound Sources"; Hazzard, D. G., Ed.; U.S. Dept. Health, Education and Welfare (FDA) 78-8048; U.S. Government Printing Office: Washington, DC, **1977**, pp. 224–239.
81. Lele, P. P. In "Ultrasound: Its Application in Medicine and Biology"; Fry, F. J., Ed.; Elsevier: New York, **1978**, Appendix I, pp. 737–741.
82. Taylor, K. J. W.; Pond, J. B. *Br. J. Radiol.* **1972**, *45*, 343.
83. Lehmann, J. F.; Herrick, J. F. *Arch. Phys. Med. Rehab.* **1953**, *34*, 86.
84. Frizzell, L. A.; Lee, C. S.; Aschenbach, P. D.; Borrelli, M. J.; Morimoto, R. S.; Dunn, F. *J. Acoust. Soc. Am.* **1983**, *74*, 1062.
85. Sommer, F. G.; Pounds, D. *Med. Phys.* **1982**, *9*, 1.
86. Gross, D. R.; Miller, D. L.; Williams, A. R. *Ultrasound Med. Biol.* **1985**, *11*, 85.
87. Williams, A. R. *Ultrasound Med. Biol.* **1977**, *3*, 191.
88. Frizzell, L. A.; Miller, D. L.; Nyborg, W. L. *Ultrasound Med. Biol.* **1986**, *12*, 217.
89. ter Haar, G.; Daniels, S. *Phys. Med. Biol.* **1981**, *26*, 1145.
90. ter Haar, G.; Daniels, S.; Eastaugh, K. C.; Hill, C. R. *Br. J. Cancer* **1982**, *45*, Suppl. V, 151.
91. ter Haar, G. R.; Daniels, S.; Morton, K. *IEEE Trans. Ultrason. Ferroelectrics Freq. Control* **1986**, *UFFC-33*, 162.

8

Biological Effects of Ultrasound in Clinical Applications

G. R. TER HAAR

Physics Department, Institute of Cancer Research, Royal Cancer Hospital, Sutton, Surrey, United Kingdom

INTRODUCTION

A discussion of the biophysics of the interactions of ultrasound with viscous media is necessarily complex. The conventional categorization of mechanisms is into those that are thermal in origin, those that arise from cavitation effects, and a third class that is neither thermal nor cavitational in origin. These mechanisms may be separated for theoretical physics considerations but in practice are difficult to isolate when biological effects are studied.

Figure 8-1 is an attempt to illustrate the problem. The ordinate represents the magnitude of an observed effect, or of a biophysical mechanism, that is, for example, increase in temperature or thermal effect or extent of cavitational activity. The abscissa shows the acoustic pressure amplitude. The occurrence of acoustic cavitation and many of the other mechanical effects produced by an ultrasonic beam are dependent on pressure amplitude, and the important parameter for thermal effects is the energy flux crossing the relevant volume, which is proportional in some cases to $(\text{pressure})^2$. It therefore seems appropriate to discuss ultrasonic biophysics in terms of pressure (usually the measured quantity for field

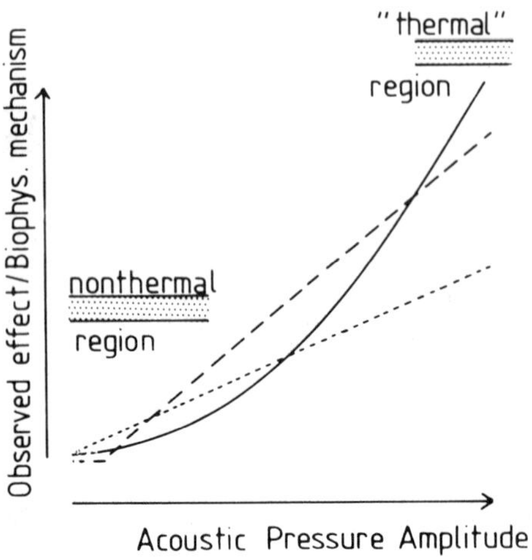

Figure 8-1. Schematic graph representing the way in which the magnitude of observed thermal (——), cavitational (– – –), and nonthermal, noncavitational (- - - -) effects vary with acoustic pressure amplitude.

characterization), only converting to intensity where appropriate. This is at odds with most existing bioeffects literature but makes sense from a physics viewpoint.

At low acoustic pressures, nonthermal mechanisms for biological change may predominate. As the pressure is increased, cavitational and thermal effects become more important until, at a high level, heating effects mask all others. No values have been put on the axes in Fig. 8-1 as they must depend on the nature of the effect observed—the appearance of a physical effect, or of a biological effect on a subcellular, cellular, or whole-organism level. In fact, the relevant values in either case are normally not known but can be expected to be highly dependent on the medium being exposed.

The approach of this chapter is to discuss the physics of the mechanisms described here, and then to put this knowledge into context for effects observed in biological tissues. As will be seen, although the physics is understood to a large extent, the link between theory and observed biological effects, which should enable prediction of bioeffects whether in terms of hazard or of benefit, is woefully inadequate at this time.

INTERACTION MECHANISMS

Heat

The energy of an ultrasonic beam is attenuated by its transit through tissue. There are two main mechanisms by which this energy is lost: absorption and scatter. Regions of different acoustic impedance within the tissue reflect the sound beam to varying extents, and scatter energy out of the primary beam to surrounding volumes. Some of the primary energy is absorbed locally (by a variety of mechanisms, including viscous shearing effects and relaxation processes), and this leads to heating. Prediction of actual temperature rises that may be obtained is difficult without a precise knowledge of the acoustic absorption coefficients and of the conduction and convection properties of the tissues involved. Blood perfusion, for example, plays an important role in determining the final temperature that is achieved, and this may depend on both the temperature and the exposure time. One can, however, make simplifying assumptions and create models that give a reasonable approximation to what may occur in plane progressive wave fields.

The acoustic pressure amplitude in the ultrasonic beam at a distance x from the point at which it is p_0 may be expressed:

$$p(x) = p_0 e^{-ax} \tag{8-1}$$

In discussion of the heating potential of an ultrasonic beam, it is more useful to rewrite this equation in terms of the energy flux crossing an area in a specified time, namely in terms of intensity ($\propto p^2$):

$$I(x) = I_0 e^{-2ax} \tag{8-2}$$

Here, I_0 is the intensity at the point $x = 0$ and a is the amplitude attenuation coefficient.

The attenuation coefficient can be expressed as the sum of the scattering coefficient a_s and the absorption coefficient α. Thus:

$$a = a_s + \alpha \tag{8-3}$$

To a first approximation, it is the magnitude of α that determines the amount of heating. Measured values of α for different tissues have found it to be between 60 and 90% of the total attenuation coefficient.[1] In general, for soft tissues, the attenuation coefficient increases approximately linearly with frequency for the frequency range used in medical ultrasound. The units most usually used for expressing absorption coefficient are nepers (Np) per centimeter (1 Np = 8.68 dB).

It is easily seen from Eq. (8-2) that the energy lost by absorption from a beam of intensity I in crossing 1 cm of tissue is $2\alpha I$. Thus, if there is no heat loss from the absorbing volume, the rate of temperature rise dT/dt is given by:

$$dT/dt = 2\alpha I/\rho C_m \tag{8-4}$$

where ρ is the density, and C_m is the specific heat per unit mass.

In order to obtain an estimate of the temperature rise, one can put representative values of α, ρ, and C_m into Eq. (8-4). Thus, for soft tissue, we may take $\rho = 1$ g cm^{-3}; $\alpha = 0.03$ Np cm^{-1}MHz^{-1}; $C_m = 4.18$ joul g^{-1} K^{-1}. This gives a temperature rise of $0.014I \times f$ K s^{-1}, where f is the frequency. Thus, for this "model" tissue, a temperature rise of 0.014K s^{-1} (0.86K min^{-1}) would be anticipated for a 1-MHz, 1-W cm^{-2} beam. Of course, in real tissue, heat would be carried out of the sample volume by convection and conduction, and the final temperature rise would be less than that predicted by Eq. (8-4).

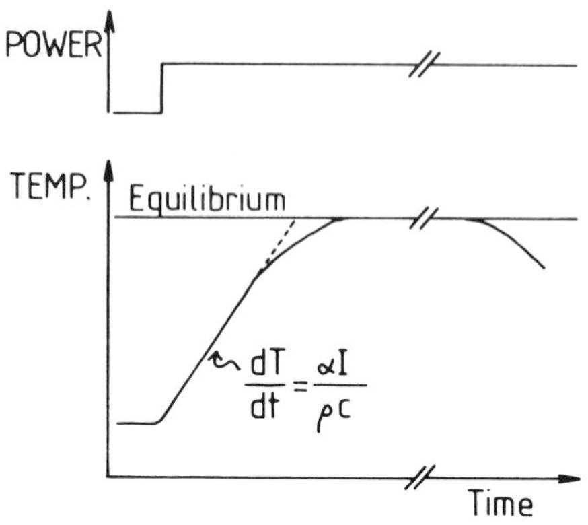

Figure 8-2. Graph showing the way in which the temperature of an exposed volume may rise and equilibrate when there is a mechanism for heat dissipation in the system. See text for details.

Figure 8-2 shows the form of the temperature rise that might be expected in an absorbing medium that has mechanisms for heat loss. The initial (linear) temperature rise is governed by Eq. (8-4), the equilibrium temperature is determined by the balance between absorbed power, and the removal of heat from the measured volume by diffusion or blood flow. In some cases, in soft tissues, if a sufficiently high level is reached ($\sim 44°C$), the temperature may be seen to drop again, despite constant power input.[2] This is most probably because of an increase in blood flow triggered by the elevated temperatures.

Nyborg[3] has considered the temperature rises that may be expected in an ultrasonic beam for a range of geometrical configurations. For example, a small region within tissue that has an absorption coefficient significantly greater than that of its surroundings may act as a point source of heat. The temperature T at a time t at a distance r from the source is then given by:

$$T - T_0 = Q(1 - erf(R))/4\pi kr \tag{8-5}$$

where T_0 is the initial temperature of the surroundings, Q is the power, k is the thermal conductivity, and R is given by:

$$R = r/2\sqrt{\rho C_m/kt}$$

The time needed for $(T - T_0)$ to reach half its equilibrium value is approximately $r^2\rho C_m/k$. The temperature rises steeply but achieves a steady-state value.

If, instead of a point source, one has a larger, spherical acoustic absorber, it can be shown that the temperature difference between the interior of the sphere ($r = 0$) and the surroundings is approximated by:

$$T - T_0 = 2\alpha It/\rho C_m \tag{8-6}$$

for short times, and:

$$T - T_0 = \alpha Ia^2/k \tag{8-7}$$

for longer times, where a is the sphere radius. It can be seen, therefore, that larger spheres can maintain a greater temperature difference than smaller ones.

Using this theory, it can be calculated that a spherical cell 8 μm in diameter in a nonabsorbing medium can maintain a temperature difference of only 0.8×10^{-6} K (W^1 cm^{-2})^1 at MHz. (This assumes $\alpha = 0.03$ Np cm^{-1}; $k = 6 \times 10^{-3}$ W cm^{-1} k^{-1}). Love and Kremkau[4] have performed a more sophisticated calculation to estimate the temperature rise that can be expected around a biological cell. They assumed a higher absorption coefficient for the nucleus than for the cytoplasm ($\alpha_{nucleus} = 0.75$ Np cm^{-1}; $\alpha_{cytoplasm} = 0.18$ Np cm^{-1}) and found that, for a cell of diameter 40 μm exposed to 1-MHz ultrasound, the temperature rise would be $\approx$0.13mK for an intensity of 1 W cm^{-2}.

These calculations show that thermally induced bioeffects are unlikely to be significant when cells are exposed to ultrasound intensities of this order when suspended in poorly absorbing liquids in vitro.

There has been some theoretical work in calculating temperature profiles that may arise from specific transducer configurations. Nyborg and Steel[5] considered in some detail the case of the field from a plane disk transducer and have calculated the maximum temperature rises that might be expected for different distances from the transducer. Other authors have calculated the influence of regions of different absorption coefficient[6,7] and different blood perfusion[8] on the temperature distributions. Similarly, focused transducers have been considered, as have arrays of either plane of focused disks.[9] There have been surprisingly few experimental studies of temperature distributions in mammalian tissues. Lehmann[10] measured temperatures in human thigh muscle and found that patients with a small thickness of muscle over bone reached their level of maximum pain tolerance earlier than those with a thicker layer. ter Haar and Hopewell[11] measured temperatures caused by contact ultrasound exposure in pig thigh muscle. Figure 8-3 shows the form of temperature profile obtained, the temperatures being taken at different depths in tissue on the beam axis (0.75 MHz, 2.0 W cm^{-2}).

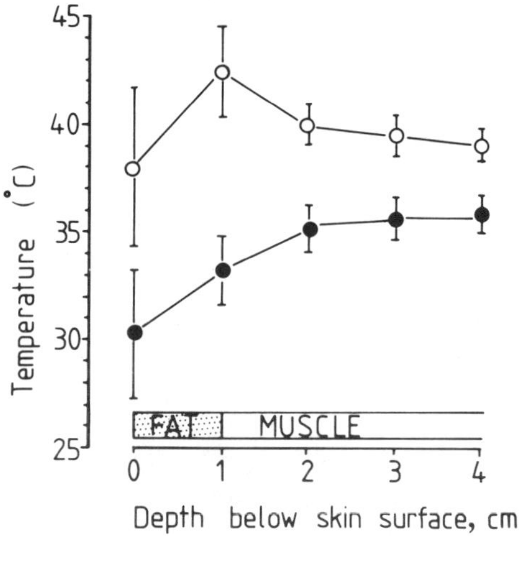

Figure 8-3. Temperature profiles measured in pig thigh muscle along the beam axis of a plane 0.75-MHz transducer in contact with the skin surface, and operating at a spatial average intensity of 2 W cm^{-2}, (● — ●) equilibrium temperatures measured before heating; (○ — ○) equilibrium temperatures arising from ultrasonic heating.

There is a variety of reasons for being interested in the temperature rise produced by the passage of an ultrasonic beam through tissue. There is the obvious question of safety: unwanted high temperatures may prove a hazard to poorly vascularized or rapidly developing tissue, for example. Although there is now a feeling that ultrasound as used in physical medicine for soft tissue injuries may be beneficial largely because of its nonthermal interaction mechanisms, the heating effect at some therapeutic intensity levels may not be insignificant. The potential for the use of ultrasound hyperthermia in cancer therapy has meant that it is important to be able to predict temperature distributions both in normal and in malignant tissues for any transducer configuration that may be used. In connection with this latter application, the effect that nonlinearities in ultrasonic propagation may have on the temperature distribution produced by a large-aperture, highly focused transducer have been calculated.[12] These calculations showed that the absorbed power density could be increased by a factor of 1.5–2.0 for regions near the beam focus because of the absorption of the high-frequency components. This has been confirmed experimentally in fresh liver samples.[13] Haran and Cook,[14] however, have shown that there is little distortion from a sine wave for plane wave propagation.

Cavitation

Detailed discussion of acoustic cavitation and its biological consequences may be found elsewhere in this book (see Chapters 1 and 7). As is suggested from Fig. 8-1, it is impossible to separate cavitational effects from other mechanical effects or from thermal effects, and so they are discussed briefly in this chapter for completeness.

There is some confusion in the definitions of cavitation that exist in the literature. In this chapter, cavitation is taken to be the formation and activity of gas- or vapor-filled cavities (bubbles) in a medium exposed to an ultrasonic field.

Commonly, two types of bubble activity are distinguished—stable cavitation and transient (collapse) cavitation. In reality, it may be difficult to separate these. Stable cavities oscillate in response to the ultrasonic pressure field. The bubble radius varies about an equilibrium value, and the cavity exists for a considerable number of acoustic cycles. Acoustic streaming and high shear stresses may be associated with such stable cavitation activity.

Collapse (or transient) cavities oscillate in an unstable manner about their equilibrium radius, grow to several times this equilibrium size, and collapse violently. This activity takes place over a few cycles of the sound field. High temperatures and pressures may be associated with collapse and may have important practical consequences of energy transfer to other forms, such as emission of light and the formation of reactive chemical species, as discussed in Chapters 4, 5, and 6.

It is generally accepted that "thresholds" exist for acoustic cavitation. Neppiras,[15,16] for example, has defined two thresholds, one for the occurrence of stable cavitation, and one for transient cavitation. For example, in a continuous-wave or tone-burst field, these thresholds may be defined in terms of the parameters P_A (acoustic pressure amplitude), P_0 (ambient pressure), R_0 (initial bubble radius), and ω (angular frequency of the ultrasonic radiation).

Experimental determination of cavitation thresholds is difficult, as the value depends on a large number of parameters, both of the acoustic field and of the irradiated sample, and also on the detection method. As an example, it has been pointed out[17] that published values for water at 1 MHz range from 1 W cm^{-2} to 2.7×10^3 W cm^{-2}. The behavior of cavitation threshold values may be summarized as follows:

1. Cavitation activity increases with increasing intensity. Historically, activity has been correlated with intensity, but a more meaningful physical statement is that activity increases with increasing acoustic pressures.
2. Cavitation threshold increases with increasing frequency.
3. Cavitation threshold increases with increasing ambient pressure.
4. Cavitation threshold decreases with increasing gas content of the sample.
5. Cavitation threshold decreases with increasing sample temperature.
6. Cavitation threshold increases with increasing sample viscosity.

Other Mechanical Effects

The passage of an ultrasonic wave through tissue causes movement of the particles that make it up. The particle is displaced from its equilibrium position by an amount known as the displacement amplitude (ξ_0) and then oscillates about its starting position with gradually decreasing amplitude until it comes to rest. There is also a velocity U_0 and acceleration a_0 associated with this motion.

The displacement amplitude ξ_0, particle velocity amplitude U_0, particle acceleration amplitude a_0, and peak pressure amplitude P_0, and intensity r are related by the equations

$$r = \mathrm{P}_0^2/2\rho c$$

$$\xi_0 = \mathrm{P}_0/\omega\rho c$$

$$U_0 = \mathrm{P}_0/\rho c$$

$$a_0 = \omega \mathrm{P}_0/\rho c$$

where ρ is the density of the propagation medium, c the velocity of ultrasound in that medium, and ω the angular frequency ($=2\pi f$).

For an intensity of 1 W cm^{-2}, at a frequency (f) of 1 MHz.

$$U_0 = 12 \text{ cm}^{-1}$$

$$\xi_0 = 0.018\ \mu\text{m}$$

$$a_0 = 71 \times 10^3 \text{ g cm}^{-2} \quad (g \text{ is the acceleration of gravity})$$

$$\rho_0 = 1.8 \text{ atm}$$

The mechanical effects that may occur have been discussed in some detail in Chapter 2. These include radiation pressure, acoustic streaming, shear stresses associated with fluid streaming, and bubble oscillations. It is not intended in this chapter to discuss the physics of these in detail, but to discuss their implications for the biological situations that may arise during medical use of ultrasound. Four different configurations are considered here:

1. *A liquid confined within fixed boundaries*—for example, plasma within a blood vessel or the cytoplasm within a plant cell. Where a fluid flows along a fixed boundary, a velocity gradient is set up, and shear stresses may be created. At sufficiently great intensities, acoustic streaming can occur within the fluid: this may increase the velocity gradient, and therefore the shear stresses.
2. *Where structures are constrained within their surroundings*—for example, lysosomes, mitochondria within the cytoplasm of cells. A structure that has a density different from that of its surroundings is subject to a cyclic displacement force that, for a spherical object, is equal to the product of its volume, the density difference, and the acceleration. For most relevant biological systems, structural inclusions have much the same density as their surroundings and the effect is small. If an object lies within a fluid but is attached to a fluid–solid boundary, it is subject to the shear stresses created in the liquid and may move within it if sufficiently "floppily" bound.
3. *Structures free to move within a field*—for example, erythrocytes, platelets, bubbles. If a structure within a field has an acoustic impedance different from that of its surroundings, it may be moved by radiation pressure and is most likely to be pushed up against a boundary. Similarly, if a gas bubble exists in the vicinity, an attractive force may be set up that can draw the structure toward it.
4. *Structures free to move within a field in the vicinity of an acoustic reflector*—for example, erythrocytes in blood vessels lying over bone. The main acoustic reflectors to be found within the body are bone and gas. Where a continuous-wave field exists, there is the possibility of standing wave formation. Where standing waves are set up, small structures with density greater and compressibility less than that of the liquid in which they are immersed move toward positions of velocity maxima. Gas bubbles smaller than resonant size move to the pressure maxima, whereas those larger than resonant size move to the pressure minima.

It is important to consider these interaction mechanisms in the light of commonly used clinical exposure conditions. Table 8-1 lists current exposure parameters drawn from a number of sources.[18] Pulse-echo diagnostic ultrasound employs pulses of microsecond duration that have high peak acoustic pressures [maxima of 7.4 MPa ($\approx$73 atm) have been measured in the positive peak, and of 3.9 MPa (38 atm) in the negative peak]. Therapeutic ultrasound (as used in physical medicine) may be used in the continuous mode and involve greater average powers, but peak pressures of only ± 0.25 MPa $\approx$ 2.5 atm are found. From our knowledge of ultrasound biophysics it should be possible to predict which interaction mechanisms will become important under specific exposure conditions.

Table 8-1. Exposure Parameters for Medical Ultrasound Exposures

Use	Frequency (MHz)	Duty factor	Average power (mW)	Spatial peak, temporal peak intensity (W cm^{-2})	Acoustic pressure (MPa)[a]
Diagnosis					
Pulse-echo	1–20	0.001	0.1–80	0.036–1100	0.1–7.4 P^+ 0.1–3.9 P^-
Doppler fetal heart monitor (CW)	2–4	1	5–30	0.001–0.03	0.003–0.015 $P^{\pm}$
Vascular doppler (pulsed)	5–10	0.01–1	0.01–40	0.001–20	0.003–0.38 $P^{\pm}$
Therapy					
CW	0.75–3	1	0–15,000	0–10	0–0.25 $P^{\pm}$
Pulsed	0.75–3	0.2	0–3,000	0–10	0–0.25 $P^{\pm}$

[a] P^+, positive pressure amplitude; P^-, negative pressure amplitude.

Table 8-2 lists the interaction mechanisms already discussed in this chapter, and also the different types of exposures encountered clinically. A temperature rise always occurs when energy is deposited in tissue, and therefore what is listed in Table 8-2 is a "significant" thermal effect, which is taken here to mean a temperature rise that produces a biological effect. This is, in itself, a subjective assessment but, in broad terms, this means of the order of degree temperature rises as opposed to millidegrees.

Thermal effects are likely only to be significant where high average powers are involved, that is, for continuous-wave or tone-burst exposures. In order to

Table 8-2. Mechanisms of Interaction and Their Likelihood of Occurrence[a]

	Continuous-wave exposures; "therapy," "Doppler"	Tone-burst "therapy" pulses (ms duration)	Short pulses "diagnosis" (μs duration)
"Significant" thermal effects	++	+	—
Cavitation	++	+	?
Radiation pressure	++	+	?
Streaming	++	+	?
Standing waves	+	—	—
Nonlinearity	+	+	+

[a] Probability of occurrence: ++, most likely, if conditions right; +, possible, if conditions right; ?, unknown; —, unlikely to occur.

activate nuclei, cavitation requires a combination of negative acoustic pressures greater than a threshold value and sufficient time for gas to diffuse into the nucleus to give bubble growth. It has been shown that continuous and tone-burst exposures can give rise to bubble formation,[19] and it seems probable that "diagnostic" microsecond pulses are of sufficient amplitude to also allow bubble growth.[20,21] The relevance of the other biophysical mechanisms to the different types of medical exposure conditions is less clear cut. Although radiation pressure and acoustic streaming may occur during short, microsecond, pulses, these phenomena seem unlikely to cause significant biological effect. They are most likely to be of importance in tone-burst or continuous-wave fields. The formation of a standing wave requires that incident and reflected waves coincide in space and time. This is only really likely when continuous waves are employed. Very little is known about the biological consequences of nonlinear wave propagation, but it can be shown that, when the pressure amplitude is of sufficient magnitude, nonlinear waveforms can be detected in water. It is not entirely clear how significant this may be in an absorbing medium such as human "soft" tissues. If it does occur, however, it is possible that an "enhanced" thermal effect may be detected.[12,13]

EFFECTS SEEN IN VITRO

Many "first line" studies looking for biological change, or therapeutic benefit are carried out in vitro. That is, cells and tissues are exposed to ultrasound while being maintained in a liquid suspension medium in a sample holder. This type of experiment can undoubtedly be valuable, but interpretation of results and their extrapolation to living animals are very complex.

Ultrasonic irradiation of cells in aqueous suspension media may create more cavitation activity than is expected in "solid" tissues in vivo. It can generally be ascertained whether or not cavitation is the primary cause of effects by repeating experiments at increased ambient pressures or varying acoustic exposure parameters, or by using a detection system that observes bubble activity and allows identification of any possible correlation between cavitation and biological effect. Another major pitfall in the execution of in vitro experiments lies in the design of the sample holder and exposure tank. Poor design can lead to unpredictable field configurations that may give unexpected hot spots or standing waves.

However, if in vitro experiments are undertaken with a proper understanding of the physics and biophysics involved, then useful results may be obtained that may give information not easily available from in vivo studies.

In vitro experiments have probably proved most useful for the study of cell membranes, for looking at genetic effects, for looking at functional changes in different cell types, and for identifying physical mechanisms that produce bioeffects.

A number of observables have been used to monitor the effect of ultrasound on cells. They do not all have immediate relevance for effects that are to be expected from medical ultrasound exposures but may shed light on mechanisms. The endpoints studied can be divided into those showing "gross" effects, such as cell lysis, loss of reproductive ability, and damage to cellular ultrastructure, and those showing more subtle effects, such as altered growth patterns, chromosomal alteration, and functional changes.

Cell Lysis and Loss of Reproductive Ability

There is extensive evidence that ultrasonic exposure of cells in suspension may lead to cell lysis, that is, complete disruption of the cell. Cavitation has been shown to be the major cause of this effect.[22,23] It is not clear as to whether or not ultrasound produces lysis in the absence of cavitation. Where lysis occurs, it is an immediate consequence of the ultrasonic exposure, not a delayed effect. Cells in mitosis may be more susceptible than those in other stages of the cell cycle.[24]

In general, cells surviving ultrasonic treatment without lysis go on to divide in the same way as untreated controls.[23,25] A common measure of effect by external agents, such as ultrasound, is the clonogenic assay, which measures the cells' ability to divide and produce viable progeny. Cells that are exposed to ultrasound while being maintained at temperatures above normal appear to suffer a loss of reproductive ability[26,27] over and above those subjected to heat alone. This appears to result from an ultrasonic interaction that is nonthermal and noncavitational in origin.[28]

Ultrastructural Changes

Many aspects of cellular ultrastructure have been studied following ultrasonic irradiation. A variety of changes has been seen, many of them not lethal.

Changes in the extracellular membrane caused by ultrasonic exposure may show up as an alteration in permeability to ion transport. For example, Chapman[29] has shown that ultrasound can induce sublethal changes in the plasma membrane of thymocytes. An immediate decrease in potassium content was found following ultrasonic exposure. Changes in membrane permeability to sodium and potassium ions have also been found (cited by Dyson[30]).

Electron microscopy of cells following ultrasonic irradiation at therapy intensities has shown changes in a variety of organelles. Mitochondria seem particularly susceptible. Damage to lysosomes has been seen following treatment of intact tissue, although it is not clear whether this is a direct or an indirect result of the ultrasound exposure.[31–33]

Where cavitation has been implicated as a damage mechanism, dilated rough endoplasmic reticulum and some other irregular lesions have been seen in addition to membrane and mitochondrial damage.[34]

In general the cell nucleus appears to survive ultrasonic exposures unscathed, the only lesions being seen being slit-like vacuoles at the nuclear membrane.[35,36] Watmough et al.[36] have suggested that intracellular cavitation microbubbles may grow from nuclei at nuclear, mitochondrial, and granular endoplasmic reticulum membranes, and this may lead to damage at these sites.

DNA and Genetic Effects

Almost all searches for genetic damage have been carried out in vitro. It appears that, with the exception of cases in which there is appreciable heat shock, currently used medical ultrasonic exposures do not lead to genetic effects.[37] Even in extreme high-intensity, cavitating, ultrasonic fields, changes in DNA were found only when there was appreciable heating or free-radical formation.[38]

Considerable effort has been exerted in looking for ultrasonically induced chromosomal alterations and sister chromatid exchanges. The vast majority of the existing evidence is that, up to intensities in the region 100 W cm^{-2} (spatial, and temporal peak) no chromosomal damage is produced. (For a comprehensive review, see Ref. 39.) Sister chromatid exchange (SCE) analysis has frequently been used as an assay for potential mutageniticity of external agents.[40] One report that ultrasound might be able to produce SCE in vitro[41] prompted a flurry of publications, the majority showing negative results even at intensities up to 5 W cm^{-2} (spatial peak, temporal average).[42]

Functional Changes

One advantage of in vitro studies that may be exploited is the ability to isolate different cell types and expose them to ultrasound. In this way, changes in normal cell function may be studied. It has been found that ultrasound may stimulate or inhibit cell function.

An example of a process that may be stimulated by ultrasound is protein synthesis by fibroblasts. It has been shown that therapeutic levels of ultrasound may increase the amount of collagent produced.[34] This effect has been shown in part to arise from cavitation activity. It has also been shown to increase blood platelet aggregation in vitro. This is though to arise from cavitation effects in the plasma which disrupt a small fraction of the platelets, causing them to release aggregating agents into their surroundings, thus leading to aggregation.[43]

Summary of In Vitro Work

As was pointed out in the introduction to this section, it is difficult, and perhaps dangerous, to extrapolate results obtained in vitro to an in vivo situation. However, such in vitro work can be useful, and it should be noted that a result obtained in such a system represents a real biological effect whose relevance in vivo requires careful consideration. Using in vitro methods, new biological responses for study in vivo may be found and fundamental interactions may be identified.

EFFECTS SEEN IN VIVO

The main body of in vivo experimental work has been carried out in small rodents, i.e., rats and mice. Reported biological effects have been somewhat anecdotal in nature and have in general failed to answer the clinically relevant questions that are raised: does diagnostic ultrasound present any hazard to the exposed tissue? How does ultrasound produce its therapeutic benefit? Which interaction mechanisms are important in vivo?

There has been a considerable amount of work aimed at identifying genetic or teratogenic effects of ultrasound. In these experiments, the embryo or fetus is exposed and examined at a later stage. In general, no adverse effects are found unless a sufficiently high intensity is used to cause significant heating (greater than 41.5°C).[44–46] Lele[47] has shown that uterine hyperthermia can cause a number of teratological effects, fetal resorption, and growth retardation. There are a number of limitations to this type of study. Necessarily, the exposure conditions are significantly different from those pertaining for human fetal exposures. Specifically, the ultrasonic beam is wide in relation to the fetal size. Human diagnostic exposures are unlikely to give rise to a significant increase in temperature.

Other in vivo studies have been concerned with effects on soft tissues, effects on bones, and changes in blood components and vasculature.

Although almost all ultrasonic exposures entail irradiation of bone, surprisingly little has been done to study this topic. There is some evidence that, if given at the correct stage of the process, ultrasound may enhance healing of bone fractures.[48] The mechanism for this is not known. It seems that in general, the main problem that can arise when bone lies in the exposed volume is periosteal heating. A liquid medium cannot support shear waves. Soft tissue surrounding a solid such as bone may propagate shear (transverse) waves, to a limited extent, but with high attenuation coefficients. At a bone–soft tissue interface, longitudinal waves propagating in bone may be partly converted to transverse propagation and then rapidly absorbed in the soft tissues. This may lead to local heating in the periosteum, which is rich in nerve endings.

Much of the experimental work on soft tissues has been investigation of ultrastructural changes. It seems that extracellular membranes, lysosomes, and mitochondria are the cellular components most readily affected. Some of the functional changes seen may be the result of acoustic streaming. For example, effects seen when smooth muscle is irradiated with ultrasound are an alteration in contractile activity[49,50] and an inhibition of action potentials.[51] These effects can be explained in terms of changes in ion transport across cell membranes, which may result from "stirring" caused by acoustic streaming.

When whole blood is exposed to ultrasound, it seems that erythrocytes are the least sensitive component to damage. Hemolysis in vitro is normally mediated by cavitation and/or high shear stresses.[52,53] At therapeutic intensities hemolysis may occur in vivo in the presence of collapse cavitation and in regions of turbulent flow,[54] although this finding has not been verified by other investigators.[55] Adenosine triphosphate (ATP) may be released at low spatial average intensities (20–30 mW cm^{-2}) when erythrocytes are irradiated in vitro in the presence of stable gas bodies.[56]

When blood vessels lie within an ultrasonic standing wave, the erythrocytes may aggregate into bands, situated at half-wavelength intervals along the axis of the sound beam. This effect is normally reversible. In such a situation damage has been shown to the luminal aspect of the plasma membrane both in the *area vasculosa* of chick embryos[57] and in uterine vessels of the mouse.[35] The appearance and site of this damage indicates that shear stresses set up by acoustic streaming may be involved.

Where ultrasound is used in cancer treatment, the aim is usually to raise the temperature locally in the region of the tumor to temperatures in the range 41–44°C. A normal response to such heating, whether induced by ultrasound or by other means, is an increased blood flow, which may manifest itself as tissue cooling.[2] It is not clear as to whether or not a similar mechanism exists in tumors. There may also be nonthermal effects of ultrasound that enhance the cytotoxicity of the exposure to the heated tissue.[27] In addition, it is possible that there is more cavitation activity at supranormal temperatures.[58]

There have been a few human epidemiological studies concerned with the safety of diagnostic ultrasound in obstetrics. In general, these have been of too small a scale to give statistically significant results, but no adverse effects have been found.[58,59] Similarly, although the relevant studies have not been carried out, no significant adverse effects have been reported or suspected in therapeutic ultrasound, despite its long-term and widespread use.

SUMMARY

This chapter has taken the reader through a discussion of the different ways in which ultrasound may interact with tissues and has presented a brief outline of the types of biological effect that have been observed in biological systems. More complete surveys of the existing biological data may be found elsewhere (see, for example, Refs. 59–61).

It should be possible from existing information to determine which of the reported effects may be clinically relevant in terms either of benefit or of hazard. Two approaches to this topic are possible. In this chapter, a largely mechanistic approach has been chosen. Following this method, it is important to ask under which conditions heating, cavitation, streaming, and standing waves become important. To summarize the information already presented, one may state that:

1. Heating may be important where
   a. the tissue has a high protein content
   b. a high intensity continuous wave beam is used
   c. a high temporal average intensity pulsed beam is used
   d. bone is included in the heated volume
   e. vascularization is poor
2. Cavitation may be important where
   a. gassy fluids are exposed
   b. small gas-filled spaces are exposed
   c. the tissue temperature is higher than normal
3. Streaming and radiation pressure may be important where
   a. a structure has an impedance different from its surroundings
   b. a fluid is free to move relative to other structures in the region
   c. continuous-wave or tone-burst excitation is used
4. Standing waves may be important where
   a. regions of high reflection coefficient lie in the field, as for example, where there are gas or bone interfaces
   b. continuous-wave or long-tone burst excitation is used with a static transducer

Using this sort of analysis a scheme such as is shown in Table 8-2 may be drawn up.

It is clear that a considerable amount of research is still required, especially in terms of relevant biological studies backed up by good and accurate physics. The standard of the physics used in this field appears to be considerably higher than that of the biology as regards the use of modern techniques.

REFERENCES

1. ter Haar, G. R. In "Physical Principles of Medical Ultrasonics"; Hill, C. R., Ed.; Ellis Horwood/Wiley: Chichester, (1986), Ch. 12.
2. ter Haar, G. R.; Hopewell, J. W. *Phys. Med. Biol.* **1983**, *28*, 889.
3. Nyborg, W. L. "Physical Mechanisms of Biological Effects of Ultrasound"; U.S. Dept. Health, Education and Welfare Publ. 78-8062; U.S. Government Printing Office: Washington, D.C., 1977.
4. Love, L. A.; Kremkau, F. W. *J. Acoust. Soc. Am.* **1980**, *67*, 1045.
5. Nyborg, W. L.; Steele, R. B. *Ultrasound Med. Biol.* **1983**, *9*, 611.
6. ter Haar, G. R. In "Hyperthermia in Radiation Oncology"; Arcangeli, G.; Mauro, F., Eds.; Proc. of 1st Meeting of European Group. Cambridge; Masson: Paris, 1979 p. 29.
7. Chan, A. K.; Sigelmann, R. A.; Guy, A. W.; Lehmann, J. F. *IEEE Trans. Biomed. Eng.* **1973**, *BME-20*, 86.
8. Hynynen, K.; Watmough, D. J.; Mallard, J. R. *Radiat. Environ. Biophys.* **1981**, *19*, 215.
9. ter Haar, G. R.; Hand, J. W. *Br. J. Radiol.* **1981**, *54*, 459.
10. Lehmann, J. F.; Delateur, B. J.; Stonebridge, J. B.; Warren, C. G. *Arch. Phys. Med. Rehab.* **1967**, *48*, 662.
11. ter Haar, G. R.; Hopewell, J. W. *Br. J. Cancer* **1982**, *Suppl. V*, 65.
12. Swindell, W. *Ultrasound Med. Biol.* **1985**, *11*, 121.
13. Clarke, R. L.; ter Haar, G. R.; Hill, C. R.; Swindell, W. *Proc. Inst. Acoust.* (1986), *8*, 171.
14. Haran, M. E.; Cook, B. D.; *J. Acoust. Soc. Am.* **1983**, *73*, 774.

15. Neppiras, E. A. *Phys. Rep.* **1980**, *61*, 159.
16. Neppiras, E. A. *Ultrasonics* **1980**, *18*, 201.
17. Coakley, W. T. *Physiotherapy* **1978**, *64*, 166.
18. ter Haar, G. R.; Hill, C. R. In "Non-ionizing Radiation Protection"; WHO: Geneva, Chapter in preparation.
19. ter Haar, G. R.; Daniels, S. *Proc. Inst. Acoust.* (1986), *8*, 147–157.
20. Flynn, H. G. *J. Acoust. Soc. Am.* **1982**, *72*, 1926.
21. Crum, L. A.; Fowlkes, J. B. *Nature (London)* **1986**, *319*, 52.
22. Kaufman, G. E.; Miller, M. W.; Griffiths, T. D.; Claravino, V. *Ultrasound Med. Biol.* **1977**, *3*, 21.
23. Morton, K. I.; ter Haar, G. R.; Stratford, I. J.; Hill, C. R. *Br. J. Cancer* **1982**, *45*, Suppl. V, 147.
24. Clarke, P. R.; Hill, C. R. *J. Acoust. Soc. Am.* **1970**, *47*, 649.
25. Bleaney, B. I.; Blackburn, P.; Kirkley, J. *Br. J. Radiol.* **1972**, *45*, 354.
26. Li, G. C.; Hahn, G. M.; Tolmach, L. J.; Shiv, E.; Pounds, D. *Radiat. Res.* **1977**, *70*, 691.
27. ter Haar, G. R.; Straford, I. J.; Hill, C. R. *Br. J. Radiol.* **1980**, *53*, 784.
28. Morton, J. T.; ter Haar, G. R.; Stratford, I. J.; Hill, C. R. *Ultrasound Med. Biol.* **1983**, *9*, 629.
29. Chapman. I. V. *Br. J. Radiol.* **1974**, *47*, 411.
30. Dyson, M. In "Biological Effects of Ultrasound"; Clinics in diagnostic ultrasound; Nyborg, W. L.; Ziskin, M. C., Eds.; Churchill Livingstone: New York, 1985, Vol. 16, pp. 121–133.
31. Dvorak, M.; Hrazdira, I. *Z. Mikrosk. Anat. Forsch. Akad. Verlags Gesellsch.* **1966**, *4*,
32. Hrazdira, I. In "Ultrasonographia Medica, Vienna"; Vienna Academy of Medicine: Vienna, 1970, pp. 457–463.
33. Taylor, K. J. W.; Pond, J. B. In "Interaction of Ultrasound and Biological Tissues"; Reid, J. M.; Sikov, M. R., Eds.; U.S. Dept. Health, Education and Welfare, Publ. No. (FDA-8008): U.S. Government Printing Office: Washington, D.C., 1977.
34. Harvey, W.; Dyson, M.; Pond, J. B.; Grahame, R. In "Proc. 2nd European Congress on Ultrasonics in Medicine"; Kazner, E.; de Vlieger, M.; Muller, H. R.; McCready, V. R. Eds.; Excerpta Medica: Amsterdam, 1975, Excerpta Medica International Congress Series No. 363., p. 10.
35. ter Haar, G. R.; Dyson, M.; Smith, S. P. *Ultrasound Med. Biol.* **1979**, *5*, 167.
36. Watmough, D. J.; Dendy, P. P.; Eastwood, L. M.; Gregory, D. W.; Gordon, F. C. A. *Ultrasound Med. Biol.* **1977**, *3*, 2205.
37. Thacker, J. In "Biological effects of ultrasound" Clinics in diagnostic ultrasound, Nyborg, W. L.; Ziskin, M. C., Eds.; Churchill Livingstone: New York, 1985, Vol. 16, pp. 67–76.
38. Thacker, J. *Br. J. Radiol.* **1974**, *47*, 130.
39. Rott, H-D. *Ultraschall* **1981**, *2*, 56.
40. Latt, S. A.; Schreck, R. R. *Am. J. Hum. Genet.* **1980**, *32*, 297.
41. Liebeskind, D.; Bases, R.; Mendez, F.; Elequin, F.; Koenigsberg, M. *Science* **1979**, *205*, 1274.
42. Miller, M. W. *Ultrasound Med. Biol.* **1985**, *11*, 561.
43. Williams, A. R.; Sykes, S. M.; O'Brien, W. D. *Ultrasound Med. Biol.* **1976**, *2*, 311.
44. Shoji, R.; Murakami, U.; Shimizu, T. *Teratology* **1975**, *12*, 227.
45. Hara, K.; Minovra, S.; Okai, T.; Sakamoto, S. *Jpn. Med. Ultrasonics* **1977**, *4*, 256.
46. Kara, K. *Acta Obst. Gynaecol. Jpn.* **1980**, *32*, 61.
47. Lele, P. P. *Ultrasound Med. Biol.* **1979**, *5*, 307.
48. Dyson, M.; Brookes, M. In "Ultrasound 82"; Lerski, R. A.; Morley, P., Eds.; Pergamon Press: New York, 1983, pp. 61–66.
49. Talbert, D. In "Proc. Ultrasonics International 1975", IPC Science and Technology Press: Guildford, Surrey, UK, 1975, pp. 279–284.
50. ter Haar, G. R.; Dyson, M.; Talbert, D. *Ultrasonics* **1978**, *16*, 275.
51. Hu, J. H.; Taylor, J. D.; Press, H. C.; White, J. E. *Aviat. Space Environ. Med.* **1978**, *49*, 607.
52. Williams, A. R., Hughes, D. E., Nyborg, W. L. *Science* **1970**, *169*, 871.
53. Rooney, J. A. *Science* **1970**, *169*, 869.
54. Wong, Y. S.; Watmough, D. J. Proc. of Ultrasound Interaction in Biology and Med. Symposium; Reinhardsbrunn, GDR, No. 10–14, Paper C-14. **1981**.
55. Gross. D. R.; Miller, D. L.; Williams, A. R. *Ultrasound Med. Biol.* **1985**, *11*, 85.
56. Williams, A. R.; Miller, D. L. *Ultrasound Med. Biol.* **1980**, *6*, 251.
57. Dyson, M.; Pond, J. B.; Woodward, B.; Broadbent, J. *Ultrasound Med. Biol.* **1974**, *1*, 133.
58. National Council on Radiation protection and Measurements. "Biological Effects of Ultrasound: Mechanisms and Clinical implications". NCRP Report 74; NCRP: Bethesda, MD, **1983**.
59. Williams, A. R. "Ultrasound: Biological effects and potential hazards"; Academic Press: New York, **1983**.
60. World Health Organization. Environmental Health Criteria 22: "Ultrasound"; WHO: Geneva, **1982**.

Index